Last Gathering of Eagles

Germany's *Freikorps* and *Reichswehr* Air Force

1918–1920

Richard Bennett

Publisher Notes
For our aviation books, please see our website at: **www.aeronautbooks.com**.
I am looking for photographs of the less well-known German aircraft of WWI. For questions or to help with photographs you may contact me at **jherris@me.com**

Interested in WWI aviation? Join The League of WWI Aviation Historians (**www.overthefront.com**), Cross & Cockade International (**www.crossandcockade.com**), and Das Propellerblatt (**www.propellerblatt.de**).

ISBN: 978-1-953201-75-1

Cover image: Anastasios Polychronis
Profiles: Bob Pearson
Design and layout: Jack Herris
Cover design: Aaron Weaver
Digital photo editing: Jack Herris

Aeronaut Books
Books for Enthusiasts by Enthusiasts
www.aeronautbooks.com

Table of Contents

Foreword

This started innocently enough -- researching another pet project, I came across a reference to *Flieger Abteilung* 431. 431? That's not a number Germany's *Luftstreitkräfte* (Air Force) used in World War I. Was it a typo?

It was not. It was the designation for one of the postwar aviation units that supported Germany's volunteer army at home, on the Polish frontier and in the Baltics between the Armistice and the Versailles Treaty that shut down German military aviation. This volunteer army, the *Freikorps*, and its successor, the *Vorlaufige* (Provisional) *Reichswehr*, have been the subject of a number of histories, but their aviation arms have been largely ignored.

Most general histories of World War I have little to say about aviation; those that do may mention the demise of Germany's air force at the Armistice, summing it up by saying the Allies forced the Germans to surrender 1700 aircraft, especially the prized Fokker D.VIIs. They may also mention some German pilots crash-landed their aircraft, rather than leave them intact for the Allies, and as an added touch, show a photo of stacks of aircraft being readied for their funeral pyres. The implication is that German military aviation ceased to exist in November 1918 and wouldn't appear again until that country's secret training mission to Lipetsk, U. S. S. R., in 1925.

The truth, not surprisingly, is far more complex. The Armistice of 11 November 1918 might be commemorated in the West as the end of World War I, but at least four more years would pass before Central and Eastern Europe would finally enjoy peace, and an uneasy one at that. Faced with internal leftist uprisings and encroachments on its eastern borders, Germany quickly raised a volunteer force from the wreckage of its wartime army and set about restoring order and securing its frontiers. This army included an air force of over 60 tactical units and several thousand personnel, many of them the brightest lights of the old *Luftstreitkräfte*. With the Allied powers slashing their air forces to peacetime strength, defeated Germany arguably had the world's most potent air arm for most of 1919. It's time this force and its flyers receive recognition for their role in history.

Below: Scenes like this were not the immediate outcome of the Armistice. Over sixteen months would pass before the Inter-Allied Aeronautical Commission of Control had the authority to order the destruction of Germany's aviation material. (Greg VanWyngarden)

Introduction

Imagine someone handing you a jigsaw puzzle, but warning you many of the pieces are scattered around the house, under furniture, in the bottoms of drawers and Lord knows where else. Others were thrown away or lost a long time ago and will never be found. Oh, and some of the pieces? Some earlier puzzler altered their shapes or colors to fit his idea of how the final picture should look. Have fun!

That's a bit like how it has been to compile this history. Not many accounts date from the days of the Weimar Republic. Memories were fresh and revisionism hadn't yet crept into the historical narrative, but there wasn't a lot of official pride in the deeds of the *Freikorps* and the early *Reichswehr*. They may have saved the Republic from the Reds and foreign invaders, but they did so grudgingly, and at least once, had turned their backs on it in a time of need. Understandably, the Weimar Republic wasn't eager to play up their accomplishments.

The pendulum swung in the other direction some years later. After coming to power, the Nazis adopted the *Freikorps* as their forebears and glorified their accomplishments. The *Reichswehr* published a series of official histories of the era, and the popular press released numerous histories and personal accounts of soldiers and flyers who served in those confused times, complete with overblown descriptions of their emotional states, patriotic spirit and false martyrdom – you'll see this in many of the quotations in this book. These are the altered jigsaw pieces, distorted by the demands of Hitler's propaganda machine.

Undoubtedly, a great deal of objective primary material was stored in the German government's archives, but most of it seems to have been destroyed in the Second World War. So we are reduced to assembling this history from surviving bits and pieces and filling in the gaps by deduction. No document is too insignificant to ignore – a flyer's ID book or travel pass may reveal more than a memorandum from the War Ministry. Surprisingly, some of the richest accounts of Germany's activities survive in the archives of her enemies of that time -- Poland and the Baltic republics – in spite of their later subjugation by the Soviets, themselves masters at erasing or rewriting history.

This has not been a solo effort – it's a compilation of fragments discovered and curated by the community of enthusiast historians who generously shared them with me. My contribution was to assemble those pieces into an overriding narrative. Responsibility for any mistakes and misinterpretations rests on my shoulders.

For years, most World War I military histories have reflected the physical distance between actions on the ground and those in the air. Accounts have focused on the land campaigns, ignoring aviation's contributions to those activities, and aviation historians have returned the snub by detaching the flyers and air battles from the fighting below. Only in recent years have historians begun placing aerial activities where they belong – in the context of the larger war. I've tried to follow their example. Why were these flyers fighting, how did they fit into the military, and this case, political, situation, and what did they accomplish? Inasmuch as the political and military conflicts of the post-World War I period are obscure to most readers, I've also provided a brief historical context to set the stage for each section.

As this is an account of German military activities, I have mostly used place names and spellings used by the Germans in 1918–20. Following independence, Poland and the Baltic republics changed many of these names to conform to their own languages and history. For readers trying to locate these places on modern maps, Appendix 4 is a cross reference to their modern names.

During much of the period covered by this account, the *Flieger Abteilung* was the basic tactical unit of the German air force. The term, usually translated as Flying Detachment or Flying Section, appears often in this book. German documents and books are remarkably inconsistent about this term, sometimes showing it as one word, two words or two words, hyphenated. Unless it's quoted from a contemporary citation, I have chosen to use the two word version. It all means the same thing.

A final remark: I am under no illusion this book will be the last word on the subject. The eternal curse on historians is that the publication of their studies will trigger the emergence of new material that fills in factual gaps and corrects speculation that previously had to suffice. This book is no exception – I *know* it contains factual errors and unwarranted assumptions. I just don't know that they are. I expect, I *hope* – it will provoke an avalanche of clarifications, elaborations and corrections. Every carefully researched history should build on earlier efforts, and I hope this one, for all its shortcomings, will prove to be a solid steppingstone. More needs to be done. The people in this account deserve their place in the historical record. Go for it, historians – tell their story.

Richard L. Bennett
Cherry Valley, Illinois, 1 August 2023

Germany's New Air Force

The terms of the November 11, 1918 Armistice required German forces to withdraw to prewar borders and evacuate all German territory west of the Rhine and around several of that river's bridgehead cities.

The Armistice may have ended fighting with the Allies, but Germany couldn't rest. Some semblance of a military force was still required by the new German Socialist Republic to maintain domestic order, secure the country's borders against encroachments by Poland and Czecho-Slovakia and to protect its Eastern frontier from Bolshevik armies advancing out of Russia. A viable air force would be an essential part of it.

Much of the imperial army had simply disintegrated or fallen under the sway of Workers' and Soldiers' Councils, but many units retained enough officers and enlisted men to become the core of a temporary army. To this core were added volunteer military formations known as the *Freikorps*, a contraction of *Freiwillige* (Volunteer) *Korps*. These were paramilitary units deputized, essentially, to serve in the army. This reconstituted army would be organized on the structural framework of 21 *Armee Korps* areas, as it had been during the Great War.

Demobilized troops and recently graduated students, eager for a chance to fight for the *Reich*, signed up to join the *Freikorps* units as they sprang up in December 1918 and January 1919. Although many of these volunteers contracted to serve only for a limited period (six months was a common term, although it could be extended), their volunteer service time was accepted as part of their permanent service records.

Eventually, hundreds of volunteer units were formed. They had no standard size or organizational structure, their reporting relationships varied from formal military hierarchy to none at all. The largest *Freikorps* were division-sized forces, usually self-contained, led by generals, fielding their own artillery and transport units and usually, aviation detachments. The smallest were platoon-sized and sometimes smaller, often assembled and led by non-commissioned officers. In between were units of all sizes and capabilities. They became the building blocks of Germany's emergency army, slotted in with surviving official army units and other *Freikorps* formations to replicate traditional military structures. However, some of these volunteer organizations' ties to the government were tenuous, and they often broke free from the army to pursue their own campaigns. Others were primarily local militias recruited to keep order in their hometowns.

Picking Up the Pieces

There was no shortage of volunteers to staff the new air force, but assembling the aircraft, tools, spare parts and supplies, in addition to providing secure flying fields, required more financial and organizational firepower than, say, outfitting an infantry company. Consequently, the government controlled the establishment of new flying units. One thing working in its favor was that the upper command structure of the old *Luftstreitkräfte* (Air Service) was, in its early days, kept intact by the new republic. *Generalleutnant* Ernst von Hoeppner, *Kogenluft* (*Kommandierender General der Luftstreitkräfte* -- Commanding General of the Air Service) and his Chief of Staff, *Oberst* (Colonel) Hermann von der Lieth-Thomsen, continued to hold their wartime posts, now reporting to Defense Minister Gustav Noske instead of *Feldmarschall* von Hindenburg, and many other staff and command officers still held their commissions. Major Wilhelm von Haehnelt, *Kofl. Heim* 8 (*Kommandeur der Flieger Heimat* 8 - Commander of Flying Troops, Homeland District No. 8) when the Armistice was declared, was appointed *Idflieg* (*Inspektor der Fliegertruppen* -- Inspector of Flying Troops) 15 December 1918[1]. In addition, the organizational structure of the *Luftstreitkräfte* and its combat units, refined over four-plus years of war, would be the template for the new volunteer air force.

Only four days after the Armistice, the War Ministry established *Grenzschutz Ost* (Border Defense East), a military command tasked with defending Germany's frontiers with the new nations of Poland and Czecho-Slovakia. *Grenzschutz Ost* encompassed the districts of I. and XX. *Korps* (East Prussia), II. and XVII. *Korps* (Pommerania and West Prussia), and V. and VI. *Korps* (Silesia).

As part of this program, *Idflieg* issued orders authorizing two new *Flieger Abteilungen*, A and B, on 2 December, adding *FA*s C and D ten days later[2]. *Heimatschutz Schlesien Flieger Abteilung* D (Silesian Homeland Defense Flying Detachment D) was already in existence 6 December, so it appears *Idflieg*'s order was aimed at gathering existing units under its wing, rather than creating new ones. It also suggests ad hoc units were popping up along the Polish frontier without waiting for orders from above. None of this was premature; on 18 December, Polish insurrectionists seized control of much of Posen province, and along with it, portions of the II., V. and VI. *Korps* districts.

Effective 1 January 1919, the old wartime organization of homeland (*Heimat*) aviation troops was restructured, with units now reporting to the *Armee Korps* districts in which they were based. Each *Korps*

Armee Korps	Former *Heimat* District	Headquarters	*Kofl.*
I.		Königsberg	*Hpt.* Hans Kumme
II.		Stettin	*Hpt.* Wolf Thieme
III.		Berlin	*Hpt.* Erich (?) Stein
IV.	*Heim* 4	Magdeburg	*Major* Robert (?) Herber
V.	*Heim* 2	Posen	*Hpt.* Gustav Walter
VI.	*Heim* 3	Breslau	*Hpt.* Bruno Zühlke
VII.		Münster	*Hpt.* Behrends
IX.		Altona	*Hpt.* Paul Genée
X.	*Heim* 6	Hannover	*Hpt.* Jasper (?) von Oertzen
XI.		Kassel	*Hpt.* Percy Baron von Ascheberg
XII.	*Heim* 5	Dresden	*Hpt.* Walther von Teubern*
XIII.		Stuttgart	*Hpt.* Alfred Ruff
XIV.		Karlsruhe	*Hpt.* Ernst Steinkopf-Hartig
XVII.	*Heim* 1	Danzig	*Hpt.* von Liebermann
XVIII.	*Heim* 7	Frankfurt (Bad Nauheim)	*Hpt.* Franz Geerdtz
XX.		Allenstein	*Hpt.* Meinhard (?) Fenske
Gardekorps	*Heim* 8		*Major* Wilhelm Grade

Table title: ***Armee Korps Kofl.* as of 1 January 1919**

Armee Korps VIII., XV., XVI. and XXI. had been disbanded, as they had been situated in the demilitarized Rhineland or in Alsace-Lorraine. XIX. A.K., headquartered in Leipzig, Saxony, was merged into XII. A.K.
* Appointed by *Idflieg*. The Soldiers' Council at Grossenhain named *Ltn.* André as *Kofl.* It's doubtful he was accepted by *Idflieg*.[3]

command staff would be assigned a *Kommandeur der Flieger (Kofl.)* to oversee those aviators.

It soon became obvious the number of air units eventually required could outstrip the alphabet, so the original four *Abteilungen* were redesignated *FA* 400 through 403, respectively, as part of their adoption into the new *Reichs* air force. By 8 January 1919, *Flieger Abteilungen* 400 through 426 had been authorized, although few of them were operational[4]. Four days later, the survivors of *Eiserne Flieger Abteilung* 8, which had evacuated Kurland with the Red army on its heels, arrived in Insterburg, East Prussia. It would be reconstituted as *FA* 427. By 1 February, *FA*s 428 through 432 had followed. On 23 April, the former *FA* des *Ostpreussisches-Freiwilligenkorps* was redesignated FA 433, followed by *FA* 434 and 435. The former kingdoms of Bavaria, Saxony and Württemberg, in keeping with the structure of the old imperial army, had their own independent commands, along with their own smaller air forces. Initially, their activities were limited to their home states. In addition, German naval troops raised one flying unit that served alongside the *Reichs* units.

Prussia's new *FA* 400 series air units were to be allocated to three commands:

1. *Grenzschutz Ost*, discussed above,
2. *Oberfehlshaber Ost* (Supreme Commander East), the former occupation authority for northwestern Russia, including its Baltic territories. After the collapse of the surviving German units in Estonia and Kurland, *Ober Ost* evacuated to Königsberg 2 January 1919 and was wound down on the 14th[5]. Its resources and responsibilities, along with those of *Armee Oberkommando* (abbreviated *AOK*) 8, were transferred to *Grenzschutz Ost*, which took on the added task of blocking the imminent Red advance into East Prussia. When this threat faded, *Grenzschutz Ost*'s responsibilities were extended to German forces supporting Baltic nationalists striving to free their new countries from Russia. In reality, this covered Germany's barely hidden agenda of replacing Russian domination of the area with its own.
3. *Abteilung* Lüttwitz, commanded by *General der Infanterie* Walther von Lüttwitz, responsible for maintaining order in Berlin and its vicinity. Lüttwitz's

Above: A typical photo of the personnel of an unidentified *Freikorps Flieger Abteilung* and its Halberstadt CL.IV, 1908/18. The divide between officers and enlisted personnel was largely erased under the new republic, and these units were noted for their cohesiveness. (Ed Ferko Collection, 04AEF-B04-F03-PB003, Folder 3, Box 4, Aschoff Series, History of Aviation Archives, Special Collections and Archives Division, Eugene McDermott Library, The University of Texas at Dallas.)

Kofl. was future *Luftwaffe* General *Hpt.* Hugo Sperrle, previously commander of four *Flieger Abteilungen,* most recently *FA(A)* 249. Soon, this command's responsibility would expand to all parts of Germany not under the jurisdiction of *Grenzschutz Ost* or the Bavarian, Saxon or Württemberg governments.

Idflieg's budgeting plans anticipated each *Abteilung* would have 6 to 9 aircraft and a headcount of about 60 men, depending on its operational requirements [6]. The manpower projection turned out to be too modest.

Drawing up a list of aviation units was the easy part -- supplying them and finding them secure homes would be a different matter. Workers' Councils had seized control of most factories producing military goods, and nothing would leave the shipping docks without their say-so. To free up the equipment it needed for its new volunteer army, the government would either have to get the Councils' cooperation or kick them out of the factories and restore their deposed management. Most Councils were not cooperative, especially after they saw the new *Freikorps* forming under leaders of the old imperial army. Ironically, the socialist government eventually saw that its survival depended on taking back control of the military's resources, and that required breaking the grip of the workers who had supported the revolution.

The government faced the same problem on the military side. Shortly after assuming power, it decreed military units would be governed by the newly established Soldiers' Councils, now authorized to exercise veto power over their officers. As in Russia a year earlier, this idealistic experiment collapsed in the face of reality. Troops were generally unwilling to commit to any hard duty, let alone mortal risk, and their formations ceased to be useful military units. Commanders struggled to maintain any semblance of discipline, military equipment and supplies disappeared into the Black Market, and prostitutes and other camp followers often enjoyed free access to the bases.

This was the situation in many of Germany's aviation schools and *Flieger Ersatz Abteilungen* (Aviation Replacement Sections), abbreviated *FEA*. The *FEA*s were large depots where personnel, aircraft and spare parts were assembled for dispatch to frontline units, and they were the logical locations for rebuilding the air force. In late 1918 and early 1919, one of the main tasks for the new *Freikorps* units was to regain

Right: Wartime photo of *FEA* 11, Brieg, Silesia. It birthed *FA*s 400 and 401 and was the operational base of *FA*s 402 and 405. (Paul Rempe)

control of these and other facilities needed to put the air force back on its feet. Typical of the challenge they faced was the takeover of *Flieger Ersatz Abteilung* 3 by *Generalmajor* Maercker's *Freiwillige Landesjägerkorps.* Following its participation in putting down left-wing uprisings in Berlin, the *Landesjägerkorps* traveled to Weimar to secure the city for the upcoming National Assembly that would draft the new republic's constitution. That done, it was ordered to Gotha in mid-February to regain control of the city and its valuable aviation facilities:

"The situation was even worse in *Flieger-Ersatz-Abteilung* 3. It had 150 NCOs and 1,100 men, about 500 of whom were on leave. The entire "troop" lived in the city, although the airfield had accommodation for 800 men. The chairman of the Soldiers' Council, Sergeant Fliedner, was the sharpest Spartakist. He had almost 100 men behind him. *Hauptmann* Genée had recently been appointed leader of the *Ersatz-Abteilung* by the War Department. The Soldiers' Council did not recognize him, however, but asked the previous commander, *Hauptmann* Bisantz, an insubordinate officer himself, who had not carried out the order to return to his regiment, to be the commander. The *Flieger-Ersatz-Abteilung* 3 was particularly important because it managed very large holdings (including 400 aircraft), which it withheld from the government of the *Reich*. All of their vehicles were in the service of the *USP*[7]. An order of the XI. *A. K. (Armee Korps* -- Army Corps*)*[8] on February 13th ordered the Abteilung to be dissolved."[9]

The sprawling *Gothaer Waggonfabrik* (Gotha Wagon Factory, manufacturer of Gotha aircraft) abutted *FEA* 3, and gaining control of one facility required taking both. Gotha would prove a tough nut to crack -- only after a daylong firefight involving machine guns and field artillery, would the *Freikorps* secure the airfield and factory complex at a price of 48 killed, wounded and missing. Spartacist losses were said to have run into the hundreds[10].

The *Freiwillige Landesjägerkorps* accomplished its goal, as did the other *Freikorps* units throughout Germany. By the middle of 1919, most cities and military facilities had been restored to government control.

Most of the locations chosen for reestablishment of the air forces were in the eastern half of Germany, close to the threats to the country and far from prying Allied eyes. *Flugplatz* (military airfield) Döberitz, on the western outskirts of Berlin, was the birthplace of six *Flieger Abteilungen* -- 419, 420, 421, 422, 423 and 432. These six detachments would be assigned to *Abteilung* Lüttwitz, responsible for maintaining order within Germany. *FEA* 11 in Brieg, Silesia, birthed *FA*s 400 and 401. These were destined for Grenzschutz Ost's

Generalkommando VI. A. K. (General Command VI. Army Corps), headquartered in Breslau.

Three *Flieger Abteilungen* were raised in *Flieger Beobachter Schule* (Aviation Observer School), Königsberg, East Prussia (408, 409, and 424, all installed in Lithuania) and four in *FEA* 8, Graudenz, West Prussia (410, 411, 412 and 435 for *Grenzschutz Ost*). In all, fifteen *FEA*s, training schools and airfields launched the combat units of the *Reich*'s new air force [11].

Integration into the *Reichswehr*

Although the *Freikorps* had performed valuable service to the *Reich*, its organizational structure was less than ideal, containing too many ad hoc units of varying sizes, many too small to make useful contributions to the country's defense. Some of them were only loosely tied to the army's command structure, following their own agendas, such as Rudolf Berthold's *Eiserne Schar* (Iron Host), which operated more as a roving vigilante band than an integral part of Germany's armed forces. It went where Berthold, not the army high command, decided it should go.

To impose some order on the situation, the National Assembly enacted the *Gesetz über die Bildung einer vorlaüfige Reichswehr* (Law on Formation of a Provisional *Reichswehr*) on 6 March 1919. This act, which took effect six days later, authorized the government to organize a new army from the *Freikorps* and remnants of the old imperial army. Existing *Freikorps* units might be retained as is, combined with others, or disbanded altogether to suit the needs of the provisional army.

The law replaced the old *Armee Korps* structure with an organization based, eventually, on 44 brigades (1 through 43, plus a short-lived 50) [12] and incorporated the formerly independent forces of Bavaria, Saxony and Württemberg. Two types of brigades were formed -- large and small.

Apart from the Bavarian units, which underwent a great deal of consolidation, most *Flieger Abteilungen* were retained, practically unchanged, and assigned directly to one of the new brigades. They were renamed *Reichswehr Truppen-Fliegerstaffeln* (Troop flying squadrons), numbered 1 - 38 and 50 after the brigades to which they were attached, or *Artillerie-Fliegerstaffeln* (Artillery flying squadrons), numbered 101 - 130. For example, *Freiwillige Flieger Abteilung* 422 became *Truppen-Fliegerstaffel* 15, *FA* 423 was redesignated *Artillerie-Fliegerstaffel* 116, and two Bavarian units, *Freiwillige Flieger Abteilung* Fürth and *Flieger Abteilung* Schmalschläger, were merged to form Bavarian *Truppen-Fliegerstaffel* 21a, reporting to the 21st Brigade. With one exception, only large brigades had air units, so there are gaps in the numbering sequence.

Functionally, *Artillerie-Fliegerstaffeln* appeared to be tasked with reconnaissance and artillery support, while the *Truppen-Fliegerstaffeln* focused more on short-range reconnaissance and tactical ground support. Because of the lack of significant aerial opposition in most areas, there were no dedicated single-seater units, but some of the *Staffeln* carried a few fighters for protection. As time went on, several units' duties became more diverse, and they became the aerial equivalent of a Swiss Army knife, their aircraft inventory including C., CL., D. and even, in a few cases, J.- type aircraft. About 60% of the units were *Truppen-Fliegerstaffeln*.

The new *Staffeln* continued to operate officially until 30 September 1919, when, as dictated by the Versailles Treaty, the provisional *Reichswehr* was to be replaced by the 100,000 Man *Reichswehr*, and military aviation would cease to exist. In reality, many of the aviation units continued to operate well into 1920.

The *Reichswehr* reorganization took several months to accomplish. Most of the new brigades officially came into existence in June; the 400-series *Flieger Abteilungen* took on their new identities about the same time.

For some *Flieger Abteilungen*, this change ran head-on into the Versailles Treaty. Germany's acceptance of the Treaty meant acceptance of the new border with Poland, ending, for the time being, the threat of war and many of the plans for expanding *Grenzschutz Ost*. Many of the new brigades and their air detachments in *Grenzschutz Ost* became redundant, and their new unit numbers now served mainly as bookkeeping identities until they were disbanded. For a few others that remained in the Baltic, the conversion never officially went into effect. A good example was *FA* 424. When its presence was no longer required in Lithuania, it moved on 10 May 1919 to the airfield at Gut Weitenfeld, near Alt-Auz, Kurland. There it sat idle in anticipation of becoming *Truppen-Fliegerstaffel* 37 when its new parent, 37. Brigade, would be set up in June in West Prussia. Plans changed, however, and by the time the brigade was finally established in August, *FA* 424 was already in the process of joining the West Russian Army of Colonel Bermondt-Awaloff, becoming its 3rd *Plastun* Aviation Division. Nevertheless, contemporary documents occasionally referred to it as *Truppen-Fliegerstaffel* 37.

Flieger Abteilung 425 remained at Aleksotas, across the Niemen River from Kaunas, until 7 July, when it, too, moved north to Gut Weitenfeld. As early as June, however, it had been redesignated *Truppen-Fliegerstaffel* 1, anticipating a transfer to 1. Brigade, headquartered in Königsberg. In any event, this never came to pass--with the arrival of *FA* 433 on 28 July, the two units were merged to form *Flugpark* Kurland. *FA* 425 was officially disbanded 14 August by *AOK Nord* [14]. The

Projected *Vorlaufige Reichswehr Fliegerstaffeln* and Their Parent Brigades			
***Rw* Fliegerstaffel**	**ex-*FA***	***Rw* Brigade**	**Brigade Hq**
Artillerie-Fliegerstaffel 101	427 & Fick (427)	1	Königsberg
Truppen-Fliegerstaffel 1[1]	425, 433	1	Königsberg
Artillerie-Fliegerstaffel 102	414	2	Stettin
Truppen-Fliegerstaffel 2	415	2	Stettin
Artillerie-Fliegerstaffel 103	432	3	Potsdam
Artillerie-Fliegerstaffel 104	419	4	Magdeburg
Artillerie-Fliegerstaffel 105	408	5	Glogau
Truppen-Fliegerstaffel 5	407	5	Glogau
Artillerie-Fliegerstaffel 106	402	6	Breslau
Truppen-Fliegerstaffel 6	403	6	Breslau
Artillerie-Fliegerstaffel 108[2]	416, 417	8	Graudenz
Truppen-Fliegerstaffel 8	400	8	Oppeln
Artillerie-Fliegerstaffel 109	435	9	Schwerin
Truppen-Fliegerstaffel 9	434	9	Schwerin
Artillerie-Fliegerstaffel 115	420	15	Berlin
Truppen-Fliegerstaffel 15	422	15	Berlin
Artillerie-Fliegerstaffel 116	423	16	Weimar
Artillerie-Fliegerstaffel 117	412	17	Danzig
Artillerie-Fliegerstaffel 120	410	20	Allenstein
Truppen-Fliegerstaffel 20	404	20	Allenstein
Artillerie-Fliegerstaffel 121		21	Munich
Truppen-Fliegerstaffel 21	A1 (Krausser)	21	Munich
Truppen-Fliegerstaffel 21a	Fürth & Schmalschläger	21	Munich
Truppen-Fliegerstaffel 21b	Kriegelsteiner & Dessloch	21	Munich
Schlacht-Fliegerstaffel 21	*Schlasta* Anschütz	21	Munich
Artillerie-Fliegerstaffel 121	A2 (Häfner)	21	Munich
Artillerie-Fliegerstaffel 127	406	27	Frankfurt a. O.
Truppen-Fliegerstaffel 27	418	27	Frankfurt a. O.
Artillerie-Fliegerstaffel 129	401	29	Bernstadt
Truppen-Fliegerstaffel 29	431	29	Bernstadt
Truppen-Fliegerstaffel 30	408	30	Berlin
Truppen-Fliegerstaffel 31	429	31	Düsseldorf
Truppen-Fliegerstaffel 34	413	34	Stettin
Truppen-Fliegerstaffel 37	424	37[3]	Gross Plowenz
Truppen-Fliegerstaffel 38	426[5]	38	Suwalki
Truppen-Fliegerstaffel 50	428	z. b. V. 50[4]	Silesia
Schlacht-Fliegerstaffel 201	409	21	Munich
Truppenfliegerstaffel Bautzen	*Grenzjäger Flieger Abteilung* 1	12/19	Bautzen
Artillerie-Fliegerstaffel Grossenhain	*Grenzjäger Flieger Abteilung* 2	12/19	Grossenhain

Projected *Vorlaufige Reichswehr Fliegerstaffeln* and Their Parent Brigades (cont.)

Reichswehr designations for *FA*s 405, 411, 421, & 430 have yet to be identified. 421 may have joined *Rw.* Brigade 4.

1. FA 425, redesignated *Truppen-Fliegerstaffel* 1, never joined *Rw*-Brigade 1 in Königsberg. Instead, it merged with *FA* 433 at Alt-Auz 28 July 1919, becoming *Flugpark* Kurland. It joined Bermondt's West Russian *Armee.*
2. *Artillerie-Fliegerstaffel* 108 was transferred to Graudenz, Silesia. However, Germany's acceptance of the Versailles Treaty made that move short-lived, and on 14 August 1919, it was ordered to Jüterbog for demobilization.
3. *Rw*-Brigade 37 existed only on paper from June to 12 July 1919, and was to be based at Schwetz, West Prussia. It wasn't activated and was replaced by a second incarnation, formed in August in Gross Plowenz, West Prussia. *FA* 424 never reported to it, joining the West Russian Army instead. In a confusing move, on 10 September 1919, the General Command of the VI. Reserve Korps placed *TFst* 37 and "*TFst* 1 (*FA* 433)" under *Grufl.* 37[13]. By this time, *FA* 433 had ceased to exist, having been merged with *FA* 425 as *Flugpark* Kurland, or 1. *Flugpark*, 14 August.
4. *Rw*-Brigade z.b.V. 50 was a short-lived formation that existed between June and 12 July 1919, only on paper. *FA* 428 apparently was to become part of this Brigade.
5. *FA* 426 became *Truppen-Fliegerstaffel* 38 before joining the West Russian Army as the 2. *Plastun* Aviation Division.

combined unit remained in Gut Weitenfeld until about 10 September, when it took up quarters at Zokniai, on the outskirts of Schaulen, Lithuania, as part of the West Russian Army.

In late June or early July, the 1. *Garde-Reserve-Division*, along with the three Sachsenberg *Flieger-Abteilungen*, had returned from Kurland to Germany, arriving in Stettin. The division was redesignated *Reichswehr* Brigade 34, with *FA* 413 becoming *Truppen-Fliegerstaffel* 34. *Flieger Abteilungen* 416 and 417 were detached and merged to form *Artillerie-Fliegerstaffel* 108, based at Graudenz. Its stay there was short, because on 14 August, it was ordered to return to its birthplace in Jüterbog for demobilization [14]. Two weeks later, *Truppen-Fliegerstaffel* 34 was ordered to Insterburg to wind down.

Unit Names -- An Exercise in Confusion

Initially, postwar aviation units were organized along the same lines as the wartime German army, with the Prussian army being the dominant force. Bavaria, Saxony, Württemberg and the navy had their own independent units and naming conventions.

One of the most difficult aspects of researching the *Freikorps* era is that units were often known by two, three or more names, all within the span of little more than a year. Reconciling these is essential to fitting together the scattered bits of historical information, and this entails untangling three official and two unofficial systems of unit nomenclature:

1. *Flieger Abteilung* 400 series designations.
As described above, Prussia assigned its new *Flieger Abteilungen* numbers from 400 to 435. This was essentially an extension of the designation system used during the Great War, where fighter squadrons were numbered with one or two digits, two-seater units had numbers in the 200s and *FA*s operating in the Middle East fell into the 300s. The navy and the three former kingdoms of Bavaria, Saxony and Württemberg did not participate in this new numbering system. Only Bavaria had more than three aviation detachments, and these were usually named after their commanding officers or home bases.

The Prussian system had a further distinction -- units were known as "*Flieger Abteilung*" (Flying Detachment) or "*Freiwillige Flieger Abteilung*" (Volunteer Flying Detachment) or "*Grenzschutz* (Border Defense) *Flieger Abteilung*". Use of the word "*Freiwillige*" implied the unit had been raised independently of the army. Although some units may have recruited their personnel without army sponsorship, a large measure of War Ministry support was nearly always involved in organizing and equipping them. This distinction was not always observed in memoirs and histories, so the same unit might or might not be described as "*freiwillige*". In fact, all their personnel were volunteers, and therefore, technically "*freiwillige*". *Grenzschutz* units received that designation because their role was to defend Germany's eastern borders.

2. Provisional *Reichswehr* designations.
Although the *Freikorps* had seen Germany through the difficult early days of 1919, the government wanted a military force organized more in line with its needs and more responsive to its will. The 6 March 1919 creation of the

Provisional *Reichswehr* absorbed the *Freikorps* units into a temporary army intended to serve until the terms of the Versailles Treaty became known. Many of the *Freikorps* units went unchanged, simply assuming new designations; many were combined with other units, some were broken up, their parts distributed to other Reichswehr units, and a few were dissolved.

In this transition, most of the existing *Flieger Abteilungen* were left intact and simply redesignated as *Truppen-Fliegerstaffeln* or *Artillerie-Fliegerstaffeln*. The redesignated units derived their designations from the *Reichswehr* brigade to which they reported, *Truppen-Fliegerstaffeln* taking the brigade number, while *Artillerie-Fliegerstaffeln* added 100 to the brigade number. The 2nd Brigade's air units, for example, were *Truppen-Fliegerstaffel* 2 and *Artillerie-Fliegerstaffel* 102. Not all brigades had aviation units, so there are gaps in the numbering sequence.

For most units, implementation of this reorganization took place by the end of June 1919, but others dragged out over months, so some *Abteilungen* retained their 400-series identities well into the autumn of 1919, coexisting with those bearing Provisional *Reichswehr* designations. This situation occurred primarily in the Baltics. In addition, many documents continued to refer to detachments by their original numbers long after they received their new *Reichswehr* numbers. This may have been done to remind readers of the new *Fliegerstaffels*' origins. Whatever the reason, over one hundred years later, it creates confusion about which unit designation was the officially accepted one.

3. *Reichswehr Fliegerhorst* designations. After 30 September 1919, the German government began replacing the Provisional *Reichswehr* with the 100,000 Man *Reichswehr*, as dictated by the Versailles Treaty. In addition, it began shrinking and consolidating its aviation forces and facilities in anticipation of their final dissolution the following year. Many airfields, *FEA*s and training schools were shut down; the survivors were redesignated *Reichswehr Fliegerhorste* (military airfields), and they became the facilities to which aircraft and personnel of the former *Truppen-* and *Artillerie-Fliegerstaffeln* were assigned. There doesn't appear to be any dedicated numbering convention to identify these aircraft detachments; they were simply components of the *Fliegerhorste*. For example, the three Saxon units consolidated at Grossenhain were named *Flugabteilung der Reichswehr Fliegerhorst Grossenhain*. Other units had similar names.

4. The name the unit chose for itself. Volunteer units, whether land-based or aerial, were noted for their fierce *esprit de corps*, and most chose names that might honor their leader, the location where they originated or some description of their uniqueness. A unit number might have been sufficient for the War Ministry's recordkeeping, but the detachments themselves wanted more distinctive names. Many used their self-assigned names in everything other than official documents and correspondence, and this practice was carried over into the Provisional *Reichswehr*.

5. Unofficial descriptive designations. If all of the above weren't confusing enough, numerous authors, diarists and historians of the time chose names they believed to be descriptive of the unit, even though those names were unlikely to have been used by personnel belonging to the unit nor by any army body to which it reported. Nevertheless, they are now part of the historical record. Appendix 3 is an alphabetical cross listing of many of these names and their official equivalents.

A good example of these multiple identities is the air unit of Division Lettow, formed and led by the hero of East Africa, General Paul von Lettow-Vorbeck. Early recruiting posters sought personnel for *Jagdstaffel* (hunting squadron, a fighter unit, abbreviated *Jasta*) *Löwe* (Lion), implying it would be a fighter unit. However, it became primarily a two-seater unit, and the *Reichswehr* history refers to it, more correctly, as *Flieger Abteilung Löwe*. Some accounts also refer to it as the *Flieger Abteilung* of the Lettow Division. The government designated it *Flieger Abteilung* 422 and finally, *Truppen-Fliegerstaffel* 15, when the division was folded into the 15th Brigade of the Provisional *Reichswehr*.

Adding to the confusion, there were:

West Russian Army Units. After many of the *Freikorps* units in the Baltic mutinied against orders to return to Germany in August 1919, they joined the West Russian Volunteer Army formed by Colonel Pavel Bermondt-Awaloff. Several aviation detachments were among the mutineers and became the core of Bermondt's air force.

Above: *Freikorps* Oberland's DFW C.6991/17 and its crew. Nürnberg, May 1919. (Author's collection)

FA 424/*Truppen-Fliegerstaffel* 37 became the 3rd *Plastun* Aviation Division[15]. In addition, the 4th *Plastun* Aviation Division, composed of the 9th, 10th and 11th *Plastun* Aviation Detachments, was the reconstituted *Kampfgeschwader* Sachsenberg. It drew heavily on the personnel of the original *Kampfgeschwader*, who, after going back to Germany to be demobilized, returned to Kurland to rejoin their old leader under the Russian flag. In German documents, the units known to the Russians as *Plastun* Aviation Detachments were simply known as *Abteilungen,* using the same unit number. To confuse matters further, many German documents continued to refer to the detachments by their original *FA* 400-series or *Reichwehr Fliegerstaffel* numbers, right up to the collapse of Bermondt's Baltic adventure.

Polizei Fliegerstaffeln. When it became known the Versailles Treaty would forbid all military aviation, the Germans made a last-ditch attempt to preserve some semblance of an air force in the form of aviation police. In the words of *Hpt*. Ernst Otto Eberstein,

"The ratification of the peace treaty will probably take place soon, and then the war, so gloriously conducted, is over. With it at the same time, our military aviation. Let's hope that I can get them at least in the guise of police flying."[16]

In addition to handling airport security and monitoring air traffic, the *Polizei Fliegerstaffeln* (police flying squadrons) were expected to be used as law enforcement's aviation arm. Initially, four *Staffeln* were planned, to be drawn from the ranks of selected *Reichswehr Fliegerstaffeln*. The Allied Armistice Commission, after initially seeming to accept the idea, flatly turned it down. This triggered months of head-butting between the Allies and Germans as new *Polizei* units sprang up throughout Germany, only to be banned, in turn, by the Armistice Commission. Despite the Allies' best efforts, some units managed to survive until November 1920, months after the Versailles Treaty specified German military aviation should cease. The earliest of them drew their leaders and personnel from disbanding *Reichswehr* detachments and can reasonably be considered their direct descendants.

Cats and Dogs. As comprehensive as the various unit numbering systems were, a handful of aviation detachments fell outside them. Most of these appear to be provisional groupings not intended to have a long life expectancy. They include the *Sächsische Freiwillige Fliegerstaffel 18,*

Right: *"Rollende* Photo *Abteilung"*. The Sachsenberg's *Geschwader's* photo processing laboratory was housed in one of its railroad cars. (Ed Ferko Collection, 04AEF-B108-A40-Album 40, Box 108, Aschoff Series, History of Aviation Archives, Special Collections and Archives Division, Eugene McDermott Library, The University of Texas at Dallas.)

also known as *Freiwillige Flieger-Geschwader Staffel* 327, and *Fliegerabteilung* Schlieben, formerly *Flieger-Freikorps Horst* Hannover. With the pending dissolution of the German air force 30 September 1919, a number of temporary units and commands were set up to facilitate the demobilization. These included *Hauptflugpark*s 301 and 302 and *Jagd-Fliegerstaffel* 202 [17]. The latter was commanded by *Oblt.* Oskar *Frhr.* von Boenigk, and his command lasted only from 18 to 29 September 1919 [18]. In addition, several land-based *Freikorps* possessed one or two aircraft -- these included the Düsseldorf *Freikorps* and Bavaria's *Freikorps* Oberland, which had DFW C.V 6991/17.

Mother Units. Although the *Freikorps/Reichswehr* aviation detachments weren't direct descendants of the *Luftstreitkräfte* and *Marine Staffeln* of the 1914-18 War, some did have their roots in those units. It was usually a case of flyers reuniting with their old wartime leaders. For example,

Above: Group photo of *FA* 423/*Artillerie-Fliegerstaffel* 116, probably at the unit's home base of Weimar. (Paul Rempe)

no sooner had Sachsenberg's and Osterkamp's *Marine Jastas* demobilized at Jüterbog, than many of their personnel signed on to serve with the new *Kampfgeschwader* Sachsenberg. In a couple of instances, the transition is more direct, undoubtedly engineered by the War Ministry. *FA* 419 was established at Marburg under *Hauptmann* Creydt, final leader of *FA(A)* (*Flieger Abteilung (Artillerie) -- Artillery Flying Detachment)* 216, which had recently demobilized there. *Hauptmann d. R.* Hermann Gröckel was dispatched in January to Kowno (Kaunas), Lithuania, to set up *FA* 425, taking over the surviving aircraft and facilities of *FA(A)* 214, the only *Luftstreitkräfte* unit not driven out of the Baltics by invading Soviet forces. The most direct relationship was that of *FA* 427, built on the remnants of *Eiserne Flieger Abteilung* 8 after its retreat from Riga to Insterburg.

Unit Sizes and Structure

In spite of *Idflieg*'s early projections, *Freikorps* air units, like their land-based counterparts, didn't conform to any fixed size or structure. Far and away the largest was the Sachsenberg *Geschwader*, which arrived in the Baltic with 700 personnel and 70 aircraft. Its size was partly due to the fact that it comprised three distinct *Staffeln*. Moreover, it had to be essentially self-sufficient in a region far from home. In addition to staffing its armory, workshops, photo section, motor pool, and payroll office, the *Geschwader* had two railroad train crews and a machine gun detachment for protection.

Its aircraft inventory also seems nearly twice that required for three *Staffeln*, so it probably included replacements for machines expected to be lost in combat and accidents. Events would bear out the wisdom of this precaution.

More common were *Flieger Abteilungen* with ten to sixteen aircraft. These units were typically part of the so-called "Generals' *Freikorps*", self-contained division-sized units operating within Germany proper. This structure was also common for units assigned to *Grenzschutz Ost* on Germany's eastern borders. *FA* 423, part of *Generalmajor* Maercker's *Freiwillige Landesjägerkorps*, based in Weimar, is a typical and, fortunately, well documented example. On 15 April, it had 19 aircraft (thirteen LVG C.VI plus six Fokker D.VII) and sixteen pilots. A group photo of the *Abteilung* shows 126 men, probably most or all of its full strength.

Freiwillige Flieger-Abteilung Krug is a good example of a unit at the small end of the spectrum. This Bavarian unit had 3 officers, 15 NCOs and 26 enlisted men [19].

Organizational structure, likewise, varied from unit to unit. *Abteilungen* not closely integrated into larger military units had to handle most of their own administrative functions, where smaller units could depend on their

Right: Recruiting poster for the *Garde-Kavallerie-Schützen-Korps*, including Division Lettow. *Jagdstaffel* Löwe is listed fifth from bottom in the left-hand column. This poster also recruits for an aviation unit called *Flieger- Komp.*(anie) Heintz. It may have been combined with *Jasta* Löwe into *FA* 422. (*Das Buch der deutschen Freikorpskämpfer*)

Reichswehr

Freiwillige!

Kommt zum

Garde-Kavallerie-Schützen-Korps

Werbezentrale: Nürnberger Straße 70—71
(Deutsches Künstlertheater)
(Heuduck)

Garde-Kavallerie-Schützen-Division
Werbezentrale: Berlin-Wilmersdorf, Uhlandstr. 155
Ecke Ludwigkirchstraße

Dragoner-Regt. 8, Berlin W 15, Uhlandstraße 155
Lehr-Inf.-Regt., Steglitzer Str. 29, Charlottenschule
Freikorps Lützow, Berlin W, Bellevuestraße 14
Husaren-Regt. 11, Berlin W 15, Uhlandstraße 155
Ulanen-Regt. 5, Berlin W, Uhlandstraße 155
Kürassier-Regt. 4, Charlottenburg, Hochschule für Musik
Jäger-Regt. z. Pferde 2, Berlin SW 11, Bernburger Straße 3
Jäger z. Pferde 6, Charlottenburg, Knesebeckstraße 4
Detachement Küntzel, Friedenau, Rheingaustraße 7, Berlin, Flottwellstraße 3, Zimmerstraße 62, Luisenstraße 31a
Freikorps von Klewitz, Charlottenburg, Kantstraße 182, Berlin W, Kaiserallee 1—12, Joach.-Friedr.-Gymnasium
Freikorps Schleswig-Holstein, Berlin W 15, Uhlandstraße 155. Wünsdorf, Baracke 27
3. Garde-Feld-Artillerie-Regt., Berlin, Scharnhorststraße 23
Schw. Feldhaubitz-Batl. Braune, Schöneiche bei Zossen
Brigade von Taysen, Berlin W, Potsdamer Str. 22a, Königgrätzer Str. 105, Wilhelmstr. 102 (Verw. d. Palais des Prinzen Friedrich Karl von Preußen)
1. Pionier-Komp., Alt-Glienicke, Gastwirtschaft von Ebel, Friedrichstraße 66, Berlin W, Uhlandstr. 155
Staffelstab und Kolonnen, Marienfelde, Berliner Straße, Ecke Dorfstraße, Berlin W, Uhlandstr. 155
Fernsprecher- und Funkerabteilung, Berlin W 15, Uhlandstraße 155
Radfahrer-Komp., Berlin W 15, Uhlandstraße 155

Division Lettow
Werbezentrale: Charlottenburg, Hotel Fürst Bismarck
(Am Knie) Berliner Straße 160

Schutztruppen-Regt. 1, Grunewald, Bismarckallee 29
Schutztruppen-Batterie 1, Grunewald, Bismarckallee 29
Schutztruppen-Pionier-Abtlg. 1, Schlachtensee
2. Marine-Brigade, Wilhelmshaven und Zossen, Stammlager
3. Marine-Brigade, Kiel-Wyk, Deckoffizierschule, und Berlin, früheres Königl. Schloß
Eskadron Krossa, Neu-Glienicke b. Potsdam
Radfahrer-Kompanie Redlich, Kohlhasenbrück bei Potsdam
MG.-Abtlg. von Elterlein, Schmargendorf, Schützenhaus
Freiw. Feldartl.-Regt. Nr. 15 (Regt. Osiander), Jüterbog, Neues Lager
Flak-Abtlg. v. Lettow, Wannsee, Kleine Seestraße 3
Schwere Feldhaubitz-Abtlg. v. Lettow, Jüterbog, Neues Lager
Pionier-Abtlg. v. Lettow, Wannsee, Kleine Seestr. 3
Minenwerfer-Komp. v. Lettow, Wannsee, Kleine Seestraße 3
Div.-Nachrichten-Kommandeur v. Lettow, Wannsee, Kleine Seestraße 3
Jagdstaffel Löwe, Bork b. Potsdam
Staffelstab Div. v. Lettow, Jüterbog, Fußart.-Schießschule
Sanitätskomp. v. Lettow, Jüterbog, Fußart.-Schießschule
Div.-Kraftfahrtruppe, Wannsee, Königstraße 68
Leichter Panzerkraftwagen-Zug, Wannsee, Königstraße 68
Sturm-Lehr-Regt., Nürnberger Straße 70/71
MG.-Scharfschützen-Korps Prey, Berlin W, Kurfürstenstraße 97
Sturmbataillon Schmidt, Berlin W, Nürnberger Straße 63
Radfahrer-Komp. v. Koeller, Charlottenburg, Kantstraße 78
Eiserne Eskadron, Berlin W, Kurfürstendamm 229, hpt. r.
Flieger-Komp. Heintz, Charlottenburg, Hardenbergstraße 27 (altes Café Josty am Zoo)
Pionier-Bataillon, Wannsee
MW.-Sturmdetachement Heuschkel, Charlottenburg, Kantstraße 6
Sanitäts-Komp., Grunewald, Delbrückstraße 20
Flakabteilung, Lichterfelde, Garde-Schützen-Kaserne, Zimmer 104
Eisenbahnbau-Komp. Soller, Grunewald, Delbrückstraße 20
Freikorps Schwarze Jäger, Berlin W 15, Uhlandstraße 155
Staffelstab GKSK., Nürnberger Straße 63
Funker-Abteilung GKSK., Berlin NW, Alt-Moabit, Meierei C. Bolle
Fernsprech-Abteilung GKSK., Berlin NW 62, Perleberger Straße 26

Aerzte:
Korpsarzt, Berlin W 15, Nürnberger Straße 3 (Hof links)

Zeitfreiwillige

Berlin W 50, Kurfürstendamm 243
Technische Zeitfreiwillige: Technische Abteilung, Lichterfelde, Moltkestraße 52

parent organization for those services. *Flieger-Abteilung* Krug, mentioned above, was probably one of these. With only 44 personnel, it must have been dependent on its parent organization for day-to-day administrative services like financial, medical and clerical support.

Fortunately, a few organizational records have survived. Someone had the foresight to photograph a wall poster breaking down *FA* 423's organization, which was divided into eight sections, also known as *Abteilungen*:

The poster can be translated:

***Abt.* I. Administrative**

Ia. Scheduling, personnel quarters, promotions, incoming mail, handling of *Abteilung* correspondence, motor vehicle driving licenses & regulations

Ib. Organizational correspondence, pay, rosters, passes, travel tickets & ID documents

Ic. Court cases, recruitments and discharges & estates of deceased

Abt. II. Finance Cash management, accounting, payroll, food, clothing and equipment, personnel affairs & craftsmen.

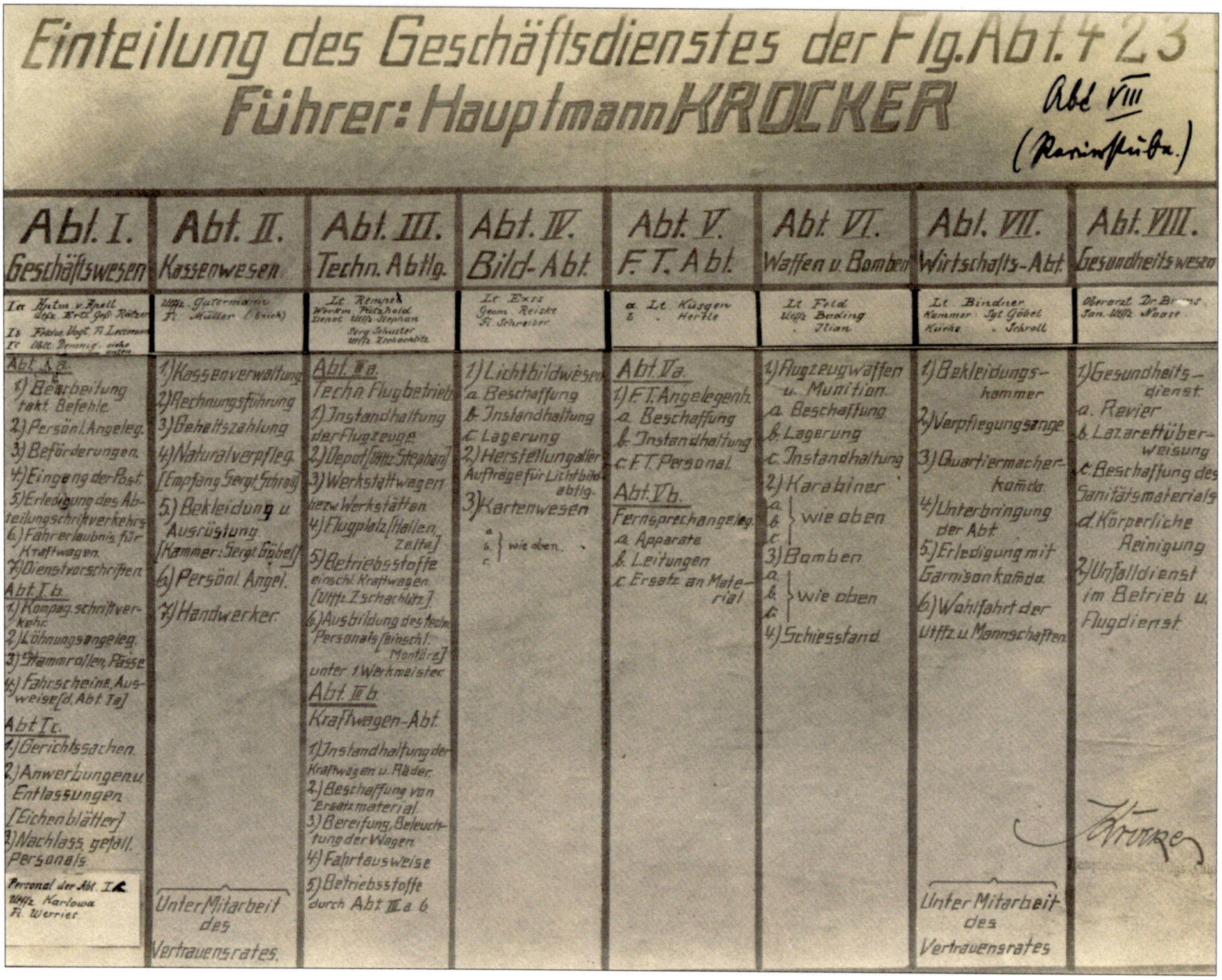

Above: *FA* 423 poster delineating the unit's sections and their responsibilities. (Paul Rempe)

***Abt.* III. Technical IIIa.**

Technical flight operations - aircraft maintenance, warehouse, workshops & repair vehicles, hangars, camping, fuels and lubricants & training of technical personnel

IIIb. Motor vehicles - vehicle and motorcycle maintenance, replacement parts procurement, vehicle tires and lighting, travel ID passes and operating materials

***Abt.* IV. Photography**

1. Photograph production
2. Fulfillment of orders for photographs
3. Map production

***Abt.* V. Radio Telegraphy**

Va. Radio Telegraphy matters - procurement, maintenance, personnel

Vb. Telephone matters - equipment, cables & replacement of material.

***Abt.* VI. Weapons & Bombs**

1. Aircraft weapons & munitions - procurement, storage & maintenance.
2. Carbines - procurement, storage & maintenance.
3. Bombs - procurement, storage & maintenance
4. Shooting range

***Abt.* VII. Economic**

1. Uniform storage
2. Dining matters
3. Quartermaster commandant
4. *Abteilung* housing
5. Dealings with garrison commandant
6. Elective travel for NCOs & enlisted men

***Abt.* VIII. Health Care**

1. Health service - district, hospital transfers, procurement of medical supplies & personal hygiene
2. Accident service in operations and flight activities

Above: *Kraftwagen Abteilung* (Motor Pool) of *FA* 426 -- Aircraft may have ruled the skies, but only with a great deal of support from the people on the ground. (Author's collection)

This document is especially valuable, because it provides an insight into the size and complexity of the organization and number of personnel required to keep a dozen or so aircraft in the air. It shows the organizational structure of the *Freikorps* and *Vorlaufige Reichswehr* air units was based on that of the *Flieger Abteilungen* of the Great War.

Overview of Aviation Operations and Aircraft

None of the aviation detachments were free-lancers; all were assigned to support ground-based forces. They were to find themselves engaged in a much different war than they had just left. Gone were the swirling fighter dogfights, because little or no aerial opposition existed. The brutal artillery duels of the Great War were also a thing of the past, and with them, most artillery spotting. The flyers were tied almost exclusively to their brothers on the ground, providing observation and tactical support in the Baltic and along the Polish Front.

Within the *Reich*, the flyers were used principally for reconnaissance, dropping propaganda leaflets over Red-controlled cities, courier service and transporting critical mail to avoid struck postal facilities. They were rarely allowed to fire their guns in anger, lest they alienate the general population. More often, they would zoom around, loosing a few shots into the sky (euphemistically called "demonstration flights") to remind their opponents of their ability to inflict harm. Given these tactical requirements, the distinction between *Artillerie-* and *Truppen -Fliegersatffeln* seems to be meaningless.

A few air-to-air engagements took place on the Polish Front and in the Baltic; otherwise, the flyers' eyes were focused on the ground, and with good reason -- the great majority of German aircraft and crew losses were due to ground fire. For the most part, the mission dictated the choice of aircraft. Two-seaters -- observation and ground assault machines -- made up the bulk of the *Freikorps* and Provisional *Reichswehr* aviation units' strength. A few single-seat fighters were carried on strength by most units, but they were

Left: The LVG C.VI saw wide use in the air units of the *Freikorps* and Provisional *Reichswehr*. This shot of an *FA* 424 pilot posing with one provides an excellent look at its nose details, including its Astra propeller. (Ed Ferko Collection, 04AEF-B128, Box 128, Aschoff Series, History of Aviation Archives, Special Collections and Archives Division, Eugene McDermott Library, The University of Texas at Dallas.)

used primarily for courier service and, in the case of the Baltic and *Grenzschutz Ost*, protection of the two-seaters. This was reminiscent of 1915, when Fokker *Eindeckers* were parceled out to two-seater units to protect the slower, less maneuverable C-types.

Hundreds of aircraft served with the *Freikorps* and *Reichswehr*, but there are no complete inventories of them. Partial lists of a few units have survived, and along with photos of some of the *Abteilungen*, we can draw some general conclusions. Among two-seat general reconnaissance aircraft, the LVG C.VI seemed to be used in the greatest numbers. For close support work, the Halberstadt CL.IV was favored, followed by the Hannover CL.II, CL.III and CL.IIIa.

In spite of the Allies' demand for the handover of all of Germany's Fokker D.VIIs, enough examples remained in Germany to serve with many units, including *FA*s 410, 422, 423, 424, 431, 434, *FA* Haas and *Geschwader* Sachsenberg. Sachsenberg made extensive use of Junkers D.Is and CL.Is, in addition to numerous

Above: Halberstadt CL.IV (Rol.) 9494/18 of *FA* 406/*Truppen-Fliegerstaffel* 27. The CL.IV was widely used by the postwar *Abteilungen*, and based on photographic evidence, Roland-built examples predominated. (Author's collection)

Halberstadt CL.IV(Rol.) 9494/18
FA 406 / *Truppen-Fliegerstaffel* 27

other types. Elsewhere, only *FA*s 406, 427, 430 and 431 seem to have had any Junkers aircraft -- CL.Is and J.I armored trench strafers -- on strength. Other aircraft types known to have been used were:

AEG G.IV
Albatros B.II, C.I, C.III, C.XII, C.XV, D.III, D.Va, J.II
DFW C.V
Fokker D.VIII
Friedrichshafen G.IIIa
Halberstadt CL.II, C.V
Hannover CL.IV, C.V
LVG C.V & P.I

Above: *Lt. d. R.* Richard Kraut, *Lt. d. R.* Alfred Brocke, and *Ltn.* Heydt of *FA* 434/*Truppen-Fliegerstaffel* 9, Graudenz. Kraut was one of the rare postwar flyers to put a personal marking on his machine. His "K" on a black rectangle was inspired by his *Jasta* 63 insignia, interlaced "R" & "K" on a black background. (Author's collection)

Above: *FA* 410/*Artillerie-Fliegerstaffel* 120 is a typical example of an aviation unit that fielded an assortment of different aircraft types in the second half of 1919. In this photo, L to R, Fokker D.VIII, Pfalz D.IIIa 6014/17, Fokker D.VII (OAW) 8453/18, Hannover CL.II or CL.III, DFW C.V and LVG C.VI. (Michael Seidel)

Pfalz D.IIIa
Roland D.VIb
Rumpler C.I, C.IV
SSW D.IV
Zeppelin-Stakken R. XIV or R.XV.

The diversity of aircraft types seemed to increase in the second half of 1919; by this time new aircraft

Above: The oversize military serial repainted on this former *Jasta* 61 Pfalz D.IIIa is almost an ironclad guarantee it had done time with a training school before its career ended with *Freikorps FA* 410/*Artillerie-Fliegerstaffel* 120. The chordwise arms of the upper wing crosses extend to the leading and trailing edges, and there appears to be a chordwise white band to the left of the wing's centerline. (Michael Seidel)

production had wound down, so the flying detachments had to draw their inventory from the abundance of castoffs from demobilized *Grenzschutz Ost* units, airbases and training facilities.

Bermondt-Awaloff's West Russian Army was the dumping ground for a great number of these aircraft in the late summer and autumn of 1919, some units receiving four or five aircraft for every pilot. For Germany, it was a choice between sending them to the Baltic, where they might be of use to brethren still fighting there, or eventually yielding them to the Inter-Allied Aeronautical Commission of Control and the scrap heap.

Aircraft Colors and Markings

The great majority of the aircraft used by the *Freikorps* and *Reichswehr* units bore standard factory paint schemes without any of the colorful embellishments common on the Western Front. There's no evidence personal markings were frowned upon by the new *Reichs* government, though it's no stretch to assume the commanders of some of the *Freikorps* formations might not have been keen on having gaudily painted aircraft under their command.

A more likely reason for the plain vanilla schemes is that these were primarily 2-seater units, like the *Flieger Abteilungen* of the Great War, where a simple identifying letter or numeral might be a plane's only distinctive marking. Nevertheless, unique markings were applied to some aircraft, and this book contains color profiles of as many of these as we could find and confirm.

A large number of former training school machines were doled out to air detachments in the second half of 1919, often bringing with them outdated markings like *patée* crosses (this is the classic "iron cross" shape, where the arms flare outward from the center to the tip) and the broad-bordered transitional crosses, along with the narrow-armed *Balkenkreuz* (literally, "beam cross", the straight-armed cross that became standard in the Spring of 1918). Some trainers were uncamouflaged or had repainted fuselages and military serials. Another feature of school machines was the frequent omission of the fabric wheel spoke covers. Many of these aircraft found their way into the *Freikorps*/*Reichswehr* inventory.

Aircraft serving with Bermondt's West Russian Volunteer Army carried a mixture of markings; those attached to the German Legion and Iron Division retained standard German insignia, while those attached to the Russian components of the army, the Keller and Wirgolitsch *Korps,* were to be identified with Russian markings, although there's some doubt this was fully carried out .

Balloon Units

If the postwar military histories paid little attention

Above: Personnel of *Luftschiffer Abteilung* 2 and a freight train cargo of hydrogen cylinders at Brietzig, Pommerania. (Greg VanWyngarden)

to the flying forces, they almost totally ignored the balloon troops. Consequently, information about them is sparse.

Captive balloons were a fixture on the Western Front, where they allowed observers to spot the fall of artillery fire and to keep watch on enemy activities well behind the trench lines. The static nature of the Front allowed balloon companies to establish semi-permanent stations, so unless an offensive forced their relocation, they tended to remain in the same places for long periods of time.

Postwar, Germany's military situation was generally more mobile. The need for observation balloons diminished but didn't completely disappear, and they were used in the Baltic, on the Polish front, and even in the *Strassenkämpfe* (street fighting) in Germany's cities.

At the end of February / beginning of March, one balloon company was operating in Kurland -- *Freiwillige Ballonzug* (Volunteer Balloon Company) 102, part of the 1. *Garde* (Guards) Reserve Division[20]. It was later transferred to the General Command of the VI. Reserve Corps and replaced by a new unit, *Freiwillige Ballonzug* 108, commanded by *Rittmeister* Seer[21]. When the division was ordered to stand down in preparation for its transfer to *Grenzschutz Ost*, the balloon company was transferred to the Iron Division to make it available for the 22 May assault on Riga[22].

Feldluftschiffer (Field airship) *Abteilung* 101 served with *Grenzschutz Ost*. It was commanded by *Hpt.* Erich Werner from 11 February until 11 July 1919, when he was transferred to command *Luftschiffer Abt.* 17. Under his command were fourteen NCOs and enlisted men[23]. This seems to have been far less than the normal personnel complement of a balloon company, so they might have been the residual cadre of a unit about to be demobilized.

Another balloon unit reporting to *Grenzschutz Ost* was *Grenschutz-Feldluftschiffer* 27, based in Bentschen. It was present there 9 September. *Luftschiffer Abt.* 2 was stationed in Brietzig, Pommerania in the spring of 1920. This might have been its final station before disbanding.

Balloon units also participated in the battle of Munich. *Abteilung* von Bogendorfer and *Abteilung* Probstmayr, components of *Gruppe* West, each operated a balloon section [24].

Above: Fokker D.VII (OAW) 8448/18, *FA* 423, Weimar. Plenty of D.VIIs remained in Germany even though the Allies demanded "all of the D 7's". (Paul Rempe)

When Did the German Air Force Actually Go Out of Existence?

All of 1919's mobilization and reorganization activity raises the question of when Germany's air force was supposed to be dissolved, let alone when it actually ceased to operate. It's commonly assumed this occurred immediately after the 11 November 1918 Armistice. Although Article 5 of the Armistice terms did require Germany to surrender "1,700 aeroplanes (fighters, bombers - firstly, all D 7's and night bombing machines)" [25] to the Allies, it made no mention of the rest of Germany's sizeable aircraft inventory, much of which was equipment returning from the Front or aircraft caught in the supply pipeline from the manufacturers to the Front. In addition, there were hundreds of machines at storage depots and training schools, along with aircraft in various states of completion at the factories, committed under production orders that hadn't been cancelled. Initially, as mentioned above, much of that inventory was being kept from the government by the Soldiers' and Workers' Councils.

In addition, although the armistice required all German military forces, including air forces, to evacuate occupied territories, the west bank of the Rhine and a 30 km radius of the Rhine bridgehead cities of Cologne, Mainz & Coblenz, it made no mention of German aviation units and equipment located elsewhere. They were free to continue operating. While peace negotiations ground on at Paris, the Armistice had to be renewed three times -- 13 December 1918, 16 January 1919 and 16 February 1919. None of those renewals altered the original Armistice terms.

On 28 June 1919, Germany reluctantly signed the Versailles peace treaty. Conflicts inherent in wording between some of the Articles of the treaty created huge loopholes that allowed the German air force to continue to operate for months. For example, Article 198 limited Germany to no more than 100 unarmed seaplanes or flying boats to be used exclusively in assisting the Allies in minesweeping the North and Baltic Seas. After 30 September 1919, the date the treaty was supposed to take effect, these aircraft were to be grounded, along with the *Reichswehr* flying detachments, a much larger air force. However, this was undercut by Article 199, which did not call for complete demobilization of the air force and its

Above: This well-known photo of the "Döberitz *Jagdstaffel*", most likely *Artillerie-Fliegerstaffel* 103, dates from 15 August 1919, less than two weeks before the unit was to be disbanded. Fokker D.VIII #5 displays a tactical number characteristic of the *Marine Jagdstaffeln* in Flanders. (Greg VanWyngarden)

Above: This damaged photo of a flying club in Gräfendorf, a village in Lower Franconia, east of Frankfurt a. M., seems to show where #5 found a new home. Lucky club -- it was blessed with highly advanced equipment. Dispersion to obscure locations was a common practice to conceal aircraft from the Armistice Commission inspectors. (Greg VanWyngarden)

Facing Page: Coincidentally, this ad touted the Fokker D.VIII as *"Das ideale Sport-Flugzeug"*. Could this have been the inspiration for the Gräfendorf move? (Author's Collection)

Fokker-Werke
G.m.b.H.
Schwerin i.M.
Das ideale
Sport-Flugzeug.

Above: Last stop for the German air force -- the scrap heap. (Greg VanWyngarden)

personnel until two months after the treaty came into force. As it turned out, ratification didn't take place until 10 January 1920.

So although the Inter-Allied Aeronautical Commission of Control (IAACC) was given free rein to travel throughout Germany and inventory its aviation assets, the Commission didn't have the authority to enforce their destruction until 10 March 1920 [26]. Consequently, diminishing numbers of German military aircraft continued to operate legally until then. A handful continued past that date, survivors in a cat-and-mouse game between Germany and the Armistice Commission.

Nevertheless, the German air arm shrank steadily between 28 June 1919 and 10 March 1920 because much of it was no longer needed. Germany's acceptance of the territorial concessions in the Versailles Treaty reduced the likelihood of a border war with Poland, making much of *Grenzschutz Ost*'s forces redundant. This led to orders for the demobilization of about 25 of the *Reichswehr*'s aviation units in the late summer and autumn of 1919.

One general order, dated 28 August 1919 [27], specified demobilization sites for most of the air force's surviving units. Some *Staffeln* were wound down, as planned, within a few weeks, but others continued a twilight existence for months, occasionally locating to other bases. A good example of this was *Artillerie-Fliegerstaffel* 115, formerly *FA* 420. The August orders called for it to return to its home base at Döberitz for dissolution, customary practice for military formations. However, when an illicit trainload of surplus aircraft and engines bound for the Baltic was intercepted at Schneidemühl, a *Flugzeugmeisterei* memo [28] called on the local air unit to take custody of it. That unit? *Artillerie-Fliegerstaffel* 115. The date of the memo? 18 February 1920.

Although the number of military aircraft in use had declined sharply in the second half of 1919 and into 1920, they hadn't necessarily gone to the scrap heap. As described elsewhere, many were slipped into the Baltics to replenish Bermondt-Awaloff's West Russian air force. Others were placed in storage, with and without engines, in hopes they could be used by the *Polizei-Fliegerstaffeln*, converted to commercial use or sold to foreign countries for much-needed cash. In a session of the House of Commons 15 April 1920, Secretary of State for War Winston Churchill testified the Air Ministry estimated Germany still had 15,248 aircraft capable of being made ready for military use in a short period of time [29]. As difficult as it may be to believe, this claim agrees closely with a 30 September 1920 report by the Armistice Commission that 15,043 aircraft and 28,125 aero engines had been inventoried and surrendered for destruction following ratification of the Versailles Treaty [30].

The funeral pyres had yet to be ignited.

Endnotes - Chapter 2

1. http://www.frontflieger.de/2-kofl.html.
2. "*Bauaufstellungen* in Januar 1919, *Inspektion der Fliegertruppen Nr.* 82/1&19 *Abt.*Ia.", Charlottenburg, 9.1.19. *Deutsche Sozialiste Republik Zentralrat* Archives 48.
3. Hannes Taeger personal correspondence, 2 October 2022.
4. "*Bauaufstellungen in Januar 1919, Inspektion der Fliegertruppen Nr.* 82/1&19 *Abt.*Ia".
5. *Reichskriegsministerium, Die Rückführung des Ostheeres, Darstellungen aus den Nachkriegskämpfen deutscher Truppen und Freikorps, Erster Band*, E. S. Mittler & Sohn, Berlin, 1936, p. 154.
6. "*Bauaufstellungen in Januar 1919, Inspektion der Fliegertruppen Nr.* 82/1&19 *Abt.*Ia.
7. *U.S.P.* = *Unabhängige Sozialdemokratische Partei Deutschlands* = Independent Social Democratic Party of Germany. This was a Marxist Party that split off the mainstream Socialist Party of Germany because the latter supported Germany's participation in World War I.).
8. Gotha was located within the XI. *A. K.'s* jurisdiction.
9. Maercker, General Ludwig R. G., *Vom Kaiserheer zur Reichswehr, Ein Beitrag zur Geschichte der deutschen Revolution, Verlag von* K. F. Koehler, Leipzig, 1921, p. 115.
10. ibid., p. 370–5.
11. "*Bauaufstellungen in Januar 1919, Inspektion der Fliegertruppen Nr.* 82/1&19 *Abt.*Ia."
12. Tessin, Georg, *Deutsche Verbände und Truppen* 1918 – 1939, Biblio *Verlag*, Osnabrück, 1974, pp. 108-143.
13. *Grufl.* 37 existed within VI. *RK* as early as 20 May 1919.
14. *Armee-Tagsbefehl* 145. Ia, *Kofl., Ober-Kommando Grenzschutz Nord*, Bartenstein, 14 August 1919.
15. *Plastun* is a Russian term describing an unmounted Cossack scouting or sentry force. Its application to aviation appears to be solely due to Bermondt-Awaloff's obsession with the Cossacks.
16. "*Kriegstagebuch*" *Hpt.* Ernst Otto Eberstein, p.18.
17. *Inspektion der Fliegertruppen Tgb. Nr.* 1283/8.19 *Abt.* I. II. *Ang.*, Berlin, 28.8.1919.
18. Hildebrand, K. F., *Die Generale der deutschen Luftwaffe* 1935–1945, Biblio *Verlag*, Osnabrück, 1990, Vol. I, pp.95–6.
19. "*Flieger in den Nachkriegskämpfen*", *Reiter gen Osten, Jahrgang 7, Folge 10, Seite* 14 (October 1936).
20. *Reichskriegsministerium, Der Feldzug im Baltikum bis zur zweiten Einnahme von Riga. Januar bis Mai* 1919. *Darstellungen aus den Nachkriegskämpfen deutscher Truppen und Freikorps, Zweiter Band*, E. S. Mittler & *Sohn*, Berlin, 1937, p. 146.
21. ibid., pp. 150.
22. ibid., pp. 152, 157.
23. *Freikorps-Grenzschutz-Ost-Feldluftschiffer-Abteilung-*101-*Kommandeur-Nachlass*/362126655143?ssPageName=STRK%3AMEBIDX%3AIT&_trksid=p2055119.m1438.l2649. eBay.de, 2017.
24. *Reichskriegsministerium, Darstellungen aus den Nachskriegkämpfen deutscher Truppen und Freikorps. Die Niederwerfung der Räterschaft in Bayern* 1919, *Vierter Band*, E. S. Mittler & *Sohn*, Berlin, 1939, pp. 202, 204.
25. "Terms of the Armistice with Germany", Catalog Reference WO 106/412, nationalarchives.gov.uk. https://www.nationalarchives.gov.uk/pathways/firstworldwar/transcripts/aftermath/armistice_terms.htm.
26. Peace Treaty of Versailles Articles 159–213 Military, Naval and Air Clauses net.lib.byu.edu/~rdh7/wwi/versa/versa4.html.
27. *Inspektion der Fliegertruppen Tgb. Nr.* 1283/8.19 *Abt.* I. II. *Ang.*, Berlin, 28.8.1919.
28. *Flugzeugmeisterei Abt.* I, B. *Nr.* 3004/2.20 *Versand*, Berlin, 18.2.1920.
29. "German Guns and Aeroplanes", House of Commons Debates, 15 April 1920, Vol. 127, cc 1821, Hansard 1803–2005, https://api.parliament.uk/historic-hansard.
30. "Destruction and Delivery Aeroplanes and Engines", Dorand Report, I.A.A.C.C., Ed Ferko Collection, History of Aviation Archives, Special Collections and Archives Division, Eugene McDermott Library, The University of Texas at Dallas.

Operations within Germany

Above: Halberstadt CL.IV "Els'chen" of *Artillerie-Fliegerstaffel* 103, formerly *FA* 432. The two men lying in the grass are wearing sweaters bearing patches reading Fl J 103. This, together with their shorts and knee socks, suggests they belonged to the *Staffel's* football team

When *Idflieg* began reconstituting the air force, it parceled out the new *Flieger Abteilungen* according to perceived need. *Grenzschutz Ost* got the lion's share, about two-thirds of the new units. The balance was split between the Baltic forces and *Abteilung* Lüttwitz, the command responsible for operations within Germany proper. Initially, this amounted to 5 *Abteilungen* -- *FA* 419, 420, 421, 422 and 423. FA 432 would join them shortly. The *Freikorps* units they reported to were established in the Berlin area, with their first task clearing the capitol of revolutionary elements. That done, they would set up operations throughout Germany to be a short march to localities threatened by revolutionaries. Strikers frequently shut down rail service, often to interfere with military operations, so the *Freikorps* units were set up to travel by road to revolutionary hot spots. Their *FA*s were assigned as follows:

***FA* 419** -- Established at Marburg, Hesse, the *FA* transferred to Döberitz, where it worked up to operational status and was attached to the 3. *Landesschützen Brigade.* From late January till late February, it supported the Brigade in operations to regain control of the North Sea port cities.

***FA* 420** -- Established at Potsdam, where it joined the 7. *Garde-Kav.-Schutzen-Division.* It was still in formation at the time of the January fighting in Berlin, but took an active role in the government's campaign against the March strike and uprising.

***FA* 421** -- Based at Johannisthal and assigned to the *Deutsche Schutzen-Division.* It saw its first action in the March Berlin fighting, before being transferred to Schleissheim, Bavaria, at the beginning of May.

***FA* 422** -- Part of Division Lettow, it was formed at

Halberstadt CL.IV (Rol.)
***FA* 432 / *Artillerie-Fliegerstaffel* 103**
***Freikorps* Hülsen, Potsdam, July 1919.**

Potsdam. Originally the division was part of the *Garde-Kav.-Schutzen-Korps,* along with the 7. *Garde-Kav.-Schutzen-Division.* Based at Ludwigslust, the *Abteilung* was transferred to Bavaria late April for the battle of Munich.

***FA* 423** -- Established in Paderborn while the *Freiwilliges Landesjägerkorps* was helping put down the Berlin Spartacist rising. It moved to Luckenwalde, south of Berlin, where it worked up to operational status and then joined the *Landesjägerkorps* at Weimar.

***FA* 432** -- Formed in Potsdam and assigned to *Freikorps* Hülsen. It participated in the March fighting in Berlin.

Friekorps Hülsen also oversaw an additional unit -- *Flieger-Frei-Korps Horst* Hannover at *FEA* 5, Hannover [1], established 12 January 1919. By 1 March, it had been renamed *Fliegerabteilung* Schlieben after its commander, *Hpt.* Richard von Schlieben, former leader of *Jasta (Jagdstaffel* -- Fighter Squadron*)* 21. It was apparently intended to support government forces in the Hannover-Braunschweig vicinity. Hannover is 250 km from Berlin, an unusually long distance over which to supervise a subsidiary unit, but at that early date, Hülsen's other aviation units were all forming or engaged in the Berlin area. *FA* Schlieben may have been the only government flying formation in northwest Germany at that time.

Above: Postcard photo of Berlin's shell-pocked *Polizei-Präsidium* (police headquarters) after the 6 March 1919 fighting. This was one of the rare occasions where *Freikorps* aviators were permitted to bomb a target within Germany. (Author's collection)

Berlin and Northwestern Germany

The first order of business for the *Freikorps* was restoring order in Berlin. In the first week of January, huge crowds of Spartacists and Independent Socialists gathered in the streets, anticipating an overthrow of the Ebert government and jeopardizing the upcoming elections for the National Assembly. *Freikorps* units were forming outside the city; for the most part, their aviation units had yet to materialize, and for a brief period, the revolutionaries were masters of the skies. A few Red aircraft, based at the Adlershof - Johannisthal airfield, were busy scattering propaganda leaflets over the city. On 10 January, the first *Freikorps* detachments were turned loose, seizing control of various districts. Three days later, *Oberst* Gerstenberg's 3. *Landesschützenbrigade* took control of the Johannisthal airfield, despite the revolutionaries' threat to put the facility to the torch[2]. By the 15th, the government was firmly in control of Berlin and its suburbs, allowing the elections to take place on the 19th.

With Berlin secure for the time, General Lüttwitz was free to disperse his units to stamp out revolutionary uprisings elsewhere. Northwest Germany, with its port cities of Bremen, Bremerhaven, Wilhelmshaven, Hamburg and Emden, had been a hotbed of revolutionary activity since the Kiel mutinies that toppled the empire. Revolutionaries in control of their port facilities had a stranglehold on German commercial shipping. The ports were overdue for return to government control, and the proclamation of a Red republic in Bremen 10 January pushed it to the top of the list.

On 28 January, the Gerstenberg Division was selected to lead the attack on Red Bremen. *Hauptmann* Creydt's *FA* 419 was transferred from the 3. *Landesschützen Brigade* to provide reconnaissance and liaison service. Its flyers served with the Division throughout the campaign, beginning with dispersing warning leaflets over Bremen, followed by conducting reconnaissance during the city's conquest on 4 February. Bremerhaven and Cuxhaven were reconnoitered on 9 and 10 February, where revolutionary defenses were found to be nonexistent. During the retaking of Wilhelmshaven the night of 20 - 21 February, *FA* 419 monitored sea and land traffic into and out of the city. The campaign wound up in early March, when, following reconnaissance by *FA* 419, detachments of government forces occupied the Frisian Islands of Norderny and Borkum and rooted Spartacist workers out of Norderny's airship base [3].

No sooner had the government regained control of the North Sea port cities than violence erupted again in Berlin. The previous two months had taught the Spartacists they could not best the *Freikorps* in armed combat, so on 3 March, they called for a nationwide general strike. It was intended to be nonviolent, but revolutionary firebrands, opposed by a government determined to crush the strike by any means necessary, insured there would be bloodshed. *FA* 420, which had formed too late to participate in the January Berlin fighting, took an active role as part of the 7. *Garde-Kavallerie (Schützen)* Division. On 5 March, four of the *Abteilung*'s planes dropped ammunition, rations and bandages to *Freikorps* troops pinned down in the fighting around the city's police headquarters, which had been occupied by the Spartacists [4]. The next day they joined an all-out assault on the headquarters, dropping bombs on the building, as ground units attacked it with artillery, mortars and machine guns.

The revolutionary *Marine-Landflieger Abteilung* operated a few aircraft during this time, primarily for carrying messages and leaflet dropping. As far as is known, there were no air-to-air encounters between them and the *Freikorps* flyers. On 16 and 17 March, with Berlin back under government control, the *Freikorps* seized the remaining airfields at Neukölln and Oberspree, putting the Spartacists out of the flying business.

Southeast of Berlin, in Cottbus, revolutionaries had re-established themselves at the airfield, after having lost control of it to government troops at the end of February. They flew a few sorties, bombing local targets, until government forces retook the base.

Elsewhere, things were relatively quiet, but on 6 March, Division Lettow's *FA* 422 lost one of its star pilots, *Ltn. d. R.* Werner Preuss, former commanding officer of *Jasta* 66, with a tally of 22 victories and the *Pour le Mérite*. He died in an accidental crash at the military exercise field at Rendsburg, near Kiel, Schleswig-Holstein [5].

Maercker's Rifles and Central Germany

Historians of the period generally agree the finest of the *Freikorps* / Provisional *Reichswehr* formations was the *Freiwilliges Landesjägerkorps*. Nicknamed Maercker's Rifles after its founder, *Generalmajor* Ludwig Maercker, it was the prototype for other large *Freikorps*, and arguably, the best run and most effective of them all. It was one of the first of the large, so-called Generals' *Freikorps* and participated in crushing the Spartacist revolt in Berlin the first half of January 1919. At the time, its aviation detachment, *Flieger Abteilung* 423, with an establishment of 9 aircraft, was forming at the Paderborn *Geschwaderschule* (Squadron training school) under its leader, *Hauptmann* Victor Krocker, veteran of six wartime *Flieger Abteilungen*, most recently *FA(A)* 211.

On 22 January, the *Abteilung* moved to the Luckenwalde airfield in the Teltow district south of Berlin to complete its preparations for service. Its first loss came on the 28th, with the death of observer *Ltn.* Gerhard Pape [6]. The following month, it moved to Weimar to rejoin its parent unit, which had gone there

Above: *FA* 423 photo of the National Theater at Weimar, where the new republic's constitution was drawn up. (Paul Rempe)

30 January to secure the city. The *Reichs* government had decided to convene its National Assembly there to draft the republic's new constitution, free from the distractions and intrigues of Berlin.

No sooner had Weimar been secured when the *Landesjägerkorps* was called on to restore other areas in Thüringia to government control. Gotha, home to the sprawling *FEA* 3, headed the list. The local Workers' and Soldiers' Council had stormed the garrison armory and disarmed government troops placed there to maintain order. The *Landesjägerkorps* entered Gotha a week later, waging a daylong battle to secure it.

A memo prepared in September 1919 by the *Reichswehr Truppenamt* [7]summarizes the intense pace of *FA* 423's activities in the following weeks:

"In support of the operations against Gotha (February '19) for restoration of government authority, aircraft were employed for dropping leaflets (basis for the intervention of government troops and an appeal for restoration of peace and order).

"The 7 aircraft used arrived exactly at the moment when, in the vicinity of the post office, the inflamed revolutionary masses had assumed a highly threatening attitude toward the advancing troops. The effect on morale was extremely great. Due to the appearance of the aircraft, which dove as low as 30 meters for dropping leaflets, the terrified crowd was dispersed.

"After the ensuing occupation, the workers remained in a state of passive resistance, which manifested itself in strikes idling the means of transportation. In consequence, the airplanes had to undertake daily postal communications and the transportation of important persons between Gotha and the government in Weimar."

To facilitate these courier missions, *FA* 423 took over four Fokker D.VIIs from *FEA* 3's inventory on 22 February[8].

The *Truppenamt* memo continues:

"Halle:

The suppression of Spartacist unrest in Halle, which, at the end of February, led to the complete shutdown

Above: Two Fokker D.VIIs of *FA* 423 over the snowy landscape of central Germany. The lower machine carries a white streamer on its lower right wing. (Paul Rempe)

of Berlin - Weimar rail traffic, pursued for the purpose of making it impossible for the National Assembly in Weimar to make contact with Berlin, required the employment of aircraft. Simultaneous with the occupation of Halle by government troops on 1 March 1919, 4 aircraft appeared, which through leaflet dropping, made the public aware of the state of siege, along with other necessary proclamations. Alternative means of communicating had been rendered impossible, owing to the hostile attitude of the population against the government troops (adhesive posters with government announcements were pulled down, the posters were, moreover, torn up, [and] *Oberstleutnant* (Lieutenant Colonel) von Klüber was murdered.)

Owing to the strike there, rail transport could not be established, so the military aircraft took over postal and passenger transportation from Halle to Weimar and Berlin, thereby negating the transportation strike. The military air postal transport from Halle was maintained daily until 29 March 1919.

Magdeburg:

On 9 April '19, the government was compelled to put troops into Magdeburg, because the troop remnants of the old army, in the Spartacist wake, would not follow government orders and instituted a stark terror. At the same time, the usual railroad strike took place. Together with marching troops into Magdeburg, leaflets were again dropped, and then in the same way as in Halle and Gotha, postal and passenger transportation was initiated.

Braunschweig:

Shortly afterward, the population of the Free State of Braunschweig requested to the *Reichs* government that the entire *Land* be freed from the terror existing at that time. For preparation of the operation, which took place on 17 April '19, a reconnaissance flight was carried out from Magdeburg for surveillance of railway transportation coming from Braunschweig, because of reports received that the Spartacist leadership in Braunschweig would hinder and spoil a government intervention by blowing up rail lines. Photographing

Above: "In memory of my final days in the military. Weimar, September 1919." A glum-looking group packs up the remaining assets of *Artillerie-Fliegerstaffel* 116. (Author's collection)

and visual reconnaissance of the rail lines leading from Braunschweig revealed the reports were without foundation. Leaflets were dropped about the operation against Braunschweig on an especially large scale. The transportation system there being likewise idle, the aircraft had to maintain contact from Braunschweig to Weimar, Berlin and Hannover (X. Army Corps). Shortly before the operations against Braunschweig, revolutionary aircraft from Braunschweig tossed out great quantities of leaflets over Magdeburg with the declaration: "Down with the government."

On 2 May, *FA* 423 was officially redesignated *Artillerie-Fliegerstaffel* 116. Its duties remained unchanged, however, as the *Truppenamt* memo continues:

"Leipzig:

On 11 May 1919, on order of the *Reichs* government, action had to be taken in Leipzig. In the earliest morning hours great numbers of leaflets were dropped to inform the population of the ongoing advance of government troops. The state of siege was proclaimed by leaflet dropping. In addition, traffic with the government in Weimar and Berlin had to be initially maintained with aircraft.

Eisenach, Ruhla:

On 19 May '19 government troops marched into Eisenach and Ruhla, because the Spartacist working class there was inflicting an extraordinary terror on the city and land and was desiring to disrupt traffic along the Berlin - Weimar - Eisenach - Frankfurt road. In addition, transport of volunteers for Kurland was raided and disarmed. About 6,000 workers here stood in openly proclaimed battle against the government. For this operation, the aircraft had to drop leaflets to quickly give knowledge regarding the attack of the government and their orders to the rebellious working class, throughout not only the city of Eisenach, but also over the numerous small surrounding locations and factories.

Erfurt:

At noon, 17 June 1919, a surprise operation against Erfurt was ordered by the *Reichswehr* ministry. In Erfurt, the head of the railroad management and additional railroad officials were isolated, and therefore, the operation was idle. In a short time, the troops were concentrated in Erfurt, aircraft were sent from Naumburg and Altenburg, which delivered the orders for the operation to the troops situated there. For greater secrecy, no use was made of the telephone. On the return flight, the aircraft brought the leader

Flieger Abteilung 423/ *Artillerie-Fliegerstaffel* 116

Upon arriving at Weimar in February 1919, *Flieger Abteilung* 423 established itself at the Weimar-Lindenberg airfield, which had opened to commercial and mail service before the war. The airport hotel was converted into the unit's headquarters and officers' mess, and its four large hangars were augmented by new workshops, barracks and storage facilities. Ultimately, the airfield supported a peak complement of nineteen aircraft, thirteen LVG C.VIs and six Fokker D.VIIs. It was located within 100 km (60 miles) of any point in Thüringia, as well as Leipzig, Saxony, insuring it would be a center of activity against Red uprisings in central Germany.

In the Spring or early summer of 1919, *Artillerie-Fliegerstaffel* 116 rolled out the red carpet for Reichswehr Minister Noske's visit. Noske had previously conducted an inspection of the *Freiwilliges Landesjägerkorps*, expressing great admiration for its organization and discipline, but that was before *FA* 423 had moved to Weimar. Now it was the flyers' turn for recognition.

Above: Weimar-Lindenberg Airfield, home base of *FA* 423/ *Artillerie-Fliegerstaffel* 116. Two of the Staffel's LVG C.VI aircraft sit in front of the hangars. The building to the left of the large hangar is the prewar hotel. It housed the unit's offices and mess. (Paul Rempe)

Above: Officers and aviators of *Artillerie-Fliegerstaffel* 116. In the two front rows, L to R: Unk.. Ernst(?) Krogmann (P), *Ltn.* Ernst Exss (O), Müller (P), *Hpt.* Victor Krocker (CO), Hunninghaus (P), Unk. (O), Griedewald (P), *Lt. d. R.* Leonhard Rempe (P), Wrensch (P), *Off.-stv.* Paul Aue (P), Schwamm (P). Viereck (P), Kohne (P), Muchinow (P). Norbert Welkoborsky, wearing glasses, is the leftmost of the three men in the rear. (Paul Rempe)

Above: Flying personnel of *FA* 423. Judging from the bare trees in the background, this photo probably dates from February or March 1919. *Lt. d. R.* Leonhard Rempe is fourth from right in the front row. (Paul Rempe)

Above: Officers' mess, *FA* 423/*AFst* 116. (Paul Rempe)

Above: *Chair-Kette.* Fliers of *AFst* 116. L to R: Ozinek, Müller, Muchinow, Kohne, Ernst(?) Krogmann, Herbst, Wrensch, *Off.-stv.* Paul Aue, *Vzfw.* Hermann Hasselmann, Hunninghaus, Viereck, *Lt. d. R.* Leonhard Rempe, Schwamm, Griedewald, and Norbert Welkoborsky. (Paul Rempe)

Above: *FA* 423/*AFst* 116's home base. The airfield's prewar hangars didn't provide enough space for all the *Abteilung's* aircraft, so fourteen tent hangars were erected along the northern perimeter of the flying field. (Paul Rempe)

Above: Thirteen LVG C.VI machines are lined up for review, numbers 1 (7766/18) and 4 closest to the camera. Two white streamers are attached to their lower right wings. Behind the Fokker D.VII on the right are three additional aircraft. Two appear to be LVG C.VIs; the unmarked rudder showing behind the D.VII may belong to a Halberstadt CL.IV. (Paul Rempe)

Above: *Reichswehr* Minister Noske (center, wearing pork pie hat) looks on as *Frau* Noske peers into the gunner's cockpit of LVG C.VI 7766/18. (Paul Rempe)

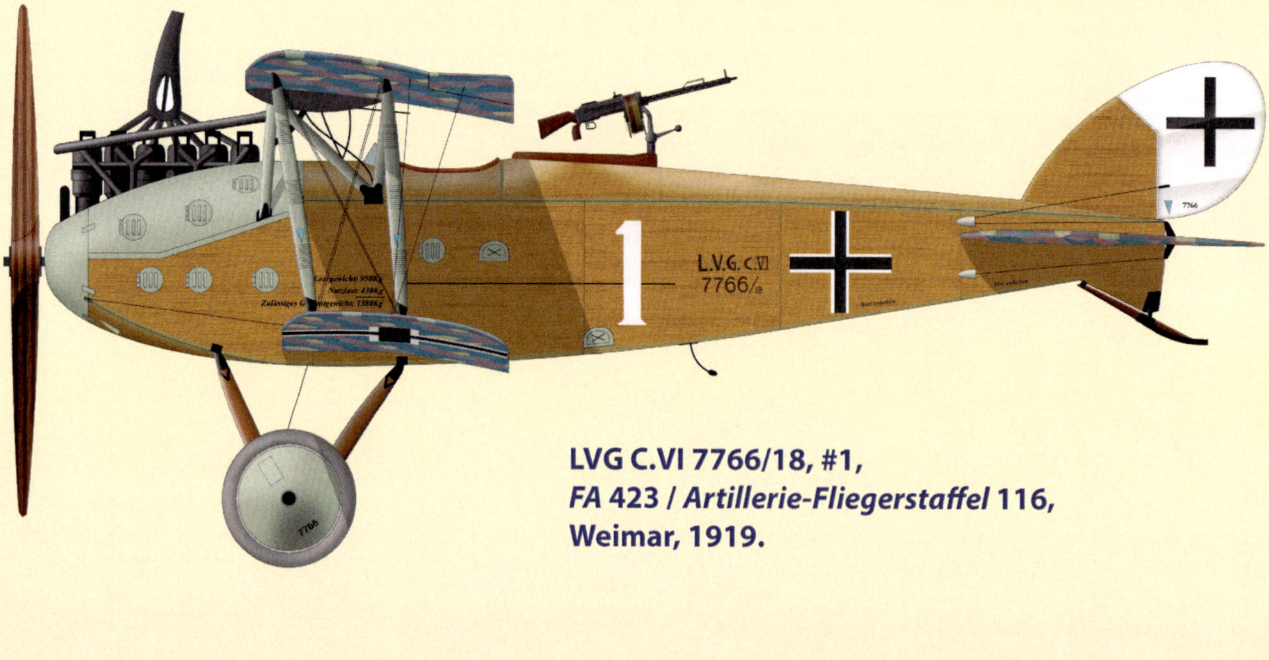

LVG C.VI 7766/18, #1,
***FA* 423 / *Artillerie-Fliegerstaffel* 116,**
Weimar, 1919.

Above: Noske inspecting *Artillerie-Fliegerstaffel* 116's flying personnel. (Paul Rempe)

Right: Paul Aue seated on Fokker D.VII F 5136/18, apparently about to conduct a wreath dropping. The ribbon on the floral wreath bears the words "*Fliegerstaffel*" and "*Maercker*". The occasion may have been Defense Minister Noske's visit. Aue wears a looped chevron, indicating his rank of *Offizier-stellvertreter* in the Provisional *Reichswehr*. (Greg VanWyngarden)

of the troops, who received his verbal instructions in Weimar. During this operation, the 6 aircraft, by dropping leaflets, explained the basis for the actions of the government in the entire Weissenfels, Naumburg, Erfurt, Eisenach and Jena railroad districts.

On the following day, the railroad workshops there called a general strike, and as notification by posters and announcements by broadcasting was impossible, the state of siege in Erfurt was announced by dropping leaflets."[9]

Following the fighting in Erfurt, *Artillerie Fliegerstaffel* 116 enjoyed a well-earned rest at Weimar, awaiting its next assignment. That never came, because Germany's 28 June acceptance of the Versailles Treaty sealed the *Staffel*'s fate. After the government briefly considered using it as the foundation of a *Polizei Fliegerstaffel*, it would be disbanded by 30 September. Three days before that deadline, the *Staffel* flew to *Fliegerhorst* Paderborn, where its aircraft were turned in and most of its personnel discharged. A sad, quiet end to an aviation unit that helped the *Landesjägerkorps* restore eight major metropolitan areas to government control in a little over 4½ months.

Operations Against Hamburg

The government's February campaign to secure Germany's North Sea ports had originally included Hamburg, but operations against that city were shelved after a coalition of moderate Socialist parties ousted the radicals from the city's administration. By June, however, the radicals had regained control, and the *Reichs* government decided to mount an attack to remove them. A corps was assembled around Division Lettow (now Reichswehr Brigade 15) to carry out the operation. Lettow's *Truppen-Fliegerstaffel* 15, which had been lent out at the end of April for the attack on Munich, had returned north and was now stationed at Blankensee bei Lübeck. On 25 June, it relocated to the airfield at Blankenese, west of the Hamburg suburb of Altona.

The operation began 1 July. *Truppen-Fliegerstaffel* 15 carried out numerous reconnaissance sorties over the districts of St. Pauli and Hammerbrook, along with the market at the *Rathaus* and the Wandsbeck train station. To avoid alienating civilians, bombing and ground strafing were forbidden. During the day, several of the *Staffel*'s planes operated out of the Fuhlsbüttel airfield on the north side of the city to minimize the time reaching their objectives. By the day's end, Hamburg had been secured [10].

By the 7th, the *Abteilung* was back at its former base at Ludwigslust, Mecklenburg-Schwerin. The same day, observer *Ltn.* Bovakht perished in a crash.

The 3rd (Löwenfeld) Marine Brigade

In addition to the Freikorps units set up under *Abteilung* Lüttwitz, three Marine Brigades were established by naval officers early in 1919. Composed of a mix of former sailors and army troops, they were assigned to Division Lettow in April 1919. One of them, the 3rd (Löwenfeld) Marine Brigade, fielded a small *Flieger Abteilung*. *Kapitänleutnant d. R.* Friedrich Christiansen, former commander of the Zeebrügge naval air station, joined the Brigade 14 April 1919 as a pilot and leader of this aviation unit, and its establishment probably coincided with his arrival. Christiansen remained with the Brigade until 21 January 1920. Interestingly, his service records state he was a pilot and corporate executive for the Caspar-*Flugzeugwerke* in Travemünde for most of that period, as well [11].

Facing Page, Below: Halb. CL. IV of 3rd (Lowenfeld) Marine Brigade.

Left: *Kragenabzeichen* (collar insignia) of the Württembergischen *Freiwilligen Abteilung* Haas.

Right: *Lt. d. R.* Leonhard Rempe, pilot and technical officer of *FA* 423/*Artillerie-Fliegerstaffel* 116. The *eichenlaub* (oak leaf) pins on his collar represent its parent unit, the *Freiwilliges Landesjägerkorps* Maercker. Variations of the oak leaf format were common *Freikorps* insignias. (Paul Rempe)

Above: *Lt. d. R.* Otto Weisshaar of *FA* Haas with his Fokker D.VII. The oak branch on the fuselage is patterned after the collar insignia of the parent *Freikorps*, württembergische *Freiwillige Abteilung* Haas. In the background is another D.VII, marked with a comet. (Tobias Weber via Hannes Taeger)

The Battle for Red Bavaria

Historical Background

Conservative Bavaria was not immune to the revolutionary fervor that swept Germany in the first days of November 1918. In Munich, an orderly peace demonstration on 7 November snowballed in a few hours to a bloodless revolution. King Ludwig III and his entourage fled the capitol, and a socialist republic, the *Volksstaat* (Peoples' State) *Bayern*, was proclaimed.

This was the first of four different moderate-to left-wing governments of varying incompetence and inability to deal with the collapsing economy, soaring unemployment, and deterioration of essential services and public order. The *Reichs* government in Berlin looked on with increasing apprehension, itching to intervene, but its hands were tied. Bavaria enjoyed *Reservatrechte* (reserve rights), dating from the formation of the German empire in 1871, that kept its military forces under the command of its government, not the *Reich*, during peacetime. In essence, it forbade *Reichs* troops from entering Bavaria without express permission, and the Bavarians, habitually suspicious of the Prussians, were in no mood to grant it.

Unlike Prussia, Bavaria had no foreign enemies at its borders, and its domestic political situation didn't require, early on at least, air power to preserve order. The new socialist government had wound down its wartime air force with remarkable efficiency. Demobilization of the Bavarian *Fliegertruppe* began in December 1918, and its wartime personnel complement had been whittled down to 302 aircrew and a handful of officers by 11 February. Further reductions were scheduled in the coming days.[12] The other side of demobilization, the disposal of surplus aircraft, equipment, and supplies, was entrusted by *Iluft* (*Inspektion der Luftfahrtruppen* – Inspector of Aviation Troops) to a newly-formed authority, *B.V.H.* (*Bayerische Verwertungsgesellschaft für Heeresgerät* – Bavarian Recycling Corporation for Army Equipment).[13]

A crisis arose 21 February 1919, when both Prime Minister Eisner and the head of a rival political party were assassinated, and the leader of the majority Social Democratic Party was gravely wounded in another attempted killing. Munich erupted in panic, the government imprisoned a number of prominent citizens as hostages against a feared right-wing coup, and leftists and rightists alike prepared for a struggle to control Bavaria. Most of the surviving government ministers fled north to the safety of Bamberg and on

17 March elected Johannes Hoffmann their new Prime Minister. That same day, *Hauptmann* Max Sorg, who had previously served with *FFA* (*Feldflieger Abteilung*) 1b, established a *Flieger Abteilung* at Kitzingen. Bavaria now had two governments. Two days later, anti-Soviet soldiers and students took control of Würzburg.

On 13–14 April 1919, a group of hard-core Spartacists, inspired by Russia's November Revolution, seized control of what was left of the Munich government, proclaiming a *Räterepublik* (Soviet Republic). In addition to Munich, Reds took control of several other cities. Elsewhere, especially in the countryside, Hoffmann's government held at least nominal control. Civil war was coming closer.

Preparations For War

Rebuffing offers of help from Berlin, the Hoffmann government assembled an army to retake Munich. That force made it as far as Dachau, northwest of the capitol, where it was routed by the Red army on 18 April. As the Red forces marched northward to meet Hoffmann's army, they added Bavaria's main wartime airbase at Schleissheim[14] to their inventory on 16 April.

Hoffmann now had no choice but to call on Berlin for help. *Reichs* War Minister Gustav Noske drove a hard bargain – he would provide an army to drive the Reds out of Bavaria, but the entire operation would be under his control, not Hoffmann's. It would be assembled from regular *Reichswehr* troops and *Freikorps* formations from neighboring states, in addition to Bavarian *Freikorps* units.

Up to this point, all the successive Bavarian governments, including Hoffmann's, had been hostile to the idea of a local *Freikorps*. Nevertheless, several had already set up shop in neighboring Thüringia and Württemberg, safely out of the reach of Bavarian authorities.

The best known of these was the *Bayerisches-Schützenkorps* (Bavarian Rifle Corps), better known as the von Epp *Freikorps* after its founder and leader, *Oberst* (Colonel) Franz *Ritter* von Epp. A bitter enemy of the leftists, von Epp had evaded arrest in Munich and made his way secretly to Ohrdruf, Thüringia, where, in mid-February, he began recruiting troops to his cause. His *Freikorps* grew quickly, and within a month, included volunteers for a *Flieger Abteilung* and *Schlachtstaffel* (ground attack squadron).

The whole operation gelled with remarkable speed; Noske had obviously anticipated Hoffmann's appeal

Above: A closer look at *Lt. d. R.* Weisshaar's oak branch-marked Fokker D.VII with him in the cockpit. (Tobias Weber via Hannes Taeger)

and begun assembling an invasion force. Less than two weeks later, government forces had entered Bavaria and prepared to encircle Munich. *Generalmajor* Oven, in charge of the entire operation, set up his headquarters in Ingolstadt, north of the capitol.

Bavaria's "White" Air Force

The quick buildup of the government invasion force didn't leave much time to assemble flyers and aircraft to support the operation, but a small air arm was cobbled together in time for the assault on Munich.

One of the earliest aviation units to take form was not even Bavarian. It was the *Flieger-Abteilung der württembergischen Freiwilligen Abteilung Haas* (Flying Section of the Württemberg Volunteer Detachment Haas), a *Freikorps* unit originally created to join *Grenzschütz Ost*, the border defense command on the Polish frontier. Aviation personnel began assembling at Böblingen in February, and the *Flieger-Abteilung* was officially formed on the 25th under *Oblt*. Robert Holtzmann, former commander of *Flieger Abteilungen* 1, 16 & 22.[15] It operated nine aircraft, including six two-seaters and three D-types. These included at least one LVG C.VI and two Fokker D.VIIs. The *Abteilung* had some features that set it apart from other *Freikorps* aviation units. It divided its component sections into eight "*Korporatsschafte*", corporate bodies, instead of the customary "*Abteilungen*" (departments). Functionally, they were probably no different. Unlike most units that placed all their aircraft under one *Abteilung*, *FA* Haas also divided responsibility for its aircraft and their support crews into four *Korporatsschafte* -- I, II, and III each were assigned two C aircraft, while *Korporatsschaft*

Lt. d. R. **Otto Weisshaar Fokker D.VII, *FA* Haas**

Comet-marked Fokker D.VII, *FA* Haas

Above: Flying personnel of Artillerie-Fliegerstaffel 121, formerly FA A2, attached to the von Epp Freikorps, pose with one of the unit's Hannover CL.IIIa ground attack aircraft. The unit insignia, a white disc on a black diamond painted over the fuselage cross, was inspired by the von Epp Freikorps' uniform sleeve insignia, a lion's head medallion sewn onto a black felt diamond. (Author's collection)

Hannover CL.IIIa, C.70XX,
***Artillerie-Fliegerstaffel* 121, *Freikorps* Epp,**
Schleissheim, Bavaria, 1919.

IV was responsible for the single-seat fighters. *Korporatsschaft* VI was the unit's motor vehicle section.

Abteilung Haas never made it to the Polish frontier, because, on 16 April, it was placed under the command of the newly-formed *Oberkommando* Ulm *der bayr. Regierungstruppen* (High Command Ulm of the Bavarian government troops), established in Ulm, Württemberg, anticipating the need to employ it in ridding Bavaria of the Reds.[16] By the 21st, *Hpt.* Sorg was its *Kofl.*

Reporting to the *Oberkommando* was *Fliegerabteilung* Franconia, headquartered at Dornstadt airfield, northwest of Ulm. Per order *OK* No. 79 op., issued 21 April, aircraft and aircrews arriving from Kitzingen would be placed under its administrative control. They and *Flieger Abteilung* Haas were instructed to obtain their fuel supplies at Donauwörth. Air units assigned to ground forces north of Munich would refuel at Ingolstadt. The order also announced that *Fliegerabteilung* Franconia would presently relocate east to Gersthofen, on the northern outskirts of Augsburg, which had just been taken from the Reds. *Flieger Abteilung* Haas would take up residence at *Lager* Lechfeld, the parade ground on the northern outskirts of Landsberg am Lech, about 50 km due west of Munich.[17]

April saw the formal establishment of the air units of the von Epp *Freikorps* at Gotha, Thüringia. On the 13th, *Flieger Abteilung* A1 came into existence. It was also known as *Flieger Abteilung* Krausser after its founding commander, *Oblt.* Friedrich *Ritter* von Krausser, who had led *Schlasta* 23b, *FA* 45b, and *Schlasta* 2 during the war. It was followed on 21 April by *FA* A2, *Flieger Abteilung* Häfner, under *Hpt.* Konrad Häfner, former commanding officer of *FA(A)* 295. Along with *FA* A1, it moved to Dornstadt to work up to combat readiness.

On the 24th, *FA* A2 and *Schlasta* Anschütz moved to Lager Lechfeld. This was apparently only a personnel move, as they found six rigged and ready-to-use aircraft waiting for the *Schlasta*. *FA* A1 didn't receive any machines and was forced to sit on the sidelines when the assault on Munich began a week later.

Flug Staffel des Freikorps Schwaben, created as a part of *Freikorps* Schwaben, arrived in Lechfeld with two Albatros C.Ias seized at Sonthofen. Along with *Flieger Abteilung* Haas and two balloon companies, they would become components of *Gruppe* West, commanded by *Generalmajor* Haas.

To facilitate communications between components of the government forces, a *Kurier-Flieger-Staffel* (Courier Aviation Squadron) was formed 21 April. It was comprised of four *Kurier-Flieger-Stationen* (Courier Aviation Stations). The headquarters station at Bamberg, the seat of Hoffman's government, had six aircraft at its disposal. *Hpt.* Ludwig Schauer, formerly an observer with *Feldflieger Abteilung* (Field Aviation Section) 6 and commander of *FA(A)* (*Flieger Abteilung* (*Artillerie*) - Artillery Aviation Detachment) 297, and now commander of the Bamberg airfield, assumed leadership of the unit. The other three stations at Munich, Würzburg, and Fürth had three aircraft apiece.[18] The Fürth Station, commanded by *Hpt.* Michael Krug, appears to have been the former *Kurier-Flieger-Station* that had been driven out of Schleissheim a few days earlier. Given that Munich and its airfields were under Red control at the time, its *Station* must have been located outside the city itself. On 25 April, the Hoffman government set up an "airmail" operation from Bamberg to Munich, scattering leaflets containing the latest news and government proclamations on the residents of the besieged capitol.[19]

Reconnaissance responsibilities over Red territory were divided between *FA* Haas and *FA* Franconia. Haas patrolled the region south of a line running southeast from Gersthofen to Rott a. Inn, east-southeast of Munich, the area assigned to *Gruppe* Haas. *FA* Franconia was responsible for the area north of a line from Haunstetten, south of Augsburg, to Fürstfeldenbrück to Rosenheim, southeast of Munich, the operational zone of *Gruppen* Friedeburg, Deetjen and Siebert. This insured an overlap of Munich, which lay between these lines.

Clipping The Reds' Wings

The Reds had the beginnings of an air arm in the form of airfields, including Gersthofen, Schleissheim and Sonthofen, along with Oberwiesenfeld, an old parade ground in Munich's Milbertshofen district; it had served as a military airfield before the new main air facility at Schleissheim was opened. They also had few aircraft and a handful of flying crews but were never able to create a viable air force. Pilots and observers were scarce, simply because few officers joined the revolution, and even enlisted pilots were in short supply. Combat flying wasn't their main priority, anyway – as with their enemies in Bamberg, communication and coordination with other areas in friendly hands was far more important. Given that both sides were flying the same kinds of planes, aircraft type identification was no longer a reliable way to tell friend from foe. The Spartacist *Kofl* (*Kommandeur der Flieger* - Commander of Flying Troops), Parnhaft, ordered his aircraft to fly red pennants as recognition markings. Wisely, he also commanded ground troops not to fire on any aircraft, regardless of circumstances.[20]

Government intelligence from 21–22 April painted a picture of a nearly-impotent Red air force: aircraft at Schleissheim not airworthy, only two private aircraft at the Red government's disposal, fuel in short supply,

Above: One of *Flugstaffel Freikorps* Schwaben's two Albatros C.Ia aircraft, seized from the Reds at Sonthofen. (Author's collection)

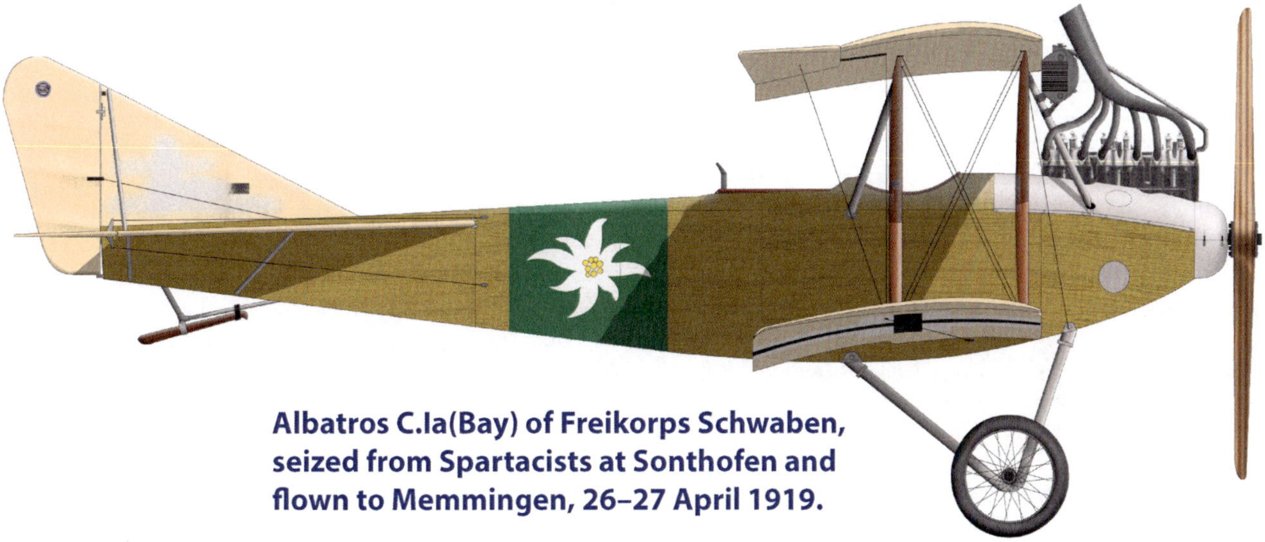

Albatros C.Ia(Bay) of Freikorps Schwaben, seized from Spartacists at Sonthofen and flown to Memmingen, 26–27 April 1919.

and stocks of bombs probably depleted.[21] Despite the meager resources opposing it, the government took the threat of a Red air force seriously enough to neutralize it. On 21 April, troops of the Probstmayr *Abteilung* occupied the Gersthofen airbase north of Augsburg. The same day, the government air forces suffered their first loss when an aviator dispersing leaflets over Oberhausen near Augsburg was hit by ground fire and forced to land. A hostile crowd nearly killed him after he landed.[22]

The night of 26–7 April, a company of troops from *Freikorps* Schwaben was ordered by *Oberkommando* Ulm to mount an operation against the Red-held airfield at Sonthofen. Intelligence had revealed the local Red commander, *Unteroffizier* Bauer, was scheduled to fly to Munich to pick up cash to pay his troops. The raiding party traveled, believe it or not, by a regularly scheduled train from Memmingen, bringing along two aircrews, pilots *Ltn.* Jehle and *Unteroffizier* Prim and observers *Ltn.* Madlener and *Ltn.* Mauler, to fly captured aircraft back to Memmingen. The raid was a complete success; by the next day, the air base had been neutralized, the Red troops' leader had been taken into custody, and the planes were in Memmingen.[23]

By the following day, government forces had advanced closely enough to Munich to neutralize the Schleissheim airbase. Its aircraft were apparently no longer serviceable anyway, due either to inadequate maintenance or to sabotage by personnel loyal to the *Reichs* government, as was later claimed.[24] The Red air force, small as it was, had been grounded.

Liberating Munich

Collectively, the forces assembled to take Munich were named *Korps* Oven, after their commander, *Generalmajor* von Oven. Authority over all Bavarian and Württemberger troops committed to the operation lay in the hands of *Generalmajor* von Möhl, acting in liaison with von Oven.

To aid communications with other units and the Military Ministry in Bamberg, a *Kurierstaffel* with two C-type aircraft was set up at an airfield near von Oven's headquarters in Ingolstadt. Led by *Oblt.* Alwin Kinkelin, a *Kasta* 36 veteran, the *Staffel* was provided command and logistical support by another newly established *Flieger Abteilung* based in Fürth and commanded by *Oblt.* Benedikt Schmalschläger, former commander of *FA (A)* 294.

Von Möhl's *Kofl* was *Hauptmann* Max Sorg, commander of the Kitzingen airfield. Sorg's *O.z.b.V.* (adjutant) was 28-victory ace *Oblt.* Robert *Ritter* von Greim, former commander of *Jasta* 34b.

By 28 April, the *Korps* had drawn a ring around Munich. It was divided into four *Gruppen*:

- *Gruppe* Nord, commanded by *Generalmajor* Siebert. Despite its name, it controlled the eastern approaches to the city.
- *Gruppe* Deetjen, under *Oberst* Deetjen, northeast of Munich,
- *Gruppe* Friedeburg, led by *Generalleutnant* von Friedeburg, to the northwest, and
- *Gruppe* West, commanded by *Generalmajor* Haas.

The same day, von Möhl was subordinated to *Korps* Oven, putting all his troops under *Reichs* control. Ultimate authority over the operation, though, was *Reichswehrminister* Noske, via von Lüttwitz's *Reichswehrgruppenkommando* I in Berlin.

Jagdstaffel Flecken, with three single-seaters, was assigned to *Gruppe* Nord. As of 27 April, it was waiting for portable hangars to be shipped from Pfaffenhofen to Ingolstadt airfield so it could take up residence there. The *Jasta* appears not have received them in time to participate in the assault on Munich.

To prepare the populace of Munich for the upcoming attack, von Oven, on 27 April, ordered all available aviation units to overfly the city in full strength, dispersing leaflets and newspapers from

Facing Page: Gruppe West's aviation forces also included two balloon companies, one led by Hpt. Gareis[27] and reporting to Abteilung von Bogendörfer, the other operated by Abteilung Probstmayr, a component of Gruppe Hierl.[28]

30 April through 3 May. Two days later, von Oven's headquarters issued more detailed orders to the aviation troops. They read like a laundry list of all the possible tasks an air force should undertake, including:

- Tactical reconnaissance, including photography.
- Transmitting intelligence and observation findings to ground forces.
- So-called "demonstration flights" (aerobatic displays and mock strafing) to intimidate Red troops in areas under imminent attack.
- Leaflet dropping.
- Ground attack with bombs and machine guns. A subsequent paragraph, however, urged extreme caution in the use of machine guns near innocent civilians and forbade dropping bombs without express orders.
- Combat against enemy aircraft. Flyers were urged to force enemy aircraft to land so their occupants could be captured. This requirement was founded on intelligence reports predicting Munich's Communist leader would flee the city by air in the event of a government victory. In fact, there were no longer any Red planes to intercept.

To distinguish themselves in the air, government aircraft were to fly two white pennants on their left wing struts. Crews were also informed Red planes could be expected to be flying red streamers.

These orders also appear to reveal a command shuffle. *Hpt.* Sorg's *O.z.b.V.* was now *Oblt.* Rauch, not Greim. On *Generalmajor* von Oven's staff, *Hpt.* Steffen held the post of *Kofl*, with *Oblt.* Heymann as his *O.z.b.V.*[29]

Oven's plan to enter Munich 4 May was upset by events. Panic had seized the Reds in the city, many of their troops were deserting, and anti-communist civilians were attacking those who remained. Worse, Red troops began killing their civilian hostages. When word of these atrocities reached von Oven, he ordered his troops to move to their jumping-off positions the evening of 1 May. They would advance into the city at noon on the 2nd. The aircraft were ordered to go into action 20 minutes before the advance, conducting their demonstration flights, dropping leaflets and firing flares, in spite of poor weather conditions. They would operate short-handed; of all the units expected to participate in the operation, only four -- *FA* Haas and *FA* Schwaben at Lager Lechfeld and *FA* A2 and *Schlachtstaffel* Anschütz, at Dornstadt – were available

Government Forces and Aviation Units Opposing Red Munich, 28.4.19

- *Korps* Oven (*Generalmajor* von Oven)
  - *Kofl.* (*Hpt.* Steffen)
    - ***Flieger Abteilung* 422** (In transit from Ludwigslust)
  - *Gruppe* Friedeburg (*Generalleutnan*t von Friedeburg)
    - ***Flieger Abteilung* 421** (In transit from Johannisthal, transferred to *Kofl.* by 9 May 1919)
    - 2. *Garde-Inf.* Division
    - *Hessische - Thüringisch - Waldecksches Freikorps*
    - *Freikorps* Görlitz
  - *Gruppe* Deetjen (*Oberst* Deetjen)
    - ***Flieger Abteilung* 409** (In transit from Lithuania)
    - *Freikorps* von Lützow
    - *Kavallerie-Schützen-Kommando* 14
    - 2. Marine Brigade
- *Generalmajor* von Möhl
  - *Kofl* (*Hpt.* Sorg)
  - *Gruppe* West (*Generalmajor* Haas)
    - *Abteilung* von Seutter
      - ***Fliegerabteilung* Haas** (*Oblt.* Robert Holtzmann)
    - *Abteilung* Graeter
    - *Bayr. Schützenkorps* (*Oberst* Franz *Ritter* von Epp)
      - ***Schlachtstaffel* Anschütz**
      - ***Flieger Abteilung* A2 Häfner**
    - *Freikorps* Schwaben
      - ***Flieger Abteilung* Schwaben**
    - *Freikorps* Liste
    - *Abteilung* von Bogendörfer
      - ***Ballonzug*** (*Hpt.* Gareis)
    - *Freikorps* Werdenfels
    - *Gruppe* Hierl
      - *Abteilung* Schaaf
      - *Abteilung* Probstmayr
        - ***Ballonzug***
  - *Gruppe* Nord (*Generalmajor* Siebert)
    - ***Jagdstaffel* Flecken** (3 D.- A/C. Awaiting hangars at Ingolstadt)
    - *Untergruppe* Denk
      - *Abteilung* Heinzmann
      - *Abteilung* Schad
      - *Abteilung* Voithenleitner
      - *Abteilung* Schaaf (*Gruppe* C)
      - *Freikorps* Oberland

and fully operational. These were all part of *Gruppe* West.

Apart from a few brief, but intense, battles, Red resistance evaporated, and organized combat essentially petered out by the end of the day. Although light, government losses included one of its aircraft. An LVG C.VI of *FA* Haas, flying about 100 m above the city, was hit exchanging fire with Reds on the ground. It was forced to land in a garden at the edge of the Theresienwiese, site of each year's Oktoberfest. The crew torched the fallen plane, and with the aid of sympathetic local residents, escaped unharmed and returned to their unit the following day.[30, 31]

While the victorious *Freikorps* instituted a White Terror, rounding up and frequently killing anyone suspected of Red sympathies, the army lost no time installing itself in Munich and its surroundings. On the 3rd, *Kofl* moved its headquarters into the Prince Luitpold Palace.

Elsewhere, mopping up continued. Pasing, a western suburb of Munich, was occupied by *Eskadron* Kipfmuller, aided by *Hpt.* Gareis' balloon company, which had arrived the evening of 1 May.

A *Fliegerbefehl* (Aviation order) dated 5 May:

1. Placed *Hauptmann* Rutz in command of *Fliegerhorst* Schleissheim.
2. Named *Hauptmann* Willibald Spang of *FA* Haas as *Grufl.* (*Gruppenflieger*, Group Aviation Commander) of *Gruppenkommando* (Group Command) West.
3. Announced a bounty for aircraft removed from Schleissheim airfield and returned to the government while the field was under Red control. The prize was 200 Marks for a C or D aircraft and 300 Marks for a G machine. This was apparently inspired by the exploit of pilot Kuhn and *Unteroffizier* Eisberger, who flew a two-seater out of Schleissheim to Bamberg on 20 April. With Munich already firmly under government control by 5 May, this order appears to be retroactive confirmation of actions already taken[32].

The fall of Munich essentially spelled the end of Red Bavaria, although communists continued to hold on in a few outlying areas, including extreme southwest Bavaria. Kempten and Lindau had to be cleared, along with unfinished business in Sonthofen. A task force assembled under Major Hierl descended on Kempten 13 May, supported by the Gersthofen *Flieger Abteilung*. This was apparently a new unit created specifically for this operation -- it operated Hannover CL.II (Rol.) 695/18 and an Albatros C.III, probably a former training school machine. Within a short time, the area was under government control.

Additional *Flieger Abteilungen*, still in transit to Munich when the battle reached its premature end, began arriving. On 4 May, *FA* 409, recalled from Lithuania, reported to *Gruppe* Deetjen. It would be stationed at Schleissheim. *FA* 421 arrived from Johannisthal, where it had been attached to the *Deutsche Schützen* Division. Too late to join *Gruppe* Friedeberg in the fighting, it was reassigned to *Kofl*. *FA* 422, which had been transferred from the Lettow Division in Ludwigslust, Mecklenburg-Schwerin, was scheduled to arrive 4 May at Oberwiesenfeld near Milbertshofen, where it would also report to *Kofl*'s staff.

*FA*s 421 and 422 were clearly short-term loans; by 9 May, Berlin had requested they be sent to Döberitz as soon as possible.

Joining The *Reichswehr*

Once the *Reichswehr* had been invited to Bavaria, it was there to stay, and the diverse Bavarian units would soon be integrated into it. Bavaria would become the responsibility of the *Vorlaufige Reichwehr*'s 21st through 24th Brigades. The consolidated and reorganized flying detachments would report to the 21st Brigade.

This process had already begun before the attack on Munich, with *FA* A2 (Häfner) of *Freikorps* Epp receiving its new designation, *Artillerie-Fliegerstaffel* 121. Häfner remained in command.

Orders dated 12 May 1919 assigned aircraft and supplies to *FA*s 409 and 421, suggesting they had left their assets behind when moving to Bavaria. *FA* 409 would receive two DFW C.Vs, 6025/17 and 17196/17, and Fokker D.VII 6823/18. *FA* 421 received Halberstadt C.Vs 8771, 8777 and 8780/18, Rumpler *Rubild* (an aircraft equipped with a camera that produced sequential strip photos) 3977/18 and 6420/18, along with a 100 hp Albatros B.II, 1309.[33] Given the date, *FA* 421's orders to return to Döberitz appear to have been been delayed.

On 14 May, Krausser's *FA* A1 became *Truppen-Fliegerstaffel* 21, *Schlasta* Anschütz became the Bavarian *Schlachtstaffel* on the 25th, and *FA*s Krieglsteiner and Dessloch were merged to form *Truppen-Fliegerstaffel* 21b on 1 June. With Bavaria's ground transportation and communications networks securely under government control, the *Kurier-Flieger Stationen* were no longer needed. On 1 July, *Kurier Flieger-Station* Fürth was detached from the *Kurier-Flieger-Staffel* and merged with *Flieger Abteilung* Schmalschläger to form *Truppen-Fliegerstaffel* 21a. On 2 June, The Würzburg *Kurierfliegerstation* was wound down. Its enlisted personnel, including pilots *Off-Stv.* Grün and Chorbacher and *Vzfw.* Schnäbele and aerial gunners *Vzfw.* Kern and Brüstle, were ordered to Kitzingen on 4 June, where *Oblt.* Schröder would reassign them.[34] On 12 July, *Kurier-Flieger Station* Würzburg was

Above & Below: Still smoking after it was burned by its crew, an LVG C.VI of *FA* Haas lies at the edge of the Theresienwiese after being downed by Red ground fire 2 May 1919. (Author's collection)

Above: The twin steeples of Kempten's St. Lorenz Basilica form the background of this photo of the aircraft of the Gersthofen *Flieger Abteilung*, stationed there mid-May 1919. The Alb. C.III sports a blue and white cockade and tail sunburst; Hannover CL.II (Rol) 695/18 appears to have a similar cockade on its fuselage, but it is scuffed up, probably by the gunner climbing into and out of his cockpit. (Greg VanWyngarden)

Albatros C.III, Gersthofen *FA*, Kempten, May 1919

Hannover CL.II (Rol) 695/18, Gersthofen *FA*, Kempten, May 1919

Above: "Exodus from Egypt." *Hpt.* Häfner (seated at right) with *Artillerie-Fliegerstaffel* 121's NCO pilots at Schleissheim. The occasion was the *Staffel*'s dissolution 30 September 1919. Their black diamond sleeve insignia, inspiration for the *Staffel*'s fuselage marking, shows prominently. *Obergefreiter* Beck is seated second from left. (Author's collection)

reduced to three aircraft, and *Flieger Abteilung* Schauer, stationed at Bamberg, was reincarnated as *Fliegerstation* Bamberg with three aircraft. *Hpt.* Schauer remained in command.[36] On the 28th, the *Kurierflieger Staffel* and the Munich and Würzburg stations were shut down. Aircraft and equipment made surplus by this force reduction were turned over to *BVH* for disposal.

FA Haas returned to its home base at Böblingen on 14 June and began downsizing through discharges and transfers. Although its parent, *württembergischen Freiwilligen Abteilung* Haas would become the basis for the *Reichswehr*'s 13th Brigade, *FA* Haas was not part of that reorganization, finally disbanding 1 September. *Hpt.* Holtzmann was reassigned to *Fliegerhorst* Böblingen.

FA 409, having previously absorbed *Flieger Abteilung* Schlieben after it arrived from Hannover, was reformed as *Schlacht-Fliegerstaffel* 201, an interesting departure from the *Vorlaufige Reichwehr*'s system for designating aviation units. Following acceptance of the Versailles Treaty, it was scheduled for demobilization at Jüterbog.[35]

No longer required in Bavaria, *FA* 422 was returned to the Lettow Division (now *Reichswehr* Brigade 15) and its airfield at Blankensee bei Lübeck, via Döbertz.

Following ratification of the Versailles Treaty, Germany's air force began contracting, closing airfields, consolidating units, and shedding personnel, all aimed at complete disbandment on 30 September 1919.

The remaining aviation assets and personnel were consolidated at a few sites to await orders to disband. On 19 September, a memo from Bavarian *Gruppenkommando Nr.* 4 outlined, among other things, details of the plan for winding down the various aviation units in Bavaria. *Truppenflieger-Staffel* 21, *Artillerie-Fliegerstaffel* 121 and the *Schlachtstaffel* were to disband at *Fliegerhorst* (military airfield) Schleissheim, while *Truppenfliegerstaffeln* 21a & 21b were to be shut down at Fürth, Nürnberg's airfield.[37]

A 21 September memo from the *Kofl.* of *Reichswehrkommando* 4 contains an itemized list of equipment requested to outfit the units at *Fliegerhorst* Schleissheim:

- *Truppenfliegerstaffel* 21: six DFW C.V and three Fokker D.VII
- *Truppenfliegerstaffel* 21a: nine DFW C.V and three Fokker D.VII
- *Artillerie-Fliegerstaffel* 121: twelve Hannover CL. IIIa
- *Schlachtstaffel*: two Fokker D.VII,

in addition to a reserve inventory of 40 Halberstadt

Above: Pilots of *Artillerie-Fliegerstaffel* 121 on the occasion of the unit's "official" disbandment at Schleissheim, 30 September 1919. L to R: *Obergefreiter* Beck, Murizk, Preusch, Stephansky and Passelot. (Author's collection)

C.Vs and 50 spare engines.[38] This request was wildly optimistic. It may have been based on the belief the Armistice Commission wouldn't be able to implement the Versailles Treaty's 30 September edict on schedule (in fact, it wouldn't). In any event, these planes were not provided, and downsizing continued, though not as completely as the *Reichswehr* planned or the Allies expected. This scenario would be repeated throughout Germany.

On 30 September, *Iluft* merged *Truppen-Fliegerstaffeln* 21, 21a, 21b, *Artillerie-Fliegerstaffel* 121, & *Bayr. Schlasta* into the *Flugstaffel der Reichswehr-Fliegerhorst* Schleissheim. The following day, the consolidated unit began reporting to *Reichswehr Flug-Inspektion* in Berlin.[39]

This action, and the ones in the preceding weeks, reveals frequent, sudden changes in the *Reichswehr*'s demobilization plans.

Saxony

Historical Background

From the beginning of the 20th Century, Saxony had been a stronghold of labor unions and the Social Democratic Party (SPD), earning it the sobriquet "Red Saxony". Following the voluntary abdication of King Frederick August III on 13 November 1918, Saxony became a Free State within the new German Republic. Actually, it was a state containing a quasi-state; Leipzig, Dresden and their surrounding areas were dominated by Workers' Councils backing the Social Democratic and Independent Social Democratic (USPD) political parties, and they had proclaimed a Saxon Soviet Republic that, despite its name, had no functioning governmental structures and submitted to the overall authority of the Free State. It was an unusual arrangement, but in the first weeks of 1919, the two republics coexisted. The *Reichs* government, on the other hand, was uneasy with the arrangement. Unlike Bavaria, which was sheltered by its reserve rights from the uninvited entry of Reichs troops, Saxony had no such protection, and Berlin was free to send troops across its borders on the pretext of restoring order.

Saxons In Lithuania

The first Saxon air unit appears to be an ad hoc formation raised by Saxon troops that made up a significant portion of the wartime *Reichs* forces in Lithuania. Following the Armistice, they held their positions in the face of the overwhelming Red advance into the Baltic territories. A hasty reorganization of these surviving forces created the nucleus of the German army that would oppose the Bolsheviks.

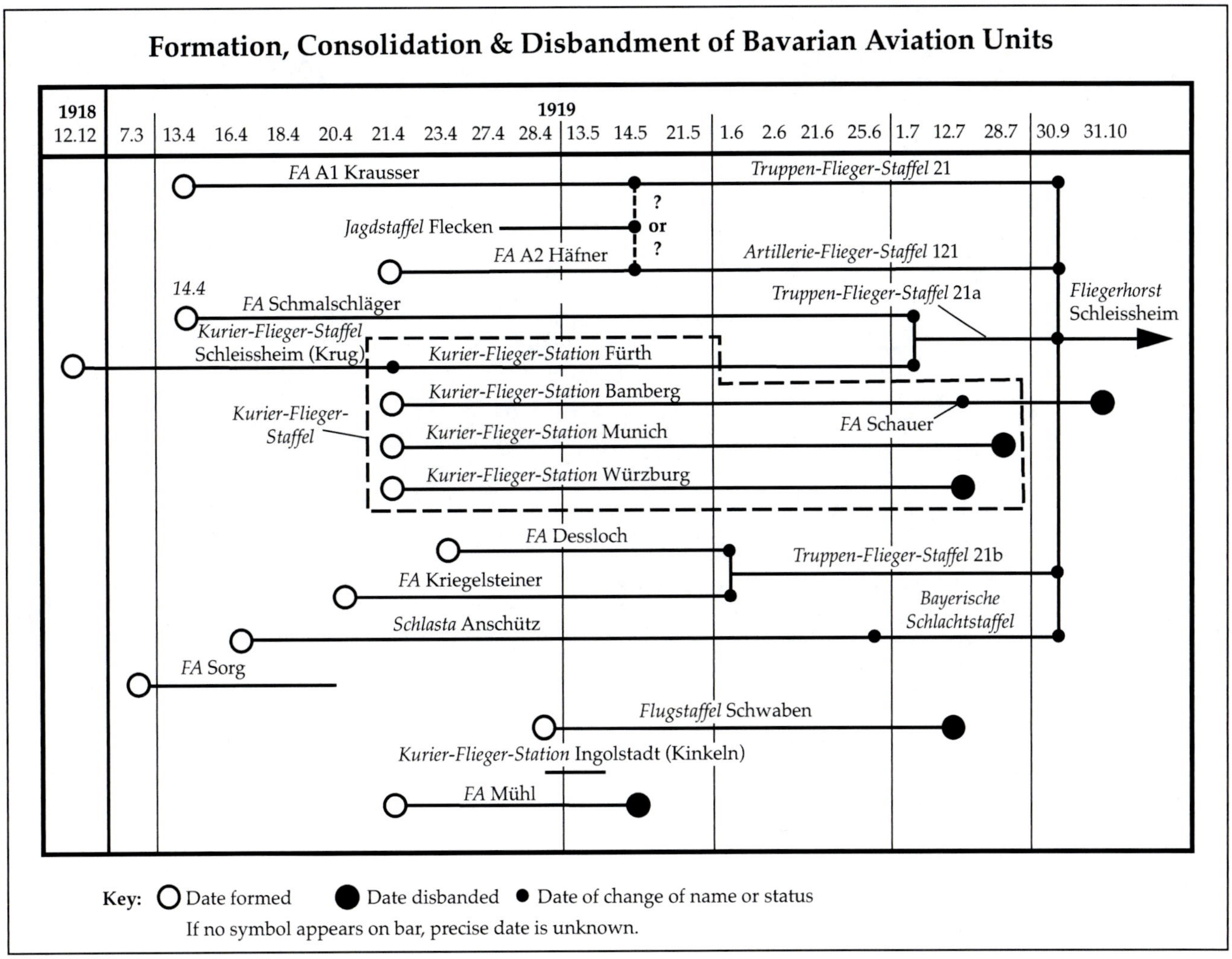

Above: By 28 April, Bavaria had formed a stunning total of fifteen aviation units to support the recapture of Munich, in addition to Württemberg's *FA* Haas and four additional units in transit. By the end of October, this unwieldy organization had been condensed into a single unit.[37]

One of these units was the 46. (*Sächs.*) *Landwehr* Division's Brigade *Süd-Litauen* (South Lithuania), and its *Kriegsgliederung* (Order of battle) listed *Freiwilligen Fliegerstaffel* 18, also known as *Freiwilligen Flieger-Geschwader Staffel* 327. It was in existence before 28 February 1919 and disappeared sometime in April.[40] Little else is known about it, including its choice of unit numbers, which don't match those of any associated units. It does not appear to have been related to *Flieger Abteilung* 214 in Kowno, because that unit had already been transformed into *FA* 425 in December 1918. Even the official *Reichswehr* history fails to mention it. It may have been absorbed into one of the 400-series *Flieger Abteilungen* that moved into Lithuania in January.

Operations Within Saxony

During the Great War, Saxony, as one of the four kingdoms of the *Reich*, had been entitled to several *Jastas* and *Flieger Abteilungen* of its own. Domestically, the kingdom was divided between two *Armee Korps* -- XII., headquartered in Dresden, with its main airfield at Grossenhain, and XIX., based in Leipzig, whose air base was at Wurzen. In the days immediately following the Armistice, *Idflieg* combined the administration of the two *Armee Korps* districts under a single *Kofl.*, *Hauptmann* Walther von Teubern. This was opposed by the Workers' and Soldiers' Council controlling the Grossenhain military airfield. From 15 January, they refused to recognize von Teubern and selected one of their own, *Ltn.* André, as *Kofl.* At one point, they even considered abolishing the position of *Kofl.* altogether. The situation at Grossenhain was similar to that at *FEA* 3, described above, and other military facilities throughout Germany -- the Workers' and Soldiers' Councils refused to recognize the authority of the central government, theft, corruption, and insubordination were rife, and the facility was

Above: *Grenzjäger Flieger-Abteilung* 1, June 1919. L to R, front row: *Ltn.* von Peterkowsky, *Oblt.* Franke, *Hpt.* Egon Doerstling (*Kommandeur*), *Hpt.* Aribert Müller-Arles, *Ltn.* Kellner, *Ltn.* v. Scheel. (Second row) *Vzfw.* Opel, *Oblt.* Wilhelmi, *Vzfw.* Paul Rothe, *Vzfw.* Otto Rühle, *Vzfw.* Metzger, *Uffz.* Geisler. (Third row) *Schirrmacher* (harness maker) Kühne, *Bordschütze* (aerial gunner) Schramm, *Uffz.* Rothe, *Vzfw.* Neitsch, *Funker* (radio operator) Möbius. (Dieter Herde via Hannes Taeger)

dysfunctional to the point of being useless. This power struggle would continue until the government and army stripped the Councils of their authority, often by force.

On 25 February, the Saxon government organized a new military arm from the remains of its wartime army and various *Freikorps* detachments. It was to field two flying units designated *Grenzjäger* (Frontier Fighter) *Flieger Abteilungen* 1 and 2. *Abteilung* 1 was commanded by *Hauptmann* Egon Doerstling, whose last wartime posting was commander of *Bogohl* (*Bombengeschwader der Oberste Heeresleitung* - Bombing Squadron of the Supreme Command) 7.[41] In March, the *Abteilung* was assigned to an airfield in Stiebitz, west of the small city of Bautzen in the southeastern part of the state. This choice appears to be driven by its proximity to Silesia and the new nation of Czecho-Slovakia. Czechs and Slovaks made up a signicant part of the population of some nearby areas of Silesia, and those regions sat on rich veins of coal, making them tempting targets for annexation by the new Czech government. It was this concern that prompted *Grenzschutz Ost* to transfer some of its forces closer to the border, including moving part of *FA* 431 to Faulbrücke bei Reichenbach on 15 March. Saxony may have anticipated it would be called on to assist in repelling a Czech invasion. A contributing factor may have had to do with the fact that Bautzen was the center of Sorb culture in Germany. Sorbs are a Slavic minority related to the Czechs and Slovaks, and the *Reichs* government may have feared they would try to secede and join their southern neighbors.

Abteilung 2 remains largely a mystery; very little documentary evidence about it has surfaced. It was formed in Grossenhain at the same time as *Abteilung* 1 and apparently remained there for its entire existence.

In February, March, and April, the rivalry between the two Socialist parties controlling Leipzig and Dresden escalated into a conflict that began to spread throughout Saxony. Strikes broke out in the mines and manufacturing industries and on the railroads. When the strikes cut rail service between Leipzig and Berlin, the *Reichs* government stepped in, fearing Saxony was planning to secede. The closest available *Freikorps* was Maercker's *Landesjäger* Brigade, which arrived

Above: Leipzig's *Hauptbahnhof* (Main Railway Station), photographed by *FA* 423. (Paul Rempe)

in Leipzig 11 May and restored government control within a week.

By 5 July, integration into the *vorlaufige Reichswehr* had taken place: *Grenzjäger Flieger Abteilung* 1 was now *Truppen-Fliegerstaffel* Bautzen, and *Grenzjäger Flieger Abteilung* 2 had become *Artillerie-Fliegerstaffel* Grossenhain. In addition, a third unit, *Fliegerkommando* Dresden-Kaditz, had been formed. No other information about it has been discovered so far, but it was probably a staff and courier detachment.

The feared Czech attack and Sorb secession did not come to pass, so *Truppen-Fliegerstaffel* Bautzen enjoyed a relatively peaceful stay in Stiebitz. On 1 October, it would return to Grossenhain, where it would be merged with *Artillerie-Fliegerstaffel* Grossenhain. The consolidated unit, *Flugabteilung der Reichswehr Fliegerhorst Grossenhain*, would remain in operation until 7 May 1920.

One of its charter personnel was 40-victory ace and *Pour le Mérite* holder, Franz Büchner, who had joined it after his previous unit, *FA* 418/*Truppen-Fliegerstaffel* 27, had disbanded.

Although the Red uprising opposing the March 1920 Kapp Putsch was centered on the industrial district of the Ruhr, strikes and street fighting broke out elsewhere in Germany, including Büchner's hometown, Leipzig. Conducting reconnaissance in a Fokker D.VII, he was brought down by ground fire from the strikers and killed on 18 March, three days after his mentor at *Jagdgeschwader* II, *Oblt*. Rudolf Berthold, perished at Harburg.[42]

The Kapp Putsch and Ruhr Uprising

On 14 March 1920, a group of right-wing militarist politicians and soldiers, backed by General von Lüttwitz and provided muscle by the Erhardt Marine Brigade *Freikorps*, seized control of Berlin and installed Wolfgang Kapp, an obscure politician and rabble-rouser, as Chancellor. When the *Reichswehr* spurned the government's request for help in putting down the coup, the government leaders fled to Stuttgart and called for a general strike. This, and the general public's lack of support, succeeded in breaking the *Putsch* within a few days, but the strike escalated into an armed Red uprising in parts of Germany, especially the Ruhr District. To restore order, the government then had to do a humiliating flip-flop and call on the army to put down the rebellion that had saved it.

Above: Halb. CL. IV of 3rd (Lowenfeld) Marine Brigade. The inscription on this photo leaves no doubt about where and when.

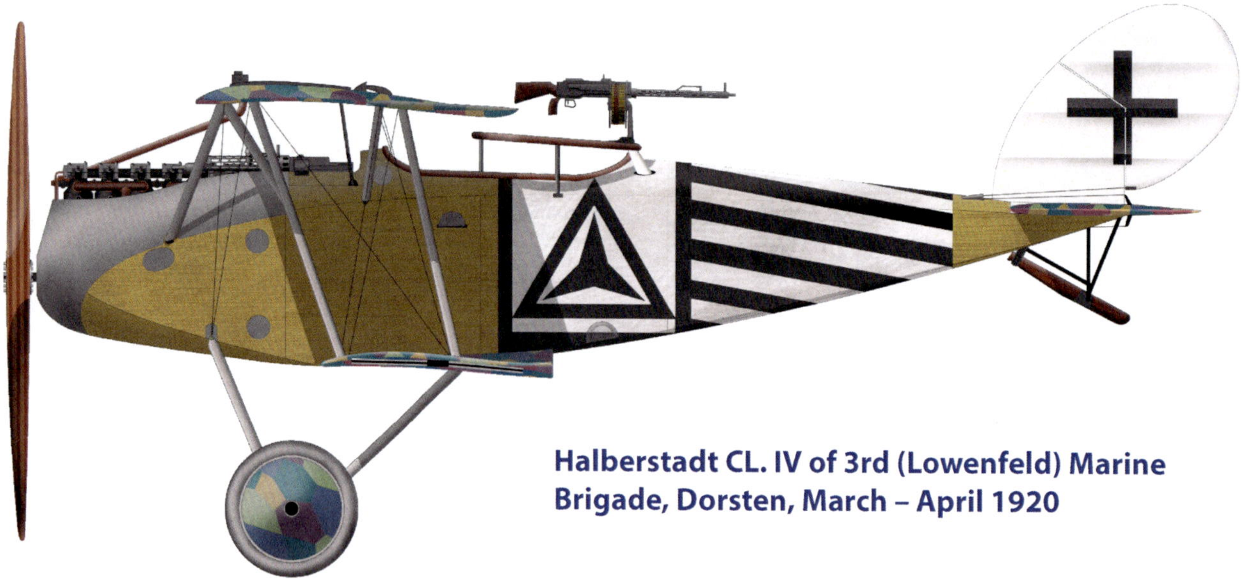

Halberstadt CL. IV of 3rd (Lowenfeld) Marine Brigade, Dorsten, March – April 1920

Bavaria remained relatively quiet during this time; the leftists and unions had been thoroughly beaten down the previous May and had no stomach for joining the revolt, so the *Reichswehr* had few reservations about sending Bavarian troops to support government forces fighting the uprising elsewhere. A draft order, dated 24 March, dispatched the von Epp *Freikorps*, now part of 21. Brigade, north to Paderborn, along with six aircraft to provide a show of force, if needed.[43] Those aircraft, drawn from the complement at *Reichswehr-Fliegerhorst* Schleissheim, were patrolling the skies over nearby Hamm and its environs by 30 March.

Other troops and aviation detachments were dispatched to the area; the 3rd Marine (Lowenfeld) Brigade was stationed in Bottrop, while at least one of the Halberstadt CL.IVs of its *Flieger Abteilung* operated out of nearby Dorsten, Westphalia.

The *Fliegerhorst* at Horst-Emscher, north of Essen,

Above: The airfield at Horst-Emscher was home to a unit believed to be involved in putting down the Red Ruhr uprising. Its lineup includes two Halberstadt CL.IV, including Roland-built 9508, and three each Fokker D.VII and LVG C.VI. (Author's Collection)

hosted an aviation unit believed to have participated in breaking the uprising. It's not known whether this was a unit based there or simply a guest, but it probably performed the usual observation, liaison, and "demonstration" flights against the strikers.

The Red uprising was put down quickly and brutally, with lopsided casualty totals in favor of the government forces. However, this success came at a cost -- among the government casualties, three flyers paid with their lives -- Franz Büchner, described above, *Ltn*. Kurt Scheibe, former commander of *FA* 407, who fell to ground fire at Halle, and *Hpt*. Rudolf Berthold, whose *Eiserne Schar* was trapped and defeated at Harburg.

Endnotes

1. *Militärpass, Flieger* Wilhelm Billerbeck.
2. "*Reichskriegsministerium, Die Wirren in der Reichshauptstadt und im nördlichen Deutschland 1918 - 1920. Darstellungen aus den Nachkriegskämpfen deutscher Truppen und Freikorps, Sechster Band (2. Reihe)*, E. S. Mittler & Sohn, Berlin, 1940, p.68.
3. ibid., pp.106 -20.
4. ibid., p.85.
5. Zuerl, Walter, *Pour le Mérite-Flieger,* Curt Pechstein *Verlag*, München, 1938, pp. 508-9.
6. Rick Duiven personal correspondence, 18 January 2000. *Aviation Awards of Imperial Germany in WW I, Volume VI -- The Aviation Awards of the Grand Duchies of Baden and Oldenburg*, states Pape was killed by ground fire during the fighting for Berlin 2 January, but *FA* 423 wasn't yet active on that date.
7. *Truppenamt* was the name given to the former operations section of the General Staff to conceal its

Fokker D.VII (OAW), *Fliegerhorst* Horst Emscher, March 1920

true purpose from the victorious Allies.
8. Rempe, Paul L., ed., *From German Cavalry Officer to Reconnaissance Pilot*, Savas Beatie, 2016, El Dorado Hills, p. 154.
9. "*Auszug eines Truppenberichtes über die Tätigkeit der Militär Flieger im Kampf um die innere Ordnung*", *Reichswehrbefelungstelle Preussen Truppen-Departement. Nr.* 124II.19. A.7.L. IV.C., *BayHStA Abt. IV Kriegsarchiv* document *Fl. Horst Gersthfn. Bd.* 13, pp.1-3.
10. *Reichskriegsministerium, Die Wirren in der Reichshauptstadt und im nördlichen Deutschland 1918–1920*, pp. 134–6.
11. Hildebrand, K. F., *Die Generale der deutschen Luftwaffe*, pp. 157–8.
12. von Hösslin, *Rittmeister a.D.* Hubert, *Die Organisation der K. B. Flieger-Truppe 1912 – 1919, Verlag der Bayerischen Kriegsarchivs*, München, 1924. p. 36.
13. Schubert, *Dipl.-Ing.* Helmut, "*Standorte und Flugfelder der Königlich-Bayerischen Fliegertruppe im Königreich Bayern*", *Blätter zur Geschichte der Deutschen Luft- und Raumfahrt* XIX., p. 36.
14. Ockel, Heinz, *Revolutionäre Volkswehr 1918-19*, Berlin, *Militärverlag*, 1968, p. 228.
15. Moore, Simone M. (tr.), *Oberst a. D.* Robert Holtzmann, "From the Balkans to the Baltic", *Over the Front*, Vol. 8, No. 3, 1993, p. 211–2.
16. "*Flieger in den Nachkriegskämpfen*", *Reiter gen Osten, Jahrgang* 7, *Folge* 10, *Seite* 14 (October 1936).
17. Pitrof, *Oberst d. Landes-Polizei a. D.* Daniel Ritter, *Gegen Spartikus in München und im Allgäu, Buchdruckerei und Verlagsanstalt* Carl Gerber, Munich, 1937, p. 64.
18. Reinhard Zankl personal correspondence, 11 October 2022.
19. Pitrof, p. 69.
20. "*Der Münchener Räte-Republik*", City Councillor Max Gerste, Munich, *Verlag der Politischen Zeitragen*, 1919, quoted in Weiss, Dr. G., *Kernpunkte der Schulorganisation*, Langensalza, Hermann Bayer & Söhne (Bayer und Mann), Herz., p. 76.
21. Pitrof, p. 65.
22. *Reichskriegsministerium, Darstellungen aus den Nachskriegkämpfen deutscher Truppen und Freikorps. Die Niederwerfung der Räterschaft in Bayern 1919, Vierter Band*, E. S. Mittler & Sohn, Berlin, 1939, p. 65.
23. Pitrof, p. 59-61.
24. *Darstellungen, Vierter Band*, p. 88.
25. *Darstellungen, Vierter Band*, pp. 195-208.
26. von Pölnitz, Götz, Emir - *Das tapfere Leben des Freiherrn* Marschall von Bieberstein, *Verlag* Georg D. Callwey, 1938.
27. *Darstellungen, Vierter Band*, p. 202.
28. *Darstellungen, Vierter Band*, p. 204.
29. *Tagsbefehl Nr.* 2, *Kommandeur der Flieger Genkdo.* Oven, Nr. Ia 12., K.H.Qu., 29.4.1919. BayHStA *Abt.*IV, *Kriegsarchiv* document *RWGrKdo* 4 Bd. 1033.
30. *Darstellungen, Vierter Band*, pp. 143, 147.
31. Hannes Taeger personal correspondence .
32. *Fliegerbefehl, Gruppenkommando West*, Munich, 5.5.19. *Landesarchiv* Baden-Württemberg, *Hauptstaatsarchiv* Stuttgart M356 *Bü* 71, *Bild* 60.
33. *Flieger Ersatz Abteilung* I *Geräteverwaltung* memo No. 33 to Abteilung. ___, *Betreff: Termin gemäss Abteilungsbbefehl vom* 10.5.19, BayHStA Abt.IV *Kriegsarchiv* document *Truppenfliegerstaffel* 21 *Rw* 827 *Bd*. 1.
34. *B. Reichswehrkommando* No 4, *Führer der Flieger* memo J. Nr.203., BayHStA *Abt.*IV *Kriegsarchiv* document *Fl. Horst Gersthfn Bd.* 20.
35. *Inspektion der Fliegertruppen Tgb. Nr.* 1283/8.19 *Abt.* I. II. Ang., Berlin, 28.8.1919.
36. *Ministerium für mil. Angelegenheiten* memo *Nr.* 80801A to *die Inspektion des Luftfahr-Wesens betreff: Stärkenachweisung für bayer. Kurier-Flieger-Station*, BayHStA *Abt.*IV.
37. *Bayr. Gruppenkommando* No. 4 memo Ia/o O *Qu. Fl.*No. 5215, 19.September 1919, Betreff: *Verringerung der Fliegertruppe*, BayHStA *Abt.*IV *Kriegsarchiv* document TR.FL.STA 21 (RW828) Bd.2.
38. *B. Reichswehrkommando* No 4, *Führer der Flieger* memo No. 5224 to *Gruppenkommando* 4 Ia/o O Qu. Fl.No. 5215 *vom* 19.9.19, BayHStA *Abt.*IV *Kriegsarchiv* document Tr.Fl.Sta 21 (RW828) *Bd*.2.
39. von Hösslin, p. 37.
40. Schroeder, *Generalmajor a. D., Die Sächsischen Freiwilligen-Truppen in Litauen 1919, Buchdruckerei der* Wilhelm und Bertha von Baensch *Stiftung* 1933, Leipzig, pp. 122-3.
41. Taeger, Dr. Hannes and Manfred Beyer, "Egon Doerstling (1890-1965) -- *Königlich Sächsischer Fliegeroffizier und General der Luftwaffe*", *Arbeitskreis Sächsische Militärgeschichte e. V., Mitteilungsblatt Nr.* 16/2006.
42. Taeger, Hannes, "Franz Büchner The Lion of Leipzig", *Over the Front*, Vol. 33, No. 2, 2018, pp. 140–6.
43. *Bayr. Reichswehr-Gru.Kdo. Nr.4 Hpt. Nr.* 13145/*Kraft Nr.* 2567, BayHStA *Abt.*IV *Kriegsarchiv* document RWGrKdo 4 Bd. 1100.

Black vs. White Eagles – Defense of the Polish Frontier

Above: Troops serving in *Grenzschutz Ost* wore a wide assortment of patches and pins with the shield of the Teutonic Knights. Pfalz D.III 8304/17 probably had the largest of them all.

Historical Background

Shortly after taking the throne of Prussia in 1740, King Frederick the Great embarked on an aggressive campaign of territorial expansion. Two wars with the Hapsburg Holy Roman Empire gave Prussia control of Silesia, with its sizeable Polish population. Between 1772 and 1795, three partitions extinguished the Polish kingdom, dividing it among Russia, Austria-Hungary, and Prussia, whose share included West Prussia and Posen, an area known to the Poles as *Wielkopolska* (Greater Poland).

The succeeding years saw the Prussians make concerted efforts to suppress the Polish language and national identity in their newly acquired territory. On the surface, they had some success, but most of the Poles simply laid low and waited for better days, which came with Germany's defeat in World War I. With the war grinding down to its conclusion, United States President Woodrow Wilson published his Fourteen Points, aspirations for how the world should be ordered after the War. Article XIII stated:

"An independent Polish state should be erected which should include the territories inhabited by indisputably Polish populations, which should be assured a free and secure access to the sea, and whose political and economic independence and territorial integrity should be guaranteed by international covenant."[1]

Although the Allies would pick apart Wilson's dreams at Versailles, there was a general understanding that Poland's sovereignty would be restored. Further, it was commonly agreed the new state would include the territory formerly held by now-Bolshevik Russia and probably the disintegrating Austro-Hungarian Empire, but no decision had been made regarding the German-held portions of old Poland. The residents of the area decided not to leave their fate in the hands of the diplomats. In the days leading up to the Armistice, they secretly began preparations to take back their homeland, activating a paramilitary force known as the People's Guard. On 18 December 1918, the Guard emerged from hiding and began taking over strategic installations and military facilities throughout the area. By the end of December, they had disarmed the German garrison in the city of Posen, but it wasn't until 6 January 1919 that they stormed the Elsenmühle airfield of *Flieger Ersatz Abteilung* 4 (*FEA* 4), a sprawling supply depot in a neighborhood west of the city

Grenzschutz Ost (Kolberg, W. Prussia)
Generalfeldmarschall Paul von Hindenburg

- *AOK Nord* (Bartenstein , E. Prussia)
  General *d. L.* Ferdinand von Quast
  Kofl. 21. Major Alfred Streccius
  - *Gkdo.* VI *R. K.*, Mitau
    - Brigade Kurland, Mitau
    - *Eiserne* Division
      - *FA* 427, Wainoden
    - *Baltische Landeswehr*
      - *FA* 433, Memel
    - 1. *Garde* Reserve Div.
      - *Kampfgeschwader* Sachsenberg (*FA* 413, 416 & 417), Wainoden
  - *Gkdo. Freiwillige Reservekorps*, Kowno
    - Brigade *Südlitauen*, Kowno
      - *FA* 424, Mariampol
      - *FA* 425, Aleksotas (Kowno)
      - *FA* 429, Wilkowischki
    - Brigade Grodno, Suwalki
      - *FA* 426, Suwalki
    - Brigade Schaulen, Schaulen
      - *FA* 408, Kielmy
      - *FA* 409, Kielmy
  - *Gkdo.* XVII. *AK*, Danzig
    Kofl. Hpt. von Liebermann
    - *FA* 411, Graudenz
    - *FA* 412, Thorn
    - *FA* 434, Thorn
    - *FA* 435, Graudenz
  - *Gkdo.* XX. *AK*, Allenstein
    Kofl. Hpt. Fenske
    - *FA* 410, Lyck
- *AOK Sud* (Breslau, Silesia)
  General *d. L.* Kurt von den Borne
  Kofl. Hauptmann Stahr
  - *Gkdo.* II AK, Stettin
    Kofl. Hpt. Wolf Thieme
    - *FA* 404, Schneidemühl
    - *FA* 414, Bromberg
    - *FA* 415, Kolmar
    - *FA* 428, Köslin
  - *Gkdo.* V. *AK*, Glogau
    Kofl. Hpt. Gustav Walter
    - *FA* 406, Glogau
    - *FA* 407, Lissa
    - *FA* 418, Glogau
  - *Gkdo.* VI. *AK*, Breslau
    Kofl. Hpt. Bruno Zühlke
    - *FA* 400, Oppeln
    - *FA* 401, Gleiwitz
    - *FA* 402, Brieg
    - *FA* 403, Breslau-Klein Gandau
    - *FA* 405, Brieg
    - *FA* 430, Breslau-Hundsfeld
    - *FA* 431, Breslau-Klein Gandau

Bases listed are those to which the units were originally posted.

Pfalz D.III 8304/17

center. Stocked with assembled aircraft, engines, and replacement parts, and furnished with workshops to maintain and repair them, it was a great windfall for the new Polish air force. Renamed Ławica, it would also become the home base for most of the Polish air force activities in Posen (now Poznan´) province. Two days after *FEA* 4 fell to the Poles, German aircraft based in Frankfurt *an der* Oder bombed its barracks.[2]

The Poles quickly organized their aviation forces in the area, establishing two squadrons, the 1. and 2. *Wielkopolska Eskadra Lotnica Polna* (Greater Poland Field Aircraft Squadrons) by mid-February. Facing them would be the *Flieger Abteilungen* of *Grenzschutz Ost*.

Grenzschutz Ost

By early January, Germany faced three threats from the East. In addition to the *Wielkopolksa* uprising in Posen, an insurgency was expected in Silesia, and the Red Army was overrunning the Baltics, driving the remains of the 8. *Armee* and *Oberfehlshaber Ost* back toward East Prussia. There was real fear the surging Bolsheviks wouldn't stop at the German border. Expectations of these developments had led to the November 15 creation of the *Grenzschutz Ost* (Border Defense East) command, led by *Generalfeldmarschall* Paul von Hindenburg, with *Generalleutnant* Wilhelm Groener as his Quartermaster General. Command headquarters was set up in Kolberg, West Prussia, and its forces, ringing Polish territory, were divided into two subsidiary commands -- *Armee Oberkommando* (abbreviated *AOK*) *Nord* (North), headquartered in Bartenstein, East Prussia, and *AOK Sud* (South), situated in Breslau. On 24 November, OHL (*Oberste Heeresleitung* - Supreme Army Command) issued a call for volunteers to join *Grenzschutz Ost*. Among the troops rushed to the area, 30 *Flieger Abteilungen* were strung along the Polish Front from Upper Silesia to East Prussia, and then into Kurland and Lithuania to help German forces fend off expected Bolshevik attacks from the East. Over 80% of the new Prussian air force had been committed to *Grenzschutz Ost*.

During the Great War, the German homeland had been divided into 21 military districts, each the base for one *A.K.* (*Armee-Korps* - Army Corps). Post-Armistice, five of them were disbanded due to the loss of Alsace-Lorraine, demilitarization of the west bank of the Rhine and the consolidation of two *Korps* commands in Saxony. The remaining sixteen districts were the building blocks for the postwar *AOK*s.

Armee Oberkommando Nord

AOK Nord covered Germany's Baltic coast from eastern Pommerania through East Prussia to the Russian frontier. It was made up of *A.K.* I, XVII and XX. When *Ober Ost* was disbanded, the army and *Freikorps* units intended for the Baltics, VI. *Reserve Korps* and the *Freiwillige Reserve Korps,* were attached to it. Ultimately, sixteen *Flieger Abteilungen* reported to *AOK Nord,* six within Germany proper, four in Kurland and six in Lithuania.

I. *Armee Korps*, headquartered in Königsberg, was assigned one flying detachment, *FA des Ostpreussisches-Freiwilligenkorps* Memel, to defend East Prussia from the Bolsheviks rolling westward through Kurland. The unit was later designated *FA* 433 and eventually seconded to the Baltic *Landeswehr*.

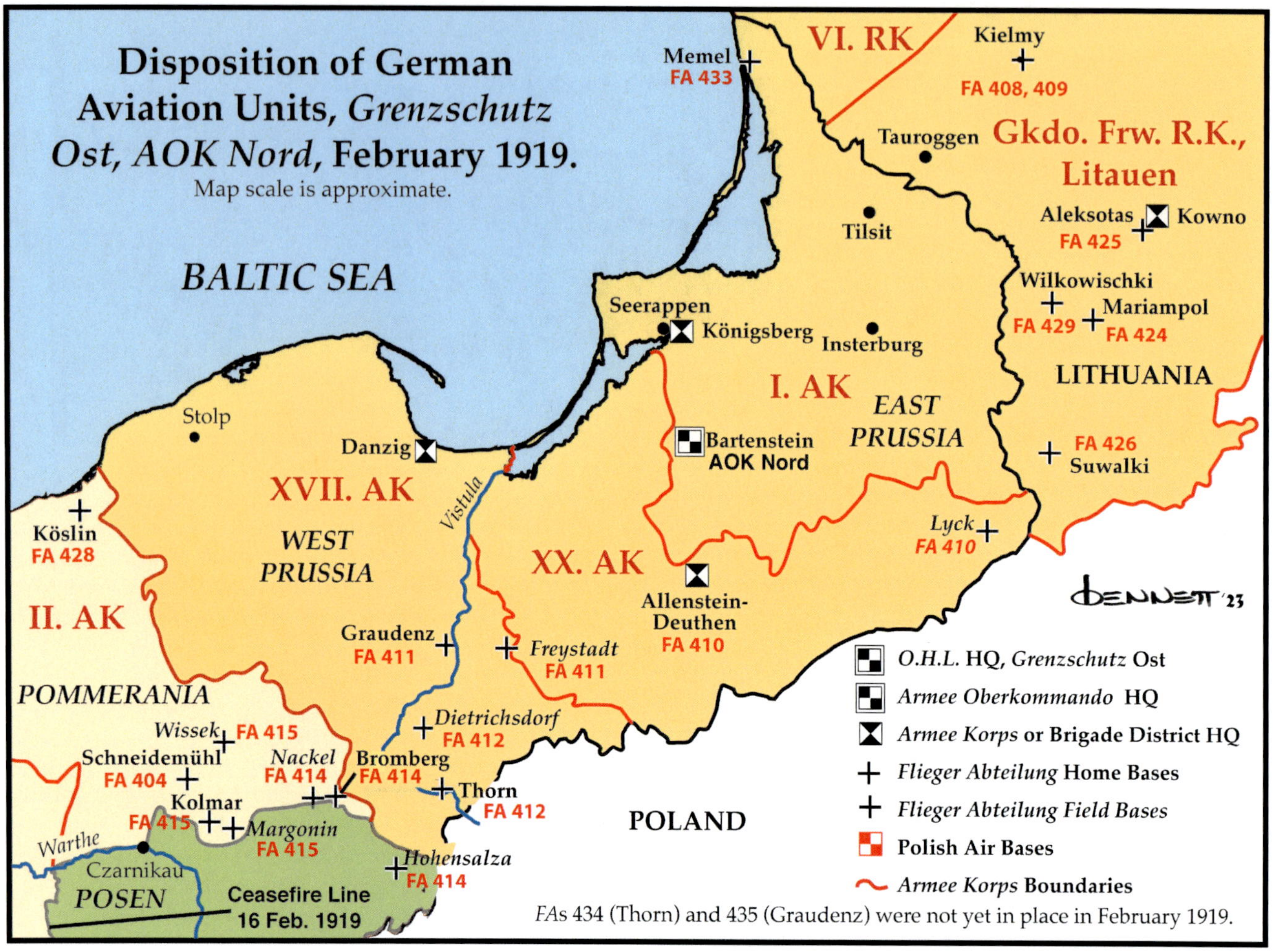

Above: Troops serving with The XVII (*Westpreussische*) *Armeekorps* often wore the evergreen tree and crossed swords insignia of the *Korps* on their collars or sleeves. This included these two aviators, members of *FA* 412 at *Fliegerhorst* Thorn. (Author's collection)

Flieger Abteilung 410/*Artillerie-Fliegerstaffel* 120

Above: No collection of German aviation unit photos seems to be complete without a shot of the *Abteilung's* officers, cigars and wine glasses in hand, at the end of what must have been a riotous dinner. This 5 May 1919 photo centers on Walter Aschoff, in civilian clothes, looking like the only sober individual in the room. Aschoff replaced *Hpt*. Koch as *Abteilungsführer* of *FA* 410/*Artillerie-Fliegerstaffel* 120. (04AEF-B01-Aschoff Series-F06-PB165, History of Aviation Collection, Special Collections and Archives Division, Eugene McDermott Library, The University of Texas at Dallas)

Flieger Abteilung 410, formed at *FEA* 8 in Graudenz, was originally intended to join the XVII. *AK*, but was ultimately assigned to the XX. *Armeekorps*, based in Allenstein. Its home base was the Deuthen airfield at Allenstein, but it soon moved east to Lyck, closer to the Polish frontier, where the action was. Following ratification of the Versailles Treaty, the *Abteilung*, now named *Artillerie-Fliegerstaffel* 120, appears to have returned to Deuthen.

Left: Staff car of *FA* 410. (04AEF-B-01-Aschoff Series-F06-PB164, History of Aviation Collection, Special Collections and Archives Division, Eugene McDermott Library, The University of Texas at Dallas)

Facing Page, Above: "*Flughalle* Deuthen". Fokker D.VII (OAW) 8453/18 is in the left foreground; other aircraft in the hangar include Rumpler C.Is, Albatros C.IIIs and LVG C.VIs. At least four iterations of the iron cross marking are in evidence here. (Michael Seidel)

Fokker D.VII (OAW) 8453/18
***FA* 410/*Artillerie-Fliegerstaffel* 120,**
***Grenzschutz* Allenstein, Deuthen, 1919**

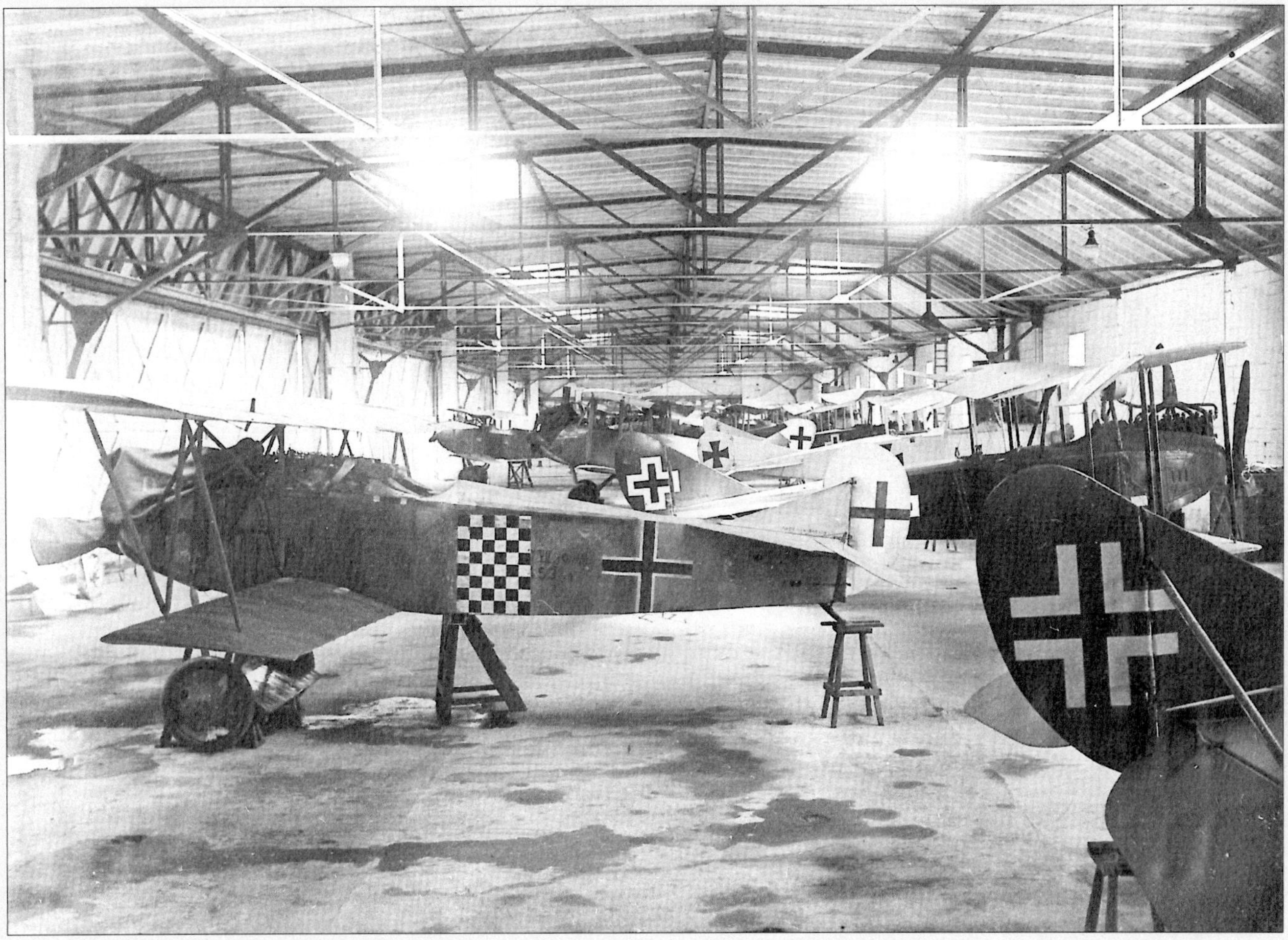

Fokker D.VII (OAW) 8453/18 outside the Deuthen hangar, along with three LVG C.VI. (Michael Seidel)

Above: "*Kopfstand*". An *FA* 410 LVG C.VI has come to grief. Crashes always provided a useful backdrop for photos to be sent to the folks back home. (Michael Seidel)

Above: Fokker D.VIII and personnel of *FA* 410. This may be the same machine portrayed in the *Abteilung's* lineup photo. If so, it has lost its wheel covers, and its fuselage and the underside of its wings appear to have been overpainted in a solid dark color. (Michael Seidel)

Above: This *FA* 410 Rumpler C.I came to a bad end. Digital manipulation of the photo reveals the wreck to be C.4041/17 (HaBra), a former training school machine. HaBra, a rarely seen abbreviation, stood for Hansa-Brandenburg. (Michael Seidel)

Above: Total writeoff. Rarely is a wreck so complete it can't be identified, but this one is a challenge. It's a two-seater with a light-colored wing covering, strongly suggesting it was a trainer before coming to *FA* 410. Little else remains to confirm its identity. (Michael Seidel)

Above: August 1919: Albatros C.XII 1911/17, marked "Beob. Sch. [Observer School] Stolp." Although Stolp was situated within *Grenzschutz Ost's* jurisdiction, it was not one of its military airbases. The Albatros bears mismatched wing panels and a mixture of cross markings -- *pateé* on the wing and *balken* on the fuselage. (Author's collection)

Above: The evergreen tree insignia on the sleeves of two of the fliers identifies them as members of the *Grenzschutztruppen des* XVII. (*Westpreussisches*) *Armeekorps*, so C.X 1911/17 was most likely a transfer to *FA* 411 at Graudenz. Its rudder and fin don't appear to carry a cross marking. (Author's collection)

Albatros C.XII 1911/17
Ex-*Beobachter Schule* Stolp
***FA* 411, Graudenz, August 1919**

Danzig-based XVII. *Armee Korps* originally had two *Flieger Abteilungen* – 411, based at Graudenz, and 412, at Thorn.

In late March or early April, Prussia formed its final two *Flieger Abteilungen*, 434 and 435. *FA* 434 was based at Thorn, while *FA* 435 operated out of Graudenz. These new units may have been created to fill in for *FAs* 411 and 412 after they transferred closer to the Polish front lines.

Armee Oberkommando Sud

AOK Sud stretched from the Baltic coast south to Upper Silesia and included. II., VI., and what was left of the *V.A.K's* territory. In all, it had fourteen *Flieger Abteilungen* under its command.

Initially, the heaviest fighting between the Germans and Poles took place in northwestern Posen, around Kolmar and Czarnikau, which changed hands three

Above: Same unit, some of the same flyers posing with a Fokker D.VII inside the hangar. A Rumpler C.IV is in the background. (Author's collection)

Above: Portion of a photo of *FA* 412's officers, NCOs and enlisted men, probably taken at Thorn. The second and third men from left in the second row appear in an earlier photo. (Greg VanWyngarden)

Left: *Ltn*. Hans Gandenberger-Moisy (O), *Lt. d. R.* Alfred Brocke, *Ltn*. Heidt (O) and *Lt. d. R.* Richard Kraut (P), posing in front of one of *FA* 434/*Truppen-Fliegerstaffel* 9's hangar doors. Three of them are sporting varieties of the *Grenzschutz Ost* insignia, a black cross on a white shield, inspired by the shield of the 13th Century Teutonic Knights. (Neal O'Connor)

Right: Officers and aircrew of *FA* 434/*Truppen-Fliegerstaffel* 9 making merry for the photographer. *Abteilungsführer Hpt*. Alfred Quade is seated third from left in the front, trying to maintain the dignity expected of a commanding officer. Observer *Ltn*. Hans Gandenberger-Moisy stands at the left of the second row, with *Ltn*. Brocke fourth from the left. Observer *Ltn. d. R.* Heidt stands at the right of the back row. (Author's collection)

Above: *FA* 434 observer *Ltn*. Heidt sees what the view is like from the front seat. Aircraft is the Fokker D.VII (OAW) of his *Abteilung* mate, *Lt. d. R.* Richard Kraut. Site is *Fliegerhorst* Thorn. (Greg VanWyngarden)

Disposition of German Aviation Units, *Grenzschutz Ost, AOK Sud*, February 1919.
BALTIC SEA
Kolberg
O.H.L.
Grenzschutz
Ost
Köslin
FA 428
POMMERANIA
Stettin
II. AK
Graudenz
FA 411
Freystadt
FA 411
Gut
Dietrichsdorf
FA 412
Wissek
FA 415
Schneidemühl
FA 404
Nackel
FA 414
Bromberg
FA 414
Thorn
FA 412
Kolmar
FA 415
Margonin
FA 415
FA 414
Hohensalza
Czarnikau
POSEN
Poznań
Lawica
2 Wielkopolska
Eskadra
Ceasefire Line
16 Feb. 1919
Meseritz
FA 406
Frankfurt
a. Oder
Oder
BRANDEN-
BURG
V. AK
Lissa
FA 407
Glogau
FA 406,
418
Rudelsdorf
Breslau
AOK Sud
FA 431
Görlitz
FA 403, 431
kl.-Gandau
FA 430
Hundsfeld
Brieg
FA 431
Faulbrücke
FA 402, 405
VI. AK
SILESIA
FA 400
Oppeln
FA 401
Gleiwitz
POLAND
CZECHO-SLOVAKIA
O.H.L. HQ, *Grenzschutz Ost*
Armee Oberkommando HQ
Armee Korps or Brigade District HQ
Flieger Abteilung Home Bases
Flieger Abteilung Field Bases
Polish Air Bases
Armee Korps Boundaries
Map scale is approximate
BENNETT '23

Above: Airfield of *FA* 406/*Artillerie-Fliegerstaffel* 127 near Meseritz. Aircraft in this photo appear to include four LVG C.VI, two Fokker D.VII, one Junkers CL.I, and one small two-seater, perhaps a Halberstadt CL.IV. (Author's collection)

Below: Personnel of Artillerie-Fliegerstaffel 127, formerly FA 406. Location is probably the unit's forward field at Meseritz. The 4-bladed " propellers" at the right rear are actually windmill vanes. (Author's collection)

Above: The Roland D.VIb was a rare bird among *Freikorps* aircraft. Like its *FA* 406 companion Halberstadt shown above, D.1535/18 has been stripped of its fuselage and tail crosses. (Author's collection)

Roland D.VIb 7535/18
***FA* 406/*Artillerie-Fliegerstaffel* 127,**
Meseritz, 1919

Above: *FA* 406's moment to clown for the photographer. Even silly pictures like this can convey a lot of information to historians; this one tells us it was probably taken at the *Abteilung's* home base in Glogau, and its inventory contained one or more SSW D.IVs. (Author's collection)

times between 5 and 8 January. *FA* 415 was especially active in this area, and one of its pilots, *Ltn. d. R.* Hans Brzenk, flying a Fokker D.VII, shot down a Polish DFW C.V between Czarnikau and Briesen.[3] This victory was Brzenk's fifth, on top of four confirmed during the Great. War. The Poles got a good measure of revenge during the fighting for Czarnikau on 8 January, when Brzenk's *Staffel* mate, *Ltn.* Max Näther, was killed by ground fire from a machine gunner sheltered in the shadows of a church.[4, 5, 6] Näther, former commanding officer of *Jasta* 62, was a 26 victory ace.[6, 7]

The volatility of the military situation in the area is reflected in the fact that four out of the five *Flieger Abteilungen* assigned to this sector changed bases at least once by June. Only one other unit on the entire Polish front moved to a different base before the signing of the Versailles Treaty.

Right: *Flugmaat* Moritz Nülle, formerly a pilot with the 2. *Marine Feld Jasta* in Flanders, joined *Truppen-Fliegerstaffel* 2 (formerly *FA* 415) on 1 July 1919 and remained with the unit until 2 january 1920. The location of his *Grenzschutz Ost* insignia was a departure from the norm. (Justin Young)

Above: Gut Dietrichsdorf, home of *FA* 412. This photo was taken 12 March 1919 by the team of pilot Körner and observer *Ltn*. Holbach. Six aircraft can be seen outside the seven tent hangars erected on the east end of the base. (Reinhard Zankl)

Fokker D.VII (OAW)
***Fl.Abt.* 418**
1919

Above: Personnel of *FA* 418/*Truppen-Fliegerstaffel* 27. L to R: *Ltn*. Gebbert, *Ltn*. Rudolf Reinau (P), *Ltn*. Richter, *Ltn*. Kuke, *Feldw*. Wiedemann, *Oblt*. Bilke, *Feldw*. Friedrich Wilhelm Morzik (P), *Feldw*. Otto Klaiber (P), *Oblt*. Oskar *Frhr*. von Boenigk (CO), *Feldw*. Bernhardt, *Oblt*. Gustav Brockmann, *Werkmstr*. (?), *Ltn*. Wilhelm Leusch (P), *Feldw*. Korfay, *Ltn*. Treuhaupt (O), *Feldw*. Friedrich, *Ltn*. Mahn (P), and *Feldw*. Grenzisa. (Rick Duiven)

Above: Fokker D.VII (OAW) of *FA* 418/ *Truppen-Fliegerstaffel* 27. The trident or pitchfork personal marking calls to mind the D.VII of *Jasta* 43's Otto Creutzmann and Josef Raesch, but neither pilot is known to have served with the *Freikorps*. (Greg VanWyngarden)

Despite fierce fighting, not much ground changed hands, so much of the significant activity took place in the air, as evidenced by Polish dispatches:[6, 8]

15 February 1919 - German planes strafe Ostrów [Ostrowo] and drop a bomb on the rail line between Ostrów & Krzeszow [Grüssau].

16 February 1919 - Fighting continued, despite a ceasefire signed in Trier. German aircraft attacked Ostrzeszow [Schildberg] and Miejska Gorka [Görchen].

18 February 1919 - A lone German aircraft attacked a train between Ostrów and Kalisz [Kalisch], killing a woman.

22 February 1919 - German aircraft strafed and dropped a few bombs on Czarnikau.

13 March 1919 - A German plane strafed Mieczkowo, while to the southwest, German aircraft patrolled over Kolnow and Pierzynami.

18 March 1919 - Polish forces observed the crash of a German plane south of their positions at Walkowice, near Czarnikau. The following day and again on the 22nd, the Germans raised an observation balloon at Kijev. In the early afternoon, eight German planes overflew Malem Grojec.

24 March 1919 - German planes circulated above Malem Grojec and Jezierzyce.

27 March 1919 - Three German planes appeared over the front near Grojecka. One strafed the town.

11 April 1919 - Significant activity by German airmen.

31 May 1919 - Intense activity by German airmen along the entire front.

Under strong pressure from the Allies, the Germans and Poles gradually came to honor the Trier ceasefire, although skirmishes and sniping continued all along the demarcation line almost until Germany's acceptance of the Versailles Treaty.

Despite the ceasefire, both sides continued to keep close tabs on their opponents. A Polish intelligence report, dated 6 July 1919 [9], presented a rundown on the German *Abteilungen* opposing them in the west section of the Front:

FA 410, stationed at Allenstein, commanded by Captain Koch, operated 3 Fokker D.VII and 6 Halberstadt C.V.

FA 411, under Captain Winkler, had 3 Fokker D.VII, 3 Halberstadt C.V, 2 Albatros C.I and one Rumpler C.IV. It was based at Freistadt, but had a secondary field at Langenau.

FA 412, stationed at Dietrichsdorf, between Culm and Culmsee, was led by Lieutenant Guteknecht [sic - Gutknecht]. No equipment information was presented.

The report noted the presence of a *Schlachtstaffel* assigned to Captain Rossbach's *Sturmabteilung* at Graudenz. Commanded by Lieutenant Lippmann, it was equipped with Halberstadt C.Vs. It also commented on a shortage of gasoline that was being addressed by the substitution of benzol, requiring the re-jetting of carburetors.

As the German government agonized over whether to accept the Versailles Treaty terms, both sides geared up for renewed fighting if the treaty were rejected. *FA* 431 completed its absorption into the *Reichswehr*, becoming *Truppen-Fliegerstaffel* 29 of the 29th brigade on 22 May. The following day, it moved to Rudelsdorf, closer to the Polish front lines, as both sides prepared for renewed fighting. It was not to be -- Germany agreed to the humiliating terms of the treaty, and the disgusted fliers returned to Breslau a few days later. Over seven weeks of inactivity followed. The only event of any significance was the crash of *Uffz.* Hassenmitter in SSW D.6178/18 on 24 July. He apparently survived, although the appearance of the wreckage suggests he had a long recovery ahead.

Finally, on 16 August, *Truppenfliegerstaffel* 29 was ordered to return to Paderborn for disbandment. Most of its personnel were discharged, but those remaining, including Eberstein, were retained for service in a planned new *Polizei Fliegerstaffel* based in Paderborn.

Role Change -- the First Silesian Uprising

Germany's acceptance of the Versailles Treaty greatly reduced the chance of further combat with the Polish army and freed Poland to transfer badly needed troops and aircraft to the east, where it was facing Lithuania and Ukrainia over disputed borders. However, it didn't guarantee peace along the German-Polish frontier. The center of the conflict moved south, to coal- and iron-rich Silesia. The majority of the population of Upper Silesia, where most of those mineral treasures lay, was Polish, and the Treaty promised a plebiscite to determine which country would eventually possess the area. In the meantime, it would remain under German control.

Inevitably, conflict would arise while both sides awaited the plebiscite. The first serious incident occurred 15 August -- *Grenzschutz* troops killed 10 Silesians in a labor dispute at a mine in Myslowitz in extreme southern Silesia. The miners responded with a general strike, and the uprising spread quickly to the towns and mines around Kattowitz.

Prior to this, the aviators in the area had to be alert to incursions by the Polish army and possible air-to-air combat against Polish flyers. Suddenly, in eight days of frantic activity, they found themselves in a new role -- flying low-level reconnaissance and ground support, hopping from town to town in support of the *Grenzschutz* troops. The *Reichswehr* finally crushed the

Flieger Abteilung 431/*Truppen-Fliegerstaffel* 29

Above: *Hpt*. Ernst Otto Eberstein (center) with some of *Truppen-Fliegerstaffel* 29's personnel. Several of them are wearing the eagle's head collar insignia of the 9. *Infanterie* Division/ *Freiwilliges Jäger-Bataillon* Schull. *Ltn*. Friedrich-Wilhelm Liebig is believed to be on Eberstein's right. *Ltn*. Käubler, formerly with *FA* 26, is on the right, arms akimbo. Käubler joined the *Staffel* 9 June 1919, so this photo was probably taken after its return to the former flight school at Breslau-Klein Gandau. The sign by the door, reading " Admission only to non-commissioned officers and flight students", reinforces this assumption. (Reinhard Zankl)

Flieger Abteilung 431 was one of the most prominent and active aviation units of *AOK Sud*. Formed at Breslau in January 1919, it was placed under the command of *Hpt*. Ermst Otto Eberstein, who had served as an observer during the Great War. As his *O.z.b.V. (Offizier zur besondere Verwendung* - Adjutant), he selected *Lt. d. R.* Alfred Lenz, whose long career as a fighter pilot on the Western Front had culminated in command of Jasta 22. On 16 January, the *Abteilung* was established at the Klein Gandau airfield in Breslau, reporting to the 9th Infantry Division.

In addition to the threat from newly independent Poland, a portion of coal-rich Silesia was being eyed by Czecho-Slovakia, which appeared to be preparing to invade it in early 1919.

To meet that threat, a portion of the 9th Division, along with *FA* 431, moved south to Faulbrücke bei Reichenbach on 15 March. At this time, the *Abteilung* consisted of 9 aircraft, 3 of them single seat fighters, along with their flight crews and about 180 support personnel. Eberstein was pleased with the move; this quiet little town was free from the distractions that tempted his men in the big city of Breslau.

The attack never materialized because a Hungarian invasion of Slovakia forced the Czecho-Slovaks to turn their attention there. This freed up part of the *Abteilung* to return to Breslau's Hundsfeld airfield 5 May.

With the approaching announcement of the Versailles Treaty terms, tensions were building along the Polish frontier. If Germany rejected the Treaty, renewed fighting was sure to break out. Eberstein's second *O.z.b.V*, *Ltn.* Martin, accompanied him, apparently leaving Lenz in command of the Faulbrücke contingent. Eberstein was not impressed by Lenz, who,

Left: *Fliegerschule* Breslau-Hundsfeld in 1917. It was the home base of *Flieger Abteilung* 430 and hosted a detachment of *FA* 431 in 1919. (Paul Rempe)

in spite of his long combat and command experience, struck Eberstein as too easygoing.[10]

FA 431 was equipped with a hodgepodge of aircraft, including a Fokker D.VII, Junkers CL.I (Jfa) 12608 and 12609/18, Ju. J.I 750/17, a Halberstadt CL.IV, LVG C.Vs and C.VIs and two SSW D.IVs, 3087/18 and 6178/18.

Above: Not the best of photos, but it shows some of the diversity of *FA* 431's aircraft -- from left to right, Junkers J.I 750/17, Halb. CL.IV and two LVGs, a C.VI and a C.V. (August Qoos album via Greg VanWyngarden)

Above: Junkers CL.I (Jfa) 12608/18 of FA 431, with *Uffz*. August Qoos in the pilot's seat. The CL.I's monoplane configuration tended to make it look small – unless humans were present to give it a sense of scale. (August Qoos album via Greg VanWyngarden)

Above: Mud bests metal -- *FA* 431 Junkers CL.I overturned on rough ground. (August Qoos album via Greg VanWyngarden)

Above: Ju. CL.I (Jfa) 12609/18 of *FA* 431, being inspected by *Hpt*. Ernst Otto Eberstein (in cockpit) and his *O.z.b.V., Ltn*. Martin, standing on the wing. An LVG C.V or C.VI is in the background.

Left: *FA* 431's inventory included this OAW-built Fokker D.VII. (August Qoos album via Greg VanWyngarden)

Right: *Uffz*. August Qoos was a pilot with *FFA* 30 and *FAs* 47, 14, and 13 before joining *FA* 431 postwar. (August Qoos album via Greg VanWyngarden)

Above: Wreck of *Uffz.* Hassenmitter's SSW D. IV 6178/18 at Breslau-Hundsfeld, 24 July 1919. (Peter M. Grosz via author)

uprising 24 August, subsequently executing around 2500 people who had participated in it.

In a follow-up report by the 117. *Infanterie-Division*[11], the flyers received high praise. The report also chronicles their activities:

"On 17 August 1919, Upper Silesia broke out in widespread insurrection fomented by Spartacists and Polish bands. Aviation was used with great success in suppressing the revolt. The opposition of the insurgents made the action of many independent bodies of troops imperative. In addition, a portion of the frontier posts, with their troop detachments, were cut off. Liaison and reconnaissance for the advancing independent troop detachments was the standing assignment for the aviators.

Analogous to the experience of the World War, the liaison between the fighting forces and their commanders in these guerilla battles was also chaotic. The aviators carried out these liaisons in the quickest way here. Through them, the brigade and the subordinate units were always advised about the situations in the rebellious districts.

For fulfilling their manifold tasks, the flyers had to go to the lowest altitudes (10 to 50 meters). From these altitudes, they were able to detect all blocked routes, felled trees, strung wires, and which uniforms were on the personnel, whether armed or not. From this, it was concluded whether they were friend or foe.

The participation in the ground fighting with their 2 to 3 machine guns is naturally highly effective. It is in the nature of things and is a well-known experience from the World War, that in every battle, the bravest infantryman in a foxhole is never more uncomfortable as under a machine gun attack from the air.

In the now-concluded fighting, the civilian threw his weapon away and hid himself in the cellar, machine gun nests erected for defense immediately went silent as soon as the flyers fired from the air.

In the following are cited individual cases of the activity of the flyers in the days from 17 to 22 August 1919.

1. Reconnaissance Activity:

a. The situation in the neighborhood of the village of Tischau was completely unclear on 17 August. One aircraft must carry out reconnaissance, it takes off and ascertains:
In the village itself, civilians in Sunday gatherings on the streets, carrying no weapons. On the platform 9 unarmed civilians. 3 km south of Tischau, on the Pless-Tischau tracks, is a locomotive with a baggage car and German soldiers. The airplane throws out a message there: "Tischau rail station at the most lightly occupied, probably not at all. Peace in the area. Following that, the column, with the aircraft escorting,

travels without fighting to the Tischau rail station. The advancing military occupied and secured it.

b. On the 18th, one aircraft carried out an order to ascertain the whereabouts of a formation of 30 German military men, who were taken captive in Nikischschacht. It located about 50 German soldiers with steel helmets leaving Myslowitz, advancing on the streets of Gieschewald. The instructions were dropped to them where to find the 30 men. The captives were freed.

c. On 22 August, under other orders, an aircraft carried out the instructions to escort advancing troop bodies on the line Ticheu [Tischau?]-Alt Berun-Neu Berun-Ticheu-Lendzin-Anhalt-Neu Gacz-Imelin and to provide them with constant reports about the places named above. The aircraft determined general peace in the cited places, some civilians in shirtsleeves, children in the streets, in Gross Chellon, 3 of our own cavalrymen, a motorcyclist travels through Neu Borun. These findings were dropped.

I. On the staff, which held the fork in the main road south of Gross Chellon. Thereupon it advances farther.

II. On an auto with officers sitting on the Alt Berun-Neu Berun road at the east exit from Alt Berun. Thereupon, the auto makes a turn and drives in the direction of Lendzin, followed by two trucks with infantry.

The plane escorts the advance and, farther south, again drops off a message that the area farther north is likewise clear. On the same morning all locations were occupied by German troops, and the rioting ended.

d. An aircraft taking off on reconnaissance finds German troops before Kattowitz and drops there their previously prepared reconnaissance report: "Aircraft receive fire from Janow and Nickischacht after 2:30. Fire was returned. At 2:45, aircraft takes no fire from Myslowitz. On the streets across from the railroad, by the second red church north of the rail station, foxholes, Front toward south. Foxholes empty. No military in Myslowitz. Civilian populace (women in bright dresses) on railway platform. No troops whatever in the Dziettzkowitz - Laurahütte area along the border.
Gieschewald and the wooded area around G. apparently peaceful, likewise Emanuelsegen. By Agatgrube and Rosalienhütte scattered gunshots. By diving down to the lowest altitudes and cutting the engine, each gunshot can be heard at once and thereby be pinpointed, whether a place was occupied or not. Where a place was occupied, with its immediate fire, the flyer became the most dangerous opponent.

2. Combat Activities:

a. On one reconnaissance flight on the 19th, one aircraft determined that individual civilians armed with guns were leaving Nikischacht toward the east. The civilians were attacked with machine gun fire, they scatter quickly, one lies still.

b. On the 18th, an aircraft recognizes, south of Rupten, a machine gun with two civilians on a straw pile; and two uniformed men as crew. Machine gun was attacked with 150 shots and silenced at once.

3. Assistance by Message Dropping
On 20th August, two aircraft dropped the following message in the courtyard of the Myslowitz mine, one of the battalions cut off by rioting, who had shot off their ammuniton, over 12,000 rounds in all, at the same time: "Laurahütte - Bogutschütz - Richthofenschacht firmly in our hands, at Eichenau-Bodzin-Schoppinitz-Janow attack takes place today if possible. Battalion should participate in this attack. Situation in Nikischchacht still unclear, Signature." Battalion laid out a recognition sign. On the basis of this message, the battalion would participate in the attack, and the surprised rioters fell back."

The German air unit described in these accounts was *Artillerie-Fliegerstaffel* 129 (formerly *FA* 401), based in Gleiwitz, west of Kattowitz. The *Staffel* suffered several losses in the fighting, the first on 19 August, when *Lt. d. R.* Weiss, an observer, was killed.

On 28 August, a plane crewed by pilot *Ltn.* Helmuth Rusche and gunner Sgt. Kieske, making a test flight, was shot down, killing both men. They were apparently the victims of friendly fire [12]. According to an item in a Warsaw newspaper [13], one of three German planes attacking Sosnowiec was hit by ground fire and crashed in nearby Myslowitz on 27 August, three days after hostilities were said to have ended. It's possible these two accounts have been conflated.

Is This Hundsfeld?

One of the most bizarre events of the German-Polish conflict took place after hostilities had officially ended. On 8 December 1919, two mechanics working in the armory of Poland's IV. Squadron at Lawica heard a plane overhead. Stepping outside, they saw a German LVG C.VI coming in to land. In German, the pilot asked them, "Is this Hundsfeld?

The mechanics replied, "This is Ławica."

The pilot, apparently unaware this was the Polish name for the Poznan (Posen) airfield, asked, "In what district is Ławica?"

"Ławica is in the vicinity of Poznan´. You are in Poland."

The pilot then told the Poles to let go of his machine so he could take off for Hundsfeld, but the mechanics, with the help of additional personnel, finally forced him to surrender the plane, and the Polish air force was richer by one LVG.[14]

The German pilot was *Ltn.* Siegfried Gussmann, a 5-victory ace formerly with *Flieger Abteilung* 3 and *Jasta* 11. How could such an experienced pilot be 175 km off course? The answer may be provided by Bartolomej Belcarz and Robert Peczkowski in their book, *White Eagles - The Aircraft, Men and Operations of the Polish Air Force:*[15]

"After the former Prussian Zone territory was incorporated into Poland, the Germans were contacted, and they were eager to sell aircraft and equipment. The aircraft would be ferried by German pilots to Polish private companies. This went on unofficially, but with Polish government consent. German pilots would simply 'lose their bearings' and land on Polish territory. After a short inspection of the aircraft they would get their payment and go back for another aircraft. At that time, it was virtually the only way to obtain aircraft for the Polish Air Force."

Endnotes – Chapter 4

1. https://www.archives.gov/milestone-documents/president-woodrow-wilsons-14-points
2. Taylor, Eileen R. (tr.), Benoist-Méchin, Jacques, *History of the German Army since the Armistice Vol. I, From the Imperial Army to the Reichswehr*, Howard Fertig, New York, 1988, p.225.
3. ibid.
4. von Salomon, Ernst, ed., *Das Buch des deutschen Freikorpskämpfer*, Wilhelm Limpert, Berlin, 1938, p. 234-6.
4. Chmielewski, S., *Dzieje Chodziezy*, 1998, Chodziez
5. Pilawski, W., *Ziemia czarnkowska w walce o wolnosc* 1918-1919, 2006, Czarnkow.
6. Mateusz Kabatek personal correspondence, April 16 & 21, 2010.
7. Zuerl, Walter, *Pour le Mérite-Flieger*, Curt Pechstein *Verlag*, München, 1938, pp. 503.
8. Pomarański, Stefan, *Pierwsza wojna polska* (1918-1920), *Zbiór wojennych komunikatów prasowych Sztabu Generalnego (za czas od* 26. XI. 1918 *r. do* 20. X. 1920 *r.), uzupełniony komunikatami Naczelnej Komendy W.P. we Lwowie* (*od* 2. XI. 1918 *r. do* 23. XI. 1918 *r.*) *i Dowództwa Głownego W.P. w Poznaniu* (*od* 11. I. 1919 *r. do* 14. XI. 1919 *r.*). *Zebrał, opracował, wstępem i skorowidzem opatrzył*, Warsaw, 1920.
9. *"Informacie o Wojskach Letniczych", Naczeine Dowodztwo W., Szef Lotnictwa*, 6 VI 1919, *Centr. Arch. Wojsk*
10. After aviation activities ended, Lenz, a Bavarian, moved to Kreuzberg, Silesia and established a local self-defense militia. Eberstein may not have approved of Lenz's command style, but there's no doubt about his willingness to serve his country.
11. *"Auszug eines Truppenberichtes über die Tätigkeit der Militär Flieger im Kampf um die innere Ordnung", Reichswehrbefelungstelle Preussen Truppen-Departement. Nr.* 124II.19. A.7.L. IV.C., BayHStA *Abt.* IV, *Kriegsarchiv* document *Fl. Horst Gersthfn. Bd.* 13, pp. 3-6.
12. http://www.theaerodrome.com/forum/showthread.php?t=64360&highlight=Artillerie+Fliegerstaffel. Photographs of a memorial stone to Rusche, placed at the site of his fall, date his death as 28 August.
13. *Kurjer Polski*, # 226, Warsaw, Sunday, 31 August 1919, p. 3
14. *Protokól*, Lawica, 9.12.19, *Centralne Archiwium Wojskowe sygn.* I.300.39.29
15. Belcarz, Bartolomiej & Robert Peczkowski, *White Eagles: The Aircraft, Men and Operations of the Polish Air Force, 1918–1939*, Hikoki Publications, Ottingham, 2001, p. 41

Baltic Campaign

German Aviation Units in the Baltic, December 1918 - 28 February 1919

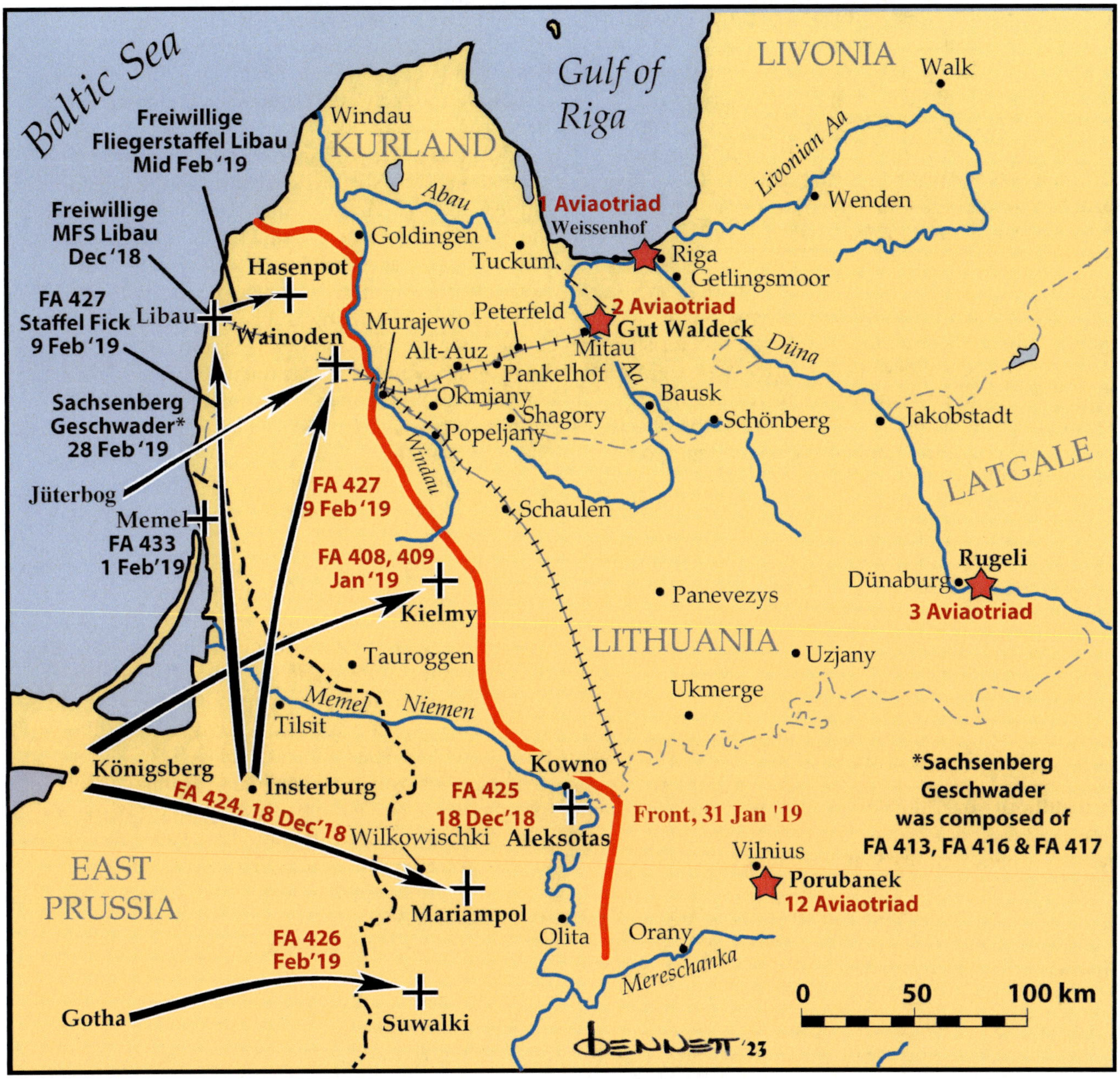

Historical Background

Germany's association with the Baltic lands goes back to the Teutonic Knights, who moved into the area in the 13th Century to spread and protect Christianity and who eventually ruled over a state incorporating most of present-day Estonia and Latvia, along with portions of Lithuania, Poland, East and West Prussia and Pomerania. Several of the area's Baltic cities were members of the Hanseatic League in the Middle Ages, and settlers emigrating east from German lands cemented ties between the two regions.

Over time, many of these settlers prospered, evolving into the estate owners, nobility and intelligentsia of the region, especially in Kurland, despite being a small minority of the overall population. Native Latvians, and to a lesser extent,

Estonians, were relegated to the middle and lower classes in the lands of their ancestors. In the 18th Century, the Baltic provinces became part of the Russian Empire, but the German elite continued to dominate the region's economic and cultural life. To many dreamers in Germany, the *Baltikum* was their idea of America's Wild West -- untamed, romantic and ripe with opportunities for those ambitious enough to strike out and seek their fortunes.

During World War I, German armies advanced through Poland into the Russian governorates of Kurland and Kowno, present-day western Latvia and Lithuania. On 3 September 1917, with Russia's Provisional Government nearing collapse, German forces occupied Riga and moved deeper into Russia. The Baltic German ruling class, apprehensive of the loss of their privileged position in post-revolutionary Russia, welcomed the *Reich*'s forces with open arms.

The November *Bolshevik* Revolution brought an end, for all intents and purposes, to Russian resistance, and a *de facto* cease-fire settled over the area. German diplomats began peace negotiations with the new Soviet government, but the Reds dragged their heels until a renewed German offensive forced them to sign the Treaty of Brest-Litovsk on 3 March 1918. The treaty ceded huge amounts of territory, including the Baltic lands, to Germany, freeing it to transfer the bulk of its *Ostfront* army to the West to participate in *Operation Michel*, the Spring offensive expected to break the Allies before American troops arrived in force. In the meantime, Germany began grooming the Baltic regions for "independence" as satellite governments.

Oberfehlshaber Ost and the Retreat from the Baltic

Real authority over the Baltics was left in the hands of *Oberfehlshaber Ost* (Supreme Command East), Germany's military government, backed by the *8. Armee*, now reduced to a skeleton force of understrength and second-line units. Its aviation presence shrank to a bare minimum -- three *Flieger Abteilungen* (*FA* 16 at Orro, Estonia, *FA* 22 at Weissenhof, across the Düna River from Riga [1], and *FA(A)* 214 at Alexsotas, near Kowno) [2]. By 11 October 1918, the German air presence in the Baltic had shrunk to such insignificance that the staff of *Kofl.* 8 was dissolved; responsibility for coordinating aviation activities with the 8. *Armee* devolved to *Hauptmann* Friedrich Erdmann, *Flivo* (*Fliegerverbindungsoffizier* - Aviation liaison officer) of *Oberfehlshaber Ost* [3]. The navy continued to maintain its *Seeflugstationen* (Naval Air Stations) at Reval, Windau, and Libau, and *Artillerie-Flieger Schule* (Artillery Flying Schools) I and II still operated at Alt-Auz and Doblen, Kurland.

The 11 November Armistice required German forces to withdraw from all territories occupied since 1914. Clause XII, however, specified they remain in occupied Russian territories until the Allies "shall think the moment suitable, having regard to the internal situation of these territories." [4] This was not an act of generosity to a defeated enemy; the Allies wanted the Germans to serve as a blocking force against anticipated Soviet expansion into the area. Britain had been chosen to lead Allied diplomatic and military efforts to nurture the newly independent Baltic republics and orchestrate their defense against the Bolsheviks. Unwilling to commit their own troops against the Reds, the British deputized the remnants of the German 8. *Armee* as their sword and shield until the new nations could provide for their own defense. This strategy was as naïve as it was cynical. Most rank-and-file German troops had no desire to do the Allies' dirty work; many had been infected with the revolutionary virus, and others simply wanted to go home. With winter approaching and violence breaking out throughout Germany, they felt their place was with their families. On 15 November, the personnel of *Seeflugstation* Reval were reported to have destroyed their aircraft and most of the supplies held by the station, with the exception of some fuel, which they sold off [5].

On the other hand, The German military leadership embraced the Allies' idea. The hidden agenda was obvious, but this enabled the Germans to pursue a hidden agenda of their own: with Allied blessing, they could continue dominating the Baltic. Some military leaders even saw it as an opportunity to rebuild the German army away from the scrutiny of the Allies and, perhaps, even restart the war against them.

The Soviets weren't party to any of this -- their agenda was to claim the Baltic lands in the name of international communism, and they weren't inclined to wait until the Allies and Germans had resolved their competing schemes. Less than two weeks after the Armistice, Bolshevik forces went on the offensive, pushing the inexperienced, understrength Estonian, Latvian, and Lithuanian armies westward. As the Red army approached, the Germans prepared to evacuate. On 17 December, with a far larger and more motivated Soviet force approaching, the 8. *Armee* left Dorpat, Estonia and withdrew south to Riga, accompanied by *FA* 16. Accidents en route cost the *Flieger Abteilung* some of its equipment. On arrival in Riga, it joined the *Eiserne* (Iron) *Flieger Abteilung*, organized 9 December by *Hauptmann* Erdmann. Erdmann had rounded up the remaining personnel and supplies of *Armee Flugpark* 8 and *Artillerie Fliegerschule Ost* I and II, combining them at Weissenhof airfield. The new *FA* operated in coordination with the *Eiserne* Brigade, a volunteer unit

Above: Two Junkers CL.Is of the *Flieger Abteilung* 417, *Kampfgeschwader* Sachsenberg, over Kurland. Other copies of this photo bear the inscription "März 1919", dating it to a two or three week period after the *Geschwader's* arrival in Kurland. (04AEF-B108-Album 40-PB061, History of Aviation Collection, Special Collections and Archives Division, Eugene McDermott Library, The University of Texas at Dallas)

Above: Detail from another photo of the lead aircraft, apparently taken during the same flight. This enlarged, digitally enhanced image reveals the squat letter "M" on the white fuselage band. The letter's significance isn't known, but might indicate this plane was the personal mount of the *Abteilungsführer, Ltn*. Bruno Majewski.

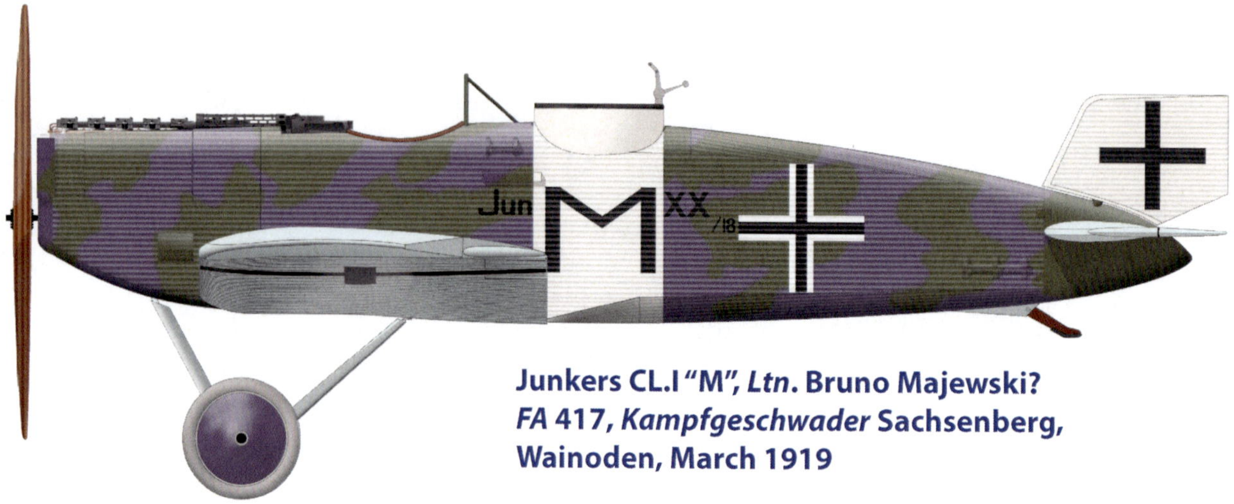

Junkers CL.I "M", *Ltn.* Bruno Majewski?
***FA* 417, *Kampfgeschwader* Sachsenberg,**
Wainoden, March 1919

made up of troops who chose to remain in Kurland and oppose the Reds. During its brief stay there, The *Eiserne FA* conducted up to three reconnaissance and bombing sorties a day against the advancing Bolsheviks, including an attack on an armored train. However, the *Abteilung*'s resources, like those of the *Eiserne* Brigade, were too scant to stop the Red advance [6].

The reinforced unit suffered its first loss to the Bolsheviks over Riga, 2 January 1919, when Johannes Busch and his observer, *Ltn.* Walter Meuer, were brought down by Red ground fire and captured. On the 21st, Busch was given a safe conduct pass back to German lines. Meuer wasn't so fortunate -- he had been tried and executed as a spy 7 January [7].

On the third, German troops left Riga. Three planes of the *Eiserne FA* departed that morning, only to run into a snowstorm on the way to their destination of Gut Paulsgnade, northeast of Mitau, with two of them being damaged on landing. Worse, some of the unit's enlisted personnel, transported to Mitau via rail, refused to disembark and instead stayed aboard the train all the way to Insterburg, East Prussia, where they deboarded and dispersed into Germany. Four days later, with Mitau threatened, the *Abteilung* abandoned its aircraft and headed west again, this time on foot, via Alt-Auz to Murajewo, where they picked up another train to East Prussia, arriving in Insterburg 12 January to rejoin the *Eiserne* Brigade [8].

With the Red Army rushing to fill the void left by the departing German army, the infant republics of Estonia, Latvia, and Lithuania were in peril of being crushed. The Soviets stopped short of overrunning Estonia, sparing Reval, to transfer the bulk of their forces to other, more pressing, fronts. This gave the Estonians breathing space to rearm and reinforce their little army with Finnish and British assistance. They also entered into a temporary alliance with White Russian forces opposing the Bolsheviks. After recapturing most of the ground they had lost, the Estonians would turn their attention south to face the Latvian Red Rifle Brigade, one of the crack units of the Red army.

Germany Returns

Lithuania was in a similar predicament to Estonia. It had to build an army from scratch as Soviet forces were advancing into its territory. Swallowing their misgivings, the new republic's leaders approached the German government for a loan to purchase arms and military equipment and to solicit volunteers to assist their army. Germany was happy to oblige, providing *Freikorps* units and remnants of the old occupation army as the nucleus of the German force on the southern Baltic Front. It was joined by the German 45. Reserve Division.

The situation in Latvia was more complicated. The Allies recognized Latvian nationalist Karlis Ulmanis as the head of the new state and provided assistance to his small army. A second Latvian military force, the Baltic *Landeswehr*, had been created by the ethnic German "Baltic barons" of Kurland, who knew a Soviet takeover would bring an end to their prosperity and privilege. They also feared life under the Ulmanis government would be little better, but, under Allied pressure, agreed to a shotgun marriage with it, hoping future events would allow them to regain the upper hand. Neither of these two armies was considered capable of defeating the Reds -- only Germany could provide the necessary resources to secure the area. The Baltic barons had been appealing to their contacts in Germany for help, but the deal clincher was Britain urging Ulmanis to seek German assistance. On 29 December 1918, the agreement was sealed. To encourage recruitment of German volunteers to fight the Bolsheviks, the Baltic barons pledged 1 million acres of land as homesteads for German volunteers who served at least six months in their service and wished to settle there once fighting had ended. It was a tempting incentive -- employment prospects in Germany were dim, and the future orientation of its government was uncertain.

The Baltic *Landeswehr* placed itself under the command of a German officer, Colonel Alfred Fletcher, and was expanded with men and equipment from Germany. The *Eiserne* Brigade was also beefed up with German volunteers and material. Renamed the *Eiserne* Division, it was now commanded by Major Josef Bischoff, a career German army officer.

From Germany, the third major component of this force, the *1. Garde-Reserve Division,* was dispatched by *Grenzschutz Ost* to Kurland after helping put down January's Spartacist uprising in Berlin. Commanded by *Generalmajor* Paul Tiede, this was a regular army unit reinforced with *Freikorps* volunteers.

All three forces were combined into the *VI. Reserve-Korps*, under the leadership of *Generalmajor Graf* Rüdiger von der Goltz. This corps would be the northern flank of the armies opposing the Bolsheviks. On 18 February, Baltic German *Hpt.* Percy Baron von Ascheberg, formerly *Kofl* of *Armee Korps* XI in Kassel, was appointed *Kofl.* of the VI. Reserve *Korps.*

South of a line running roughly southwest-to-northeast from Heydekrug, East Prussia to Bausk, Livland, in what is now mostly Lithuania, the German forces took up positions to support the forming Lithuanian army and to blunt any Red assault before it could reach East Prussia. This command, separate from von der Goltz's VI. Reserve Korps, was known as *Generalkommando zur besondere Verwendung 52* -- General Command for Special Purposes, or *G.z.b.V. 52.*

Farther south, on a line running roughly east-northeast from the Niemen River, the area was held by the *Zusammiges* (Combined) *Reserve Korps*, headquartered in Kowno. This command was later transformed into the *Freiwillige Reserve Korps*. Its right flank abutted the left flank of the Polish army.

All these commands reported to *AOK Nord* of Germany's *Grenzschutz Ost* command.

This arrangement was exactly what the Germans wanted -- they now had Allied sanction to intervene in the Baltics, protect East Prussia from a Red invasion and pursue their dream of turning the new Baltic states into satellites.

Aviation units, forming at the same time, joined the ground forces as they became combat-ready. From north to south, they were:

Units Assigned to the *VI. Reserve-Korps*

***Freiwillige Fliegerstaffel* Libau.** By January 1919, the Baltic seaport of Libau had become the center of activity for the German and *Landeswehr* forces opposing the advancing Reds, and *Freiwillige Fliegerstaffel* Libau, also known as the Libau *Fliegerstaffel*, formed in December, was part of it. This unit, operating six elderly Albatros C.IIIs and 4 seaplanes, flew in support of the Germans and their Baltic allies trying to retain a foothold in Kurland. Between 12 January, when the Iron *Flieger Abteilung* fled into East Prussia, and 9 February, when it returned as *FA* 427, the Libau *Fliegerstaffel* bore the burden of German aviation operations in Kurland [9].

Fortunately, the Libau *Fliegerstaffel* was directly opposed by Soviet 1st Latvian *Aviaotriad*, also weak and underequipped. Operating two Sopwith 1A2s out of Alt-Auz, it was able to muster only nine flights totaling 10 hours and 25 minutes from 1 January through 16 February [10]. Clearly, neither side was able to dominate the skies.

On 8 February, one of the Libau *Fliegerstaffel's* Albatroses reconnoitered Red-held Windau, collecting a few bullet holes from ground fire, while a second Albatros, flying over Libau, suffered engine failure and crashed, injuring pilot *Feldwebel* Hawelik and observer *Uffz.* Hillmann. Hawelik died from his injuries.[11]

***Staffel* Fick**, part of *FA* 427, arrived 9 February, absorbing the *Fliegerstaffel.* The *Fliegerstaffel*'s seaplanes, of limited use in the upcoming overland campaigns, remained at Libau. The combined unit, now reporting to the *Landeswehr*, continued its operations. On 13 February, pilot *Vzfw.* Jedwill and observer *Hpt. z. S.* von Gordon conducted reconnaissance over Windau.[12] Four days later, one of the *Staffel*'s planes dropped 10 small bombs on Red positions at Gut Ezern. When *FA* 433 arrived in Tuckum 23 April, *Staffel* Fick was merged into it, and the combined unit continued its association with the *Landeswehr*.

***Flieger Abteilung* 427.** Following its evacuation to Insterburg, East Prussia, on 12 January 1919 [13], the *Eiserne Flieger Abteilung* [14, 15] was reorganized into a new unit, *Flieger Abteilung* 427, again under the command of *Hpt*. Friedrich Erdmann. It would return to Kurland on 9 February with *Pour le Mérite* ace Bruno Loerzer among its personnel[16]. When *FA* 427 entered Kurland, a portion of the *Abteilung*, *Staffel* Fick, led by *Lt. d. R.* Walter Fick, had split off, taking its six aircraft directly to Libau, as described above. The main body of the *Abteilung* operated as part of the *Eiserne* Division. Taking up residence at the abandoned airship base in Wainoden 9 February, it was in action 9 days later, bombing Red field guns and their limbers in Essern [17]. On the 27th, it dropped two bombs on Murajewo [18].

The *Landeswehr* had, since December, been trying to raise its own air force in Germany, and it might

Above: Sachsenberg *Geschwader* aerial photo of Alt-Auz, looking northeast, shortly after the Reds evacuated. The former *Artillerie-Flieger-Schule* field, occupying the right half of the photo, was devoid of nearly all of its wartime structures. The large double building adjacent to the rail line was not part of the school; it was a recent addition still under construction. Gut Weitenfeld was located about 3.5 km off to the lower right of this photo. (04AEF-B108-Album 40-PB074, History of Aviation Collection, Special Collections and Archives Division, Eugene McDermott Library, The University of Texas at Dallas)

have expected its early overtures would result in it receiving the services of the Sachsenberg *Geschwader*. However, the *Geschwader* was assigned to the 1. *Garde-Reserve-Division*, leaving the *Landeswehr* high and dry, apart from *Ltn. d. R.* Fick's small *Staffel*. The commander of the *Landeswehr*'s *Stosstruppe* (shock troops), *Ltn.* Baron von Manteuffel, decided to try again, and on 14 February 1919, dispatched observer *Ltn.* Ludwig and pilot E. Hoffmann on a steamer to Pillau to press their case. They visited several officials in Germany, but couldn't get any commitments for assistance. Their final stop, on 20 February, was the Sachsenberg *Geschwader*, which was in the final stages of activation at Jüterbog. Sachsenberg was sympathetic to their request, but before any action could be taken, the *Geschwader* packed up for Kurland. Ludwig and Hoffmann accompanied them, and on arrival in Wainoden requested a transfer to Sachsenberg's unit. There they tried to facilitate the formation of a *Landeswehr* aviation unit using aircraft, personnel and other resources provided by Sachsenberg. Their efforts didn't bear fruit, but developments in East Prussia would eventually give them their aviation arm [19].

The ***FA des Ostpreussisches Freiwilligenkorps*** (East Prussian Volunteer Corps), assembled in Memel 1 February 1919 by *Hauptmann* Erich Alt, former commander of *FA(A)* 253 and *FA* 37, would eventually receive the designation *FA* 433. Originally assigned by *Grenzschutz Ost* to participate in the defense of East Prussia, it carried out its first front flight 28 February [20]. Once the Reds had been stopped in Kurland and the Germans and *Landeswehr* went over to the offensive, it was placed on loan to the *Landeswehr*. At some point, its aircraft's German crosses were overpainted with plain white rectangles to suggest they now belonged to the Latvian government forces, not Germany. As the Front moved farther to the east, the flight distance to and from Memel hampered the *Abteilung*'s efficiency, and it relocated to Tuckum 23 April, where it absorbed the Libau *Fliegerstaffel* (*Staffel* Fick). The move didn't interfere with operations; the consolidated unit conducted three bombing raids the same day, attacking shipping on the Aa and Düna Rivers [21].

Libau was also the base of a small, homegrown detachment, ***Freiwillige Flugstation der Gouvernements* Libau**.

The largest and best known of the Baltic aviation units

Kampfgeschwader Sachsenberg

Above: *Flugelschwert* (Winged sword) sleeve patch of the *Kampfgeschwader* Sachsenberg. (*Das Buch des deutschen Freikorpskämpfer*)

Above: Unidentified machine gunner of the *Sachsenberg Geschwader*, May 1919. In addition to the *Flugelschwert* insignia of the *Geschwader*, he wears the machine gunner patch on his sleeve. (Brian Lewis)

Select three *Pour le Mérite* aces with a total of 110 victories during the First World War, add to them 700 pilots, observers and support troops, equip them with 70 of the latest aircraft in Germany's arsenal, and what do you have? The *Kampfgeschwader* Sachsenberg, arguably the most potent aviation unit in the world in 1919.

By far the largest and most prominent of the *Freikorps* aviation units, the *Geschwader* was led by *Oberleutnant zur See* Gotthard Sachsenberg, who took command of the unit at the former Artillery Observer School in Jüterbog during January 1919.

Oblt. Herbert-Armin Angleroth served with the *Geschwader* and later described its formation:

"When, in December 1918, the *Reichs* government issued a call for volunteers from all the services to defend the Eastern frontier and the ethnic German population of the Baltic Provinces, the staff and men of the former Flanders *Marinejagdgeschwader* (Naval fighter squadron) were in Dessau, carrying out their demobilization orders. Gotthard Sachsenberg placed himself and the remaining loyal troops at the disposal of the *Reichs* government to continue to serve the Fatherland ...

"The War Ministry, conscious of the importance of the forthcoming assignment, placed material worth several millions at the *Geschwader's* disposal. The *Geschwader's* task was many-sided, because neither police nor constables, let alone rear echelon troops, existed in the Baltic. Those troops who constituted the Front were self-led. There was no fortified Front opposing them. With these considerations in mind, the *Geschwader* was formed as an independent unit, composed of different flying groups -- fighter *Staffel*, battle *Staffel* and reconnaissance *Staffel* -- and one machine gun company, one training detachment and later also one troop of cavalry and artillery. In order to be able to respond to the smallest possible changes in the tactical situation, the entire *Geschwader*, inclusive of personnel, equipment and workshops, was billeted in a railroad train.

"The formation of the *Geschwader* took place in Jüterbog. Its strength was established at 450 men; under the uncertain conditions of the time, the issuing of uniforms, weapons and supplies and the assignment of the men was a difficult and thankless task.

"The recruiting of volunteers went fairly smoothly. Early on, it was clearly ascertained that the Spartacists were sending large numbers of men as agitators

Above: Top Guns of the Sachsenberg *Geschwader* (L-R) -- Josef Jacobs, Gotthard Sachsenberg, and Theo Osterkamp. (Greg VanWyngarden)

into the *Geschwader* with the express purpose of undermining the spirit of the troops and rendering them unfit for the possible confrontation with Soviet Russia. Each day clothing, equipment, aircraft and auto parts, as well as weapons, were stolen and quickly sold in Jüterbog or nearby Berlin. Consequently, there was no chance for relaxation as officers and NCOs stood nightly watch. One has difficulty finding words to describe their faithful comradeship. They went about their assignments with dedication, enduring the recent great defeat and hoping at the same time that the approaching spring of 1919 would also be the Spring of the Fatherland.

In January 1919 the *Geschwader* took shape, and as would be expected, aircraft were assembled and flown, and equipment was made ready for frontline service..." [22]

Sachsenberg was joined by two other *PlM* holders, *Ltn. d. R.* Theo Osterkamp and *Ltn.* Josef Jacobs.

Osterkamp's involvement in the Baltic had already come under the scrutiny of the Spartacists, Germany's Communist party, who somehow got hold of correspondence between him and the Baltic *Landeswehr*'s agent in Germany. They published it in their newspaper, *Die Rote Fahne* (The Red Banner) on 2 January 1919 under the headline, "The Ebert Regime as Soul Sellers". The article included this letter from Osterkamp to attorney A. Volck in Berlin. Volck was apparently the *Landeswehr*'s German representative:

"Dessau, 13. 12. 18
Dear Herr Volck!
Referring to our short conference in Berlin yesterday, I offer you here my proposal in writing. I am prepared, here in Dessau, first of all, to establish a *Feldflieger-Abteilung* composed of:

- 8 C - aircraft
- 3 Troop transport trucks
- 1 Cargo truck
- Spare parts and tools
- Cost amounts to about 500,000 M.
- About 10 officers and vice deck officers [NCOs] (flying personnel)
- About 50–60 men (technical and general personnel)

and send it to Riga in private transport.
There I will place myself at the disposal of the "Iron Division" and directly subordinated to the *A. O. K.* in that place in a permanent association with them, my independent aviation unit, deploying, depending on the situation, as an observation or assault *staffel*. I myself, as well as all my people, commit ourselves to serve your venture, in the form of a disciplined military unit, as we formerly had served our Fatherland…" [23]

Osterkamp's budding private venture would soon be absorbed into the larger scheme of the Sachsenberg *Geschwader*.

Osterkamp later wrote a highly romanticized version of events in his memoir *Du oder Ich*:

Right: Officers and men of *Kampfgeschwader* Sachsenberg about to depart for Kurland. Sachsenberg is in the right center of this photo, arm in arm with the man wearing the naval jacket. His right arm rests on the shoulder of his *O.z.b.V, Oblt. z. S.* Karl Cranz. *Vzflugm.* Karl Scharon is sixth from left, *Ltn.* Eberhard Cranz is third from right in the front. *Lt. z. S.* Hellmuth Brembach is third from right, back row. (Alamy)

Right: Another departure photo of *Kampfgeschwader* Sachsenberg personnel. The man ninth from left in the back row, next to the building column, appears to be Sachsenberg. *Vzflugm.* Karl Scharon is seated on the right, front center. (Alamy)

"Machines and materials were obtained with and against the will of the authorities. Still, we were missing weapons. -- So, everything appeared to be a failure. No open requisitions, no 'detours' helped. All places were unresponsive to requests and threats. Our attempts at bribery failed on the hydra of the 'open hands'.

"In the midst of the Spartacist battle of Berlin, the Reds' password and the location of their arsenal were betrayed to us. In an all-night project with two passenger cars, hundreds of guns and cases upon cases of ammunition and pistols, as well as 50 machine guns, were taken out of this arsenal and loaded onto trucks waiting in a hidden location.

"They were wild trips.

"Over our greatcoats, the red band and the 'Louis cap' for the Red 'brothers', under the greatcoat, the uniform and *Pour le Mérite* for the government troops.

"And then, one day, everything was betrayed. --

"But we had caught wind of it and taken off before they caught us.

"The next day, our plans were displayed on the front page of *'Rote Fahne'*, and *Ltn.* Osterkamp and *Ltn.*

Left: More of Sachsenberg *Geschwader's* personnel by one of the rail cars they called home. The *Geschwader's Flugelschwert* arm patch is much in evidence here. *Internationale Eisenbahn Schlafwagen Gesellschaft* (International Railway Sleeping Car Corporation) was the German subsidiary of *Compagnie Internationale des Wagons-Lits*, operator of the famed Orient Express. Its rolling stock was commandeered by the German government when World War I broke out. (Alamy)

Sachsenberg were declared outlaws for 'selling their souls to the Kurland Barons' [24]."

This confirms Angelroth's account of the sparring that took place between the *Geschwader* and the Spartacists, but from that point on, most of this account should be taken with a grain of salt. It's one of those altered pieces of the jigsaw puzzle.

Angelroth continues:

"In early February, I was sent to Libau by our commander as head of an advance party of 4 officers and 50 men. We traveled from Jüterbog to Stettin on the railroad and from Stettin to Libau via steamer. Our transport ship was under way in good order off Danzig, when we were intercepted by English warships with searchlights, who, on examining us, ascertained that troops with weapons were on board. However, they allowed us to pass unmolested.

"The task of our advance group was to secure in whatever way possible the pending installation of the *Geschwader* in the Baltic, selecting suitable airfields and preparing for the arrival of the *Geschwader.* Further, contact with the General Staff was to be made to inform them that they would soon receive the assistance of a flying *Geschwader.*" [25]

The *Geschwader* was organized into three *Staffeln*:

***FA* 413** - *Jagdstaffel* (Fighters), flying Fokker D.VIIs and Junkers D.Is and led by Osterkamp. His second in command was 47 victory ace *Ltn.* Josef Jacobs, former commander of *Jasta* 7 [26].

***FA* 416** - *Aufklärungs-Staffel* (Reconnaissance) equipped with LVG C.VIs, DFW C.Vs, Halberstadt C.Vs, and Rumpler C-types and commanded by *Ltn.* Eberhard Cranz.

***FA* 417** - *Schlachtstaffel* (Ground support) under the leadership of *Ltn.* Bruno Majewski, formerly of *Marine Feld Jasta* I. The unit flew Junkers CL.Is and Hannover CL.IIIs.

Sachsenberg was named overall *Geschwader* commander. His *O.z.b.V* (*Offizier zur besondere Verwendung* - Adjutant) was *Oblt. z. S.* Karl Cranz, brother of Eberhard Cranz.

The gap in numbering of the *Flieger Abteilungen* is a bit curious, considering the Geschwader was formed as a single unit. Assuming *FA* numbers were assigned to units in the order they were authorized, the *FA* 413 - *FA* 416 gap suggests Osterkamp's detachment had already been approved as a standalone unit before the idea of a combined *Geschwader* was taken up *by Idflieg*. The units that were formed in the interim, *FA* 414 & 415, date from 2 and 8 January 1919, a period coinciding with the Bolshevik conquest of Riga and retreat of the Iron Brigade. These events may have driven the decision to expand Germany's aviation presence in the Baltic.

By early February, the *Geschwader* was taking shape nicely, but many positions still needed to be filled. An article in the 1 February issue of the *Berliner Volks-Zeitung* reported:

"Fitters, drivers, splicers, cobblers, tailors, carpenters, cooks, plumbers, radio operators, and other skilled personnel are still urgently needed. Volunteers who are willing to help the Fatherland can report to *Lieutnant z.S.* Cranz, Charlottenburg, Joachimsthaler Strasse 9, I, Pension Jablonsky. Subway station Zoo, travel expenses and food, 9 Marks daily allowance. Free clothing. Report by February 3, evening at the latest." [27]

Above: Airship sheds Walhalla & Walter at Wainoden. The rail sidings serving the base show clearly against the snow. (04AEF-B108-Album 40-PB064, History of Aviation Collection, Special Collections and Archives Division, Eugene McDermott Library, The University of Texas at Dallas)

By the time the *Geschwader* left Jüterbog, it had swelled to a strength of 50 officers and 650 NCOs and enlistees, backed up by a machine gun company and quartered in a passenger train. An additional train carried supplies and workshops and some of the *Geschwader*'s aircraft. The rest traveled by sea to Libau. Soviet intelligence reported the arrival of 9 aircraft at the port of Libau between 21 and 23 February. It also remarked that several Red agents were among the arriving personnel.[28]

The *Geschwader* arrived in Wainoden 28 February, joining *FA* 427 and its aircraft, which had been there since the ninth. Two huge airship hangars named Walhalla and Walther [29] provided plenty of room for *FA* 427 and the Sachsenberg *Geschwader* and its 70 factory-fresh aircraft. Railroad sidings from the nearby Libau - Schaulen line ran into the base, providing a place to park the trains housing the workshops, photo lab and living quarters.

Assigned to the *1. Garde-Reserve Division*, the Sachsenberg *Geschwader* settled into its new home. By the beginning of March, *FA* 417, the *Schlachtstaffel*, was ready to go into action. Its *Erste Flugtag* (First flight day) was the 6th, when it conducted reconnaissance in

Above: Sachsenberg and his *O.b.z.V.*, Cranz, crossing the flying field. This photo appears to date from 6 May, the *Geschwader's Erste Flugtag*. (04AEF-B008-F027-NB005, History of Aviation Collection, Special Collections and Archives Division, Eugene McDermott Library, The University of Texas at Dallas)

Above: *Vizeflugmeister* Hermann Hackbusch poses with his handiwork -- Junkers CL.I 12600/18, nose down in the snow. (AHT AL0752-015)

Below: Hackbusch wasn't the only pilot to be embarrassed on the snow-covered airfield -- the same day, *Vzfw*. Fritz Todheide bellied in his Ju. CL.I less than 100 meters away. (AHT AL0752-012)

Above: Propping a Ju.D.I at Wainoden. The airship docking rail made a useful wheel chock.

the vicinity of Okmijany and Popeljany [30].

Working up to combat readiness didn't go without incidents.

On 7 March, *Vizeflugmeister* Hermann Hackbusch nosed over Junkers CL.I 12600/18 near the hangars. Not far away, *Vzfw.* Fritz Todheide stripped off the landing gear of his CL.I. Snow on the field, estimated from these photos to be fresh, wet, and up to 2 inches deep, may have contributed to these accidents. Quite a few flyers struggled with the tricky Halb. C.V as well. Rupprecht, *Oblt.* Adolf Averdam's pilot, flipped Halberstadt C.V (Av) 6902/18, onto its back in almost the same location as Hackbusch and Todheide's mishaps. This was probably somewhat later, as no snow was present.

Mishaps notwithstanding, *FA* 417's aircraft were participating in ground attacks on Red road and rail traffic within a week.

was the ***Kampfgeschwader* Sachsenberg**, consisting of three components, *Flieger Abteilungen* 413, 416, and 417. In its day, it was the largest and best-equipped aviation unit in Germany.

The Campaign in Kurland

Although German accounts usually refer to the lines as a Front, it was certainly not one in the sense of the Western Front, where millions of men faced each other along a line about 750 km in length. The Baltic Front was about 250 km long, but the maximum combined strength of the opposing forces was no more than 75 - 100,000 men at any time. With a troop density this low, it was impossible for either side to maintain a continuous line. In most areas, it was more a string of strongpoints, with cavalry and aircraft essential for keeping watch on the gaps in between. In many ways, it reverted to the way wars had been fought in the past.

Angelroth's account describes the situation in greater detail:

"With the first flights, it was determined that the positions of the enemy were not visible and tangible, as in the West, where one continuous firing trench existed. As individual detachments, they had occupied farms and crossroads, railroad overpasses, bridgeheads and wooded cover. The places were often unrecognizable to aerial observation. Therefore, flights had to be carried out at very low altitudes, and this naturally provoked intense gunfights with rifles and machine guns. In that entire time anti-aircraft guns were not seen." [31]

The early stages of the campaign in the Baltic were characterized by mobility, and consequently, the aviation units found themselves moving frequently. They made good use of airfields and facilities that had been constructed by the Germans and Russians during the Great War. If no developed air base existed, the flyers often found a home at one of the many country estates or manors that dotted the area [32]. As on the Western Front, these properties were well suited to house aviation units, boasting large tracts of well-cleared farmland for flying fields, a comfortable manor house for headquarters and officers' lodging, and outbuildings to accommodate workshops. Many of them were owned by Baltic German families, who were usually agreeable to making them available to the flyers.[33]

The Red advance into the Baltic reached its high water mark at the end of January 1919. In the north, they had taken the port of Windau; from there, the Front ran roughly south-southeast to the outskirts of Kowno and then south into Suwalki governate, where they were opposed by Polish forces. The Red capture of Windau had been associated with one of the first atrocities that would come to characterize the bitter Red-White conflicts raging across Eastern Europe -- Bolshevik forces murdered 100 German troops who had surrendered [34]. This set the tone for the war, and the reminiscences of several German pilots reveal their visceral fear of going down and being captured behind enemy lines.

On 13 February, an aircraft of the Libau *Fliegerstaffel*, piloted by *Vzfw.* Jedwill with observer *Kapitän z. S.* von Gordon, overflew Windau to observe the results of the previous day's shelling. They found few signs of damage, but brought home several bullet holes for their troubles [35].

On the morning of the 24th, the Libau *Fliegerstaffel* lost one of its planes. Pilot *Vzfw.* Jedwill encountered engine trouble, and he and his observer, *Ltn d. R.* Fick [misspelled Tick in the article], crashed into the forest near the ammunition dump at the Libau naval port. They totaled the aircraft and several trees, but managed to walk away with only minor injuries. Weather may have played a part -- one of the Staffel's seaplanes was turned back by a heavy snowstorm and regained its harbor only with great difficulty.[36]

The same day, the Germans and *Landeswehr* finally recaptured Windau. This gave them control of the entire west bank of the Windau River to south of the Wainoden-Mitau rail line. They had now grown enough to take the offensive and prepared to launch their first major assault, Operation *Tauwetter* (Thaw). Operational orders announced:

2.) In the next few days monoplanes made of corrugated metal (painted dark green, with black crosses on a white background) will be flown at the front. The troops are to be informed about this.

3.) On *Tau[wetter]* day 2 [4 March], the airmen will shoot down several red flares as soon as they have recognized that Essern is being evacuated by the enemy in an easterly direction [37]."

Launched 3 March, the four-day advance carried the Germans and *Landeswehr* deep into Red territory. It was followed by Operation *Eisgang* (Ice flow), begun 9 March, immediately followed by *Frühlungswind* (Spring wind). Aircraft flew ahead of the advancing troops, providing valuable intelligence on the situation at the Front; on 17 March, a flyer made an emergency landing near Major Fletcher's headquarters, informing him that his *Landeswehr* had outrun its neighboring units and was in an exposed position.[38]

On 18 March, *FA* 427 and the Sachsenberg *Geschwader* moved forward along the rail line to Alt-Auz, former home of *Artillerie-Flieger-Schule I*, which had been taken by the Iron Division six days earlier. The former school's site was a far cry from what it had been three months before; nearly all its twenty-plus wartime structures had been razed, probably by the Germans departing after the Armistice, leaving only a

Above: Gut Weitenfeld, summer 1918, when it was a German training base. In February 1919, it became a shared facility of *Geschwader* Sachsenberg, which soon moved on to Peterfeld, and *FA* 427, which remained there until June, when it moved to Paulsgnade near Mitau. On 10 May, *FA* 424 arrived there, followed by *FA* 425 and *FA* 433 in July. The manor house and one outbuilding survive today as the Vitin People's House (Moshe Bukhman)

large empty field. Consequently, the Corps Command ordered that the Sachsenberg *Geschwader,* needing access to the Murajewo – Riga rail line bordering the old school field, would operate out of it, while *FA* 427 would occupy the former base of *Rfa* (*Riesenflugzeug Abteilungen* - Giant airplane detachments*)* 500 and 501 to the north. *Hauptmann* Erdmann, commander of *FA* 427, arranged for the facilities in Gut Weitenfeld, about 3.5 km to the south, to be made available to both squadrons [39].

In the first half of April, the Sachsenberg *Geschwader* was ordered forward to Peterfeld, a whistle stop on the rail line to Mitau. During the Great War, Peterfeld had been home to *Lehrabteilung* (training detachment) B of *Artillerie-Fliegerschule* 2. For the time being, *FA* 427 remained behind at Gut Weitenfeld.

The advances had carried the VI. *Reserve-Korps* to the Aa River. Mitau was taken by the Baltic *Landeswehr* 20 March, with covering support from *FA* 416 and 427. The same day, the Iron Division reached Doblen, 30 km to the west. Riga was now within striking distance, but during the ensuing two months, the opposing forces settled into a stationary war while political intrigues played out. In the meantime, the aviation units took on greater responsibility for reconnaissance and ground attack. Some notes from the logs of the *Sachsenberg Geschwader* give a general idea of the flyers' typical activities during this time:

14 March 19: Attacked enemy traffic in and around Shagory.

15 March 19: At Shagory, bombed & machine-gunned an armored train on Alt-Auz - Mitau line.

Late March 19: Reconnaissance over Düna River from Dünaburg to Riga. Angelroth & *Vzfw.* Werner Schumm brought down by ground fire. They make their way back through Red lines.

30 March 19: Sachsenberg aircraft conducts 450 km reconnaissance flight [40].

4 April 19: Pilot Prinz and observer, Jodkow (?) of *FA* 417, in Ju. CL.I 12926/18, conduct reconnaissance over Bausk and Gross Eckau [41]. *Flugmeister* Dahlmann is wounded by ground fire over Gross Eckau, returns to base [42].

Left: *Kommandeur* (Major) Hoffmann of the Baltic *Landeswehr* in a flying suit, about to board the observer's cockpit of LVG C.VI 9037/18 for a look at activities on the Front. Date is probably 19 March 1919, although the photo is dated May. (04AEF-NegB09-F26-NB002, History of Aviation Collection, Special Collections and Archives Division, Eugene McDermott Library, The University of Texas at Dallas)

Below: Hoffmann's trip might not have been a milk run – this is what's left of LVG C.VI 9037/18, presumably later the same day. (04AEF-B309-F12-PB001, History of Aviation Collection, Special Collections and Archives Division, Eugene McDermott Library, The University of Texas at Dallas)

12 April 19: Joint action with Yorck *Abteilung* -- attacked Bolshevik reinforcements in area of Schönberg.

13 April 19: Four aircraft machine-gunned and bombed Red reinforcements approaching Radziwilischki, scattering them [43].

17 April 19: Reconnaissance, locating enemy in vicinity of Rudbaren.

20 April 19: Reconnaissance of enemy positions near Durben by *Ltn.* Ludwig [44].

28 April 19: Bombing enemy columns and troop concentrations north of Bausk.

5 May 19: Bombing enemy forces at Gross Eckau, including driving a machine gun crew from its position [45].

The Red air force made its much-anticipated appearance by making several leaflet drops over Mitau on 1 April [46]; no air combat ensued. A German intelligence summary released the same day revealed just how weak the opposing Soviet air force was:

1. *Aviaotriad* at Riga / Nordeckhof: About 6 aircraft after two were lost in crashes early March.
2. *Aviaotriad* with 2 aircraft at Gut Osol. May have moved to Dünaburg 11 March.
3. *Aviaotriad* [47]

German Aviation Units in the Baltic, 1 March - 30 April 1919

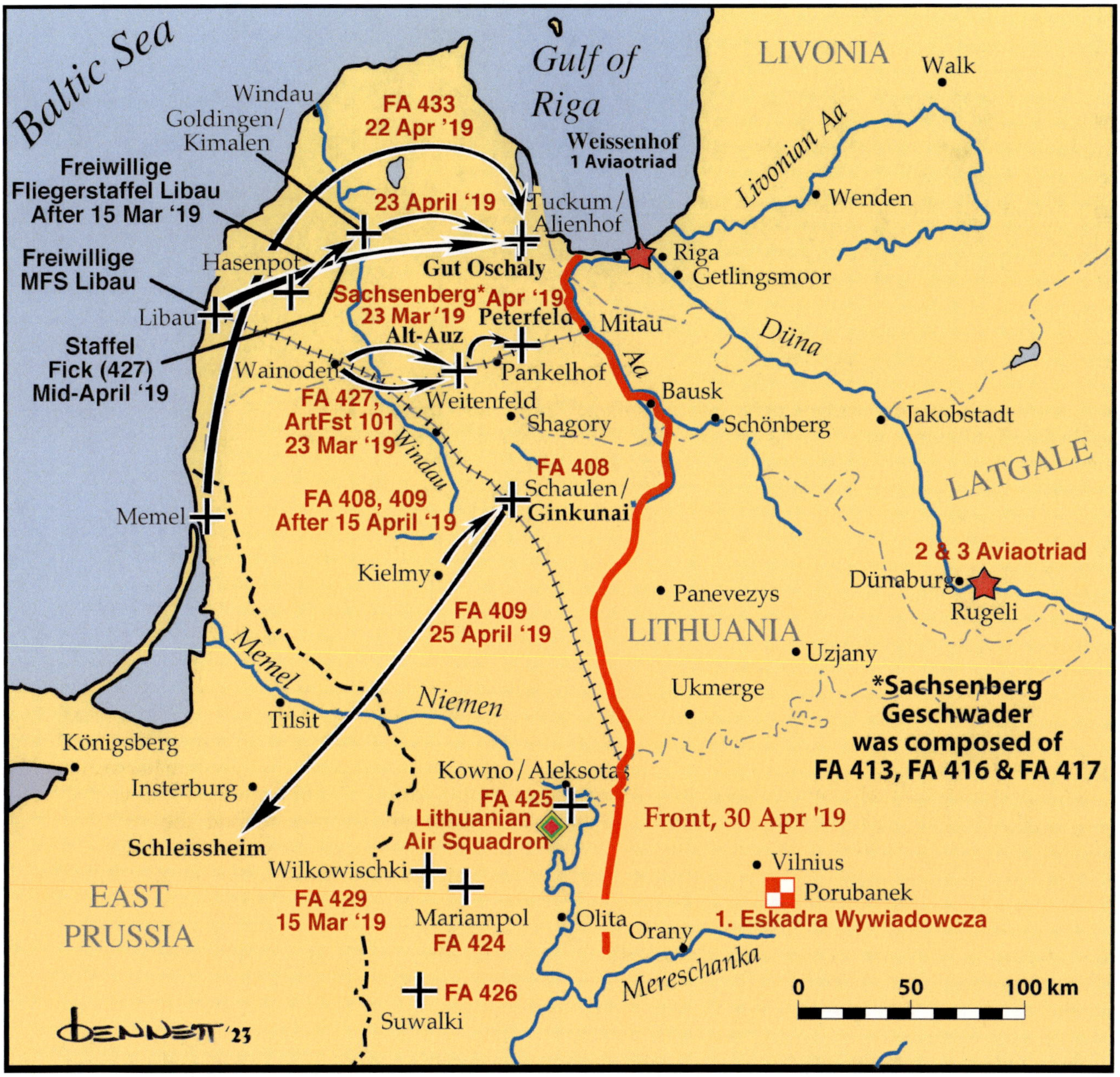

Change of Command

Political intrigue now bubbled up. If the Germans and their Baltic allies had limited themselves to military activities, their subsequent actions would probably have received wholehearted Allied support, but events in mid-April led to their eventual downfall. Per the agreement hammered out 29 December 1918, the German and *Landeswehr* forces were in the service of the Latvian nationalist government. Technically, this put General von der Goltz under the command of Latvian President Ulmanis. Ulmanis took this literally; von der Goltz regarded it as a petty formality to be ignored. Both men were dictatorial and stubborn, and it was only a matter of time before the alliance disintegrated. The success of Operations *Tauwetter* and *Eisgang* emboldened von der Goltz, who concluded

Above: In the last week of April, *FA* 433 moved to Alienhof on the outskirts of Tuckum, where it was joined by *Staffel* Fick. On close examination, the dismantled aircraft on the left, possibly an Albatros C.I or C.III, appears to have a white square painted over its fuselage insignia. The Alb. D.V on the right retains its German crosses. (*Das Buch der deutschen Freikorpskämpfer*)

his forces were now dominant enough carry out a plan hatched by the Baltic barons months earlier -- shed the cover provided by the pesky Ulmanis and his regime and pursue his long-held agenda, conquest of all of Latvia and Estonia, followed by a possible strike at Petrograd.

The breakup began 15 April, when the Pfeffer *Freikorps*, on the pretext of liberating a German officer arrested by the Latvians, disarmed Latvian troops at the Libau naval base. The following day, the Baltic *Landeswehr* arrested troops of the Latvian national army. Ulmanis evaded the Germans and took refuge aboard a transport protected by British warships anchored at Libau [48].

Rid of Ulmanis, the Baltic Barons established a new Latvian government, while von der Goltz feigned innocence of any involvement in the coup. The new president was Andreas Needra, a Lutheran pastor prominent in the *Baltendeutsch* community and willing to carry out the *Landeswehr's* bidding. The *Landeswehr,* which from its earliest days comprised a mixture of nationalities, was purged of its Latvian and Russian members and brought back to full strength with troops arriving from Germany.

The Allies, furious at this betrayal, immediately demanded the restoration of Ulmanis' government. Berlin dragged its heels, protesting it had no authority over the *Landeswehr*, technically a Latvian army, but on 8 May, it did clip von der Goltz's wings by ordering the 1. *Garde-Reserve Division* to report back to Graudenz in West Prussia. Tensions were building along the Polish ceasefire line, and *Grenzschutz Ost*'s headquarters wanted the division back. This forced von der Goltz to scale back a major assault along the entire Kurland front to a focused attack on Riga. He revised his plans, but stalled the transfer, pleading a lack of transport, while he rushed plans for his Riga operation. The attack would be spearheaded by the *Landeswehr,* allowing von der Goltz to claim this was a Latvian, not German, operation. It would be supported the Iron Division, with the 1. *Garde-Reserve Division* held in strategic reserve.

With the *Landeswehr* leading the attack, *FA* 433 needed more resources, so on 20 May, men and equipment of the idled Sachsenberg *Geschwader* were temporarily assigned to *FA* 433. In effect, the *Geschwader* became the *Landeswehr's Flieger Ersatz Abteilung*, keeping it up to strength with transfers of aircraft and their crews. As the German advance had removed the Soviet threat to East Prussia, *FA* 433 had been officially placed under the control of the *Landeswehr* and moved from Memel to Alienhof, on the eastern outskirts of Tuckum, on 23 April [49]. To reinforce the myth that the *Abteilung* was now a part of the Latvian *Landeswehr*, rather than the German army, the aircraft's German crosses were overpainted with a simple white rectangle. Because its *FA* 433 designation would continue to tie it to the German army, it was dubbed the 1. *Baltische Fliegerabteilung*. The Libau *Fliegerstaffel* (*Staffel* Fick), which had moved to Hasenpot in late January and then Kimalen, near Goldingen, in early March, arrived at the same time and was absorbed into *FA* 433.

Thus reinforced, FA 433 did yeoman work, carrying out 35 missions by mid-May, dropping almost 3000 kg of explosives on the enemy, in addition to reconnaissance and ground support. Typical of these were the activities of 28 April, when the *Abteilung* dropped 120 kg of bombs and 20 projectile mines on the Majorenhof rail station, a bridge over the Aa

Right: (L to R) Latvian Pilots Sergeant Janis Preditis, Lt. Narciss Puskelis, and Sergeant Voldemaris Jakubovs, surrounded by men of the Sachsenberg *Geschwader*. (04AEF-NegB09-F26-NB001, History of Aviation Collection, Special Collections and Archives Division, Eugene McDermott Library, The University of Texas at Dallas)

River, the Gettlingsmoor marshaling yards and ships steaming on the Aa and Düna Rivers. The following day it paid visits to Gettlingsmoor, Weissenhof airfield and traffic on the Düna River, unloading 175 kg of explosives on them, followed by strafing bridges near Neu Bilderingshof and a train near Pupe. On the 30th, its planes bombed and strafed Red troops at Karlsbad and Majorenhof.[50, 51] Although the German and *Landeswehr* aircraft enjoyed dominance of the air during this time, they were still plagued by recurrences of engine failure due to poor quality fuel. It would cost *FA* 433 two aircraft returning from a mission 2 May[52] and would haunt the Germans till the very end of their operations in the Baltic.

The Reds, with far fewer aircraft and chronic fuel shortages, failed to offer much aerial opposition, but that was no guarantee of safety for the German flyers. Ground support sorties often had to be flown at extremely low altitudes, ground fire was often fierce, and because the action occurred on the Red side of the lines, German aircrews faced the prospect of capture, and possibly, torture and execution at the hands of the Bolsheviks, as Angelroth described in his account:

"I myself was shot down on one flight over the enemy lines at the end of March. We received two hits in the engine, after which number 2 and 5 cylinders exploded and flew into our faces. My pilot, a Württemberger, *Vizefeldwebel* Schumm, was able to keep the machine under control and land in a thicket. We removed the observer's machine gun, compass and so on and succeeded in returning to Alt-Auz through the Russian lines at night."[53]

Angelroth's account also mentioned a problem that would continue to plague the Germans and their future Russian allies:

"The fuel and the oil were of such poor quality that emergency landings always had to be expected." [54]

On 10 May, Karl Cranz and his pilot, Karl Scharon, on a mission to bomb Eckau, were forced to jettison their bombs and force-land. A few weeks earlier, after an emergency landing behind enemy lines, his brother, Eberhard Cranz had to be rescued by a patrol of naval shock troops[55], and *Vzfw*. Fritz Bohne made it back to friendly territory riding bareback on a horse he commandeered from a farm field [56].

A similar fate befell *Vzfw.* Kanisch and *Uffz.* Hillmann, who described their misfortune:

"A Russian winter lay behind us. Nearly overnight, it was becoming Spring. A new *Fliegerabteilung*, under the leadership of *Hauptmann* Alt [*FA* 433], came out to us from Insterburg [sic]. The two *Abteilungen* were combined. It provided a rare respite for us, for now there were twelve aircraft crews there who could relieve one another. It was an unaccustomed existence for us, after, in the prior months, having to take off day by day on difficult winter flights in the extreme cold.

"At that time, the Aa formed the Front between our volunteer troops and the Bolsheviks. By Shagger [then known as Shagory], about 7 kilometers from Schlok, a Lettish battery stood on the Aa, which our men had made an effort to get. It was not able to be silenced by the nuisance shelling of our artillery, for following some salvoes, it regularly disappeared under a prepared cover.

Left: Neither Angelroth's nor Sachsenberg's description of the Soviet machines was entirely correct. As this photo shows, they were aluminum-doped Type 24 bis Nieuports. The cockades on the rudders and undersides of the lower wings were crudely converted to red discs (paintbrush strokes are visible), but strangely, the upper wings retained the old imperial tricolor cockades. This aircraft was N4287; the machine crashed by Hackbusch was N 4321. The third machine was N4300. (04AEF-NegB09-F26-NB003, History of Aviation Collection, Special Collections and Archives Division, Eugene McDermott Library, The University of Texas at Dallas)

Nieuport Ni.24 *bis* #4321, 1. *Aviaotriad*, Weissenhof. Surrendered to Sachsenberg *Geschwader* 9 May 1919.

German Aviation Units in the Baltic, 1 - 23 May 1919

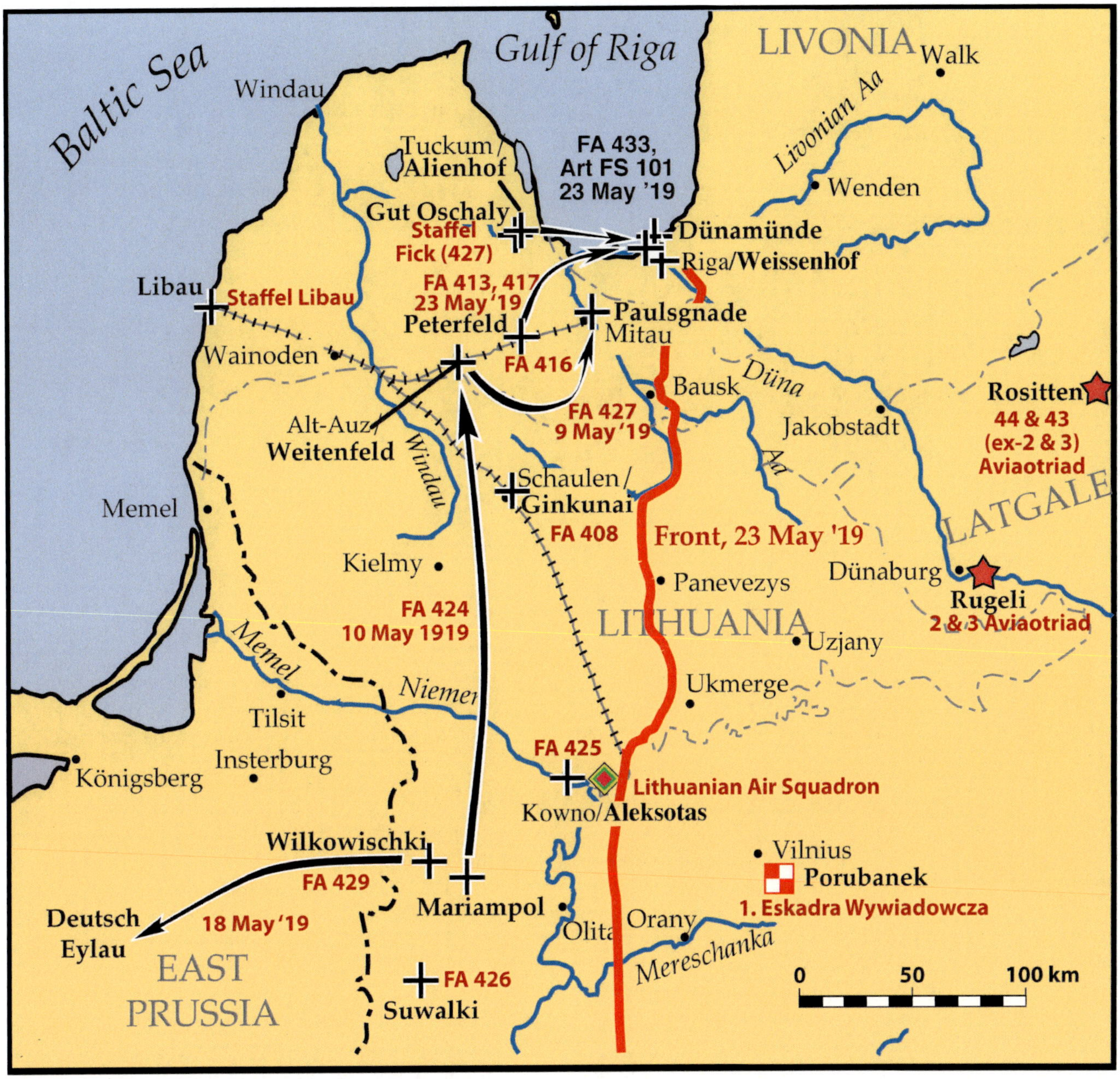

"In the first days of May, I had just obtained a Halberstadt, CL.IV, a fighting machine for assault with two fixed machine guns for the pilot, which fired through the propeller, and one machine gun for the observer. Automatically-released mines were attached to the fuselage, in addition to as many as possible stowed away on board as the machine could carry on each flight.

"One evening, the telephone rang. Shlok reports: "The battery is firing!" After a flight of exactly 88 kilometers, at 800 meters we sight our objective -- it is straight ahead near the Aa bridge, under cover. The machine descends, throttled back, the battery in our line of sight. At 400 meters, both machine guns cut loose, and the burst of shots falls in the middle of the target -- then benzene spurts in the face! Damn it! It can only be a hit in the carburetor. Both of us can hardly see any longer, so badly does it burn the eyes. Instinctively,

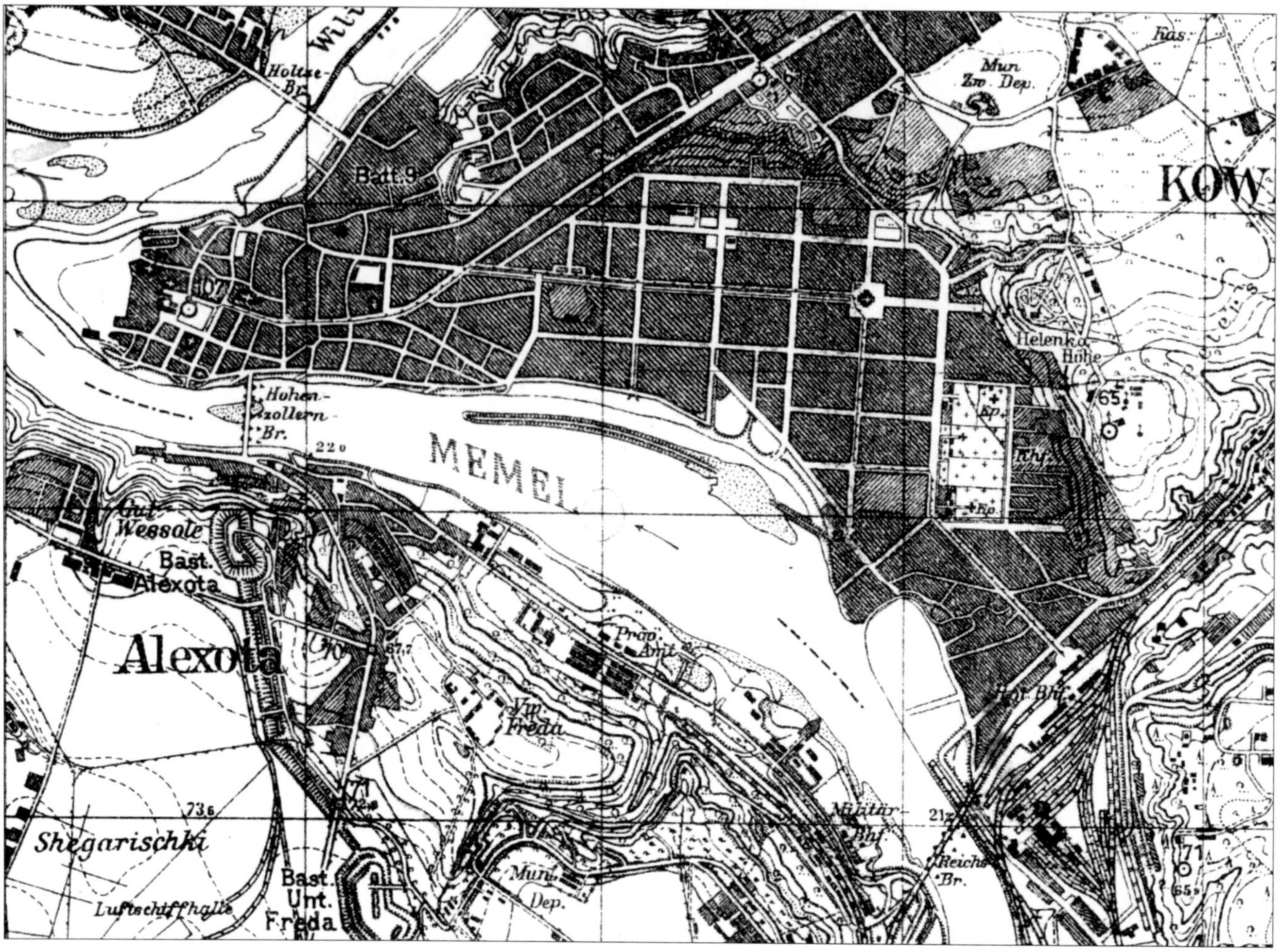

Above: Portion of a 1916 German military map showing the location of Aleksotas (identified as Alexota) airfield, due south of Kowno. The airfield, situated on the property of Gut Wesole, is believed to have been occupied by *FFA* 51 in September 1915. An army airship hangar was later constructed there, as indicated by the inscription "*Luftschiffhalle*" in the lower left corner. (Moshe Bukhman)[71]

the machine will be turned back in a wingover.

"The "long glide" now becomes useful. With a stopped engine, we glide over the Tirul Swamp back towards our lines. We must remove some of the load from the machine, or else we will not be able to reach the Aa. Lieutenant Deichmann lets the now unsecured mines fall. With a great roar, they explode on the terrain below us. He also throws our machine gun overboard. We knew not to expect any mercy on a landing. Then the river glints ahead of us. Ten meters past the Aa, we set the machine down for a swampy landing. Machine gun fire tries to reach us from the opposite bank, but we quickly became infantrymen again. By alternatively standing and ducking, we succeeded in escaping from the firing -- in the night, the machine was dismantled and transported to the airfield." [57]

This incident occurred 8 May [58].

No less a luminary than Josef Jacobs was also a victim of ground fire. Brought down in one of the swamps between Mitau and Riga 18 March, he wandered lost for a week before regaining his lines.[59]

Riga, headquarters of the Soviet forces, was only 65 km from Peterfeld and 55 km from Tuckum. To avoid Soviet retaliation against Riga's German residents, the German flyers were instructed to limit their overflights to reconnaissance and refrain from combat unless forced to engage. Fortunately, the Soviet flyers of the 1. *Aviaotryad*, operating out of the old German base at Weissenhof, usually obliged them, keeping a discreet distance. Angelroth suggested a possible reason:

"The Red Army also had its own aircraft and relatively good pilots, but the aviators were not regarded as sympathetic to the new regime; the pilots stood opposed to their new masters. Because of their convictions, we called these people 'radishes': red on the outside and white on the inside."[60]

Against this background of careful avoidance, one of the most bizarre events of the air war took place on 9 May. German narratives don't agree completely with one another, but they match on the essential events.

Above: Kielmy airfield, photographed during the Great War. It had changed little when *FA*s 408 and 409 arrived there in January 1919. The estate is now home to the Kelmė Regional Museum and Kelmė Manor. (Moshe Bukhman)

According to Kanisch:

"During our offensive flights a few days before [the 22 May attack on Riga}, an interesting episode occurred. Over Riga, we dove suddenly on two obsolete Nieuport single-seaters. They refused to fight us, moving some distance off. The next day, a machine of one of the other German *Abteilungen* encountered three of them. The pilot, preparing himself for a tough fight, stopped when he saw the enemy aircraft crews wave white scarves. He waved back and moved off in the direction of his airfield by Mitau. The three enemy fliers set down in a precise landing right alongside him. They had brought their luggage and bedding with them. They were Russian officers whom the Bolsheviks had forced to fly for them."[61]

The Germans described by Kanisch were Angelroth and Schumm in a Rumpler of *FA* 416. Angelroth picks up the narrative:

"One day, as we flew reconnaissance over Riga, we saw, to our surprise, 3 aircraft take off from the Russians' airfield and approach us. We had relatively little ammunition for a fight, having previously expended it at a freight yard. Therefore, we headed, aided by a tailwind, back to our airfield, which by that time had been moved to Peterfeld. The enemy aircraft were the Nieuport type; they were painted red. Our machine was faster, and we landed in Peterfeld. Yet scarcely had we rolled into the hangar, when the Russian aircraft also settled down quickly for landing. The field had meanwhile been placed on alert, and the barrels of our ground machine guns were pointing at the 3 enemy aircraft. How astonished we were, however, as the flyers stood peacefully alongside their machines with their hands upraised. They were made prisoners and transferred the next day to the division for interrogation. As proof of their dissatisfaction with the present regime, they wore, under their outer clothing, their old uniforms with their orders and medals, which they had put on in secret the night before taking off." [62]

Sachsenberg's reminiscence was the most detailed:

"It was ... midday, shortly before the first liberation of Riga, when suddenly, three Nieuport single seat fighters -- red stars on their wings -- came diving out

Above: Personnel of *FA* 409. Location is probably Kielmy or Ginkunai. Photo quality makes details difficult to confirm, but this appears to be another one of those shots where the subjects are having some fun with the photographer. The cyclist wears what looks to be a French Adrian helmet, and the man closest to the camera is taking a swig from a wine bottle. (Author's collection)

of the clouds toward our airfield at Peterfeld by Mitau. We immediately envisioned the same picture as we had so frequently experienced in Flanders, that of fearless English combat flyers, in squadrons of up to eighty machines, attacking our place with machine guns from an altitude of fifteen meters. -- Great alarm! The *Staffel* on duty rushed to the machines, the machine gun company manned its weapons -- clear to fire! The aircraft drop below five meters and -- land! We take the pilots prisoner, interrogate them, and it turns out the three men had received orders to drop bombs on Mitau. However, the grandparents of one of them and the sister and brother of another lived there. The flyers had only been forced to serve; under their pilots' suits they wore their old Lettish [sic] uniforms with their orders and decorations.

"With the attack on Riga close at hand, I now decided to fly back as quickly as possible with the same aircraft, without insignia modifications and with the red Soviet stars still on the wings, and attack and destroy unmolested all the bases which were vital to capturing the city and cutting off the enemy. One difficulty stood in the way. The English [sic] engines required a type of fuel that the squadron didn't carry. Hackbusch, one of the *Staffelführers*, volunteered. Hardly had he started when the engine quit, and the machine was completely wrecked. This tragic mishap was immediately used as inspiration for the salvation of the captured flyers, who would be exposed to the unbridled wrath of the Russians as soon as they found out about the surrender of the machines.

"So the crash of the Soviet machine was then photographed from all sides, and one half hour later, after the conscious, severely injured Hackbusch was to be transported out, a German two-seater took off toward Riga. The observer carried, in a droppable pouch, a letter to the commandant of Riga in which the *Geschwader*, following the custom of the Front, reported that the three aircraft had been shot down in air combat near Mitau. The flyers were slightly injured and were

Above: Fokker D.VII stunting over *FA* 424's tent hangars. (04AEF-B127-Album114-PB112, History of Aviation Collection, Special Collections and Archives Division, Eugene McDermott Library, The University of Texas at Dallas)

in captivity, their aircraft completely demolished. As evidence of this "fact" were enclosed the still-wet prints of the crash photos, on which the markings and numbers of the Lettish aircraft were recognizable. Three hours after the Lettish aircraft had taken off, a German airplane dropped the dispatch pouch, with a black-white-red streamer, over the airfield in Riga." [63]

Transcriptions of the letter survive in the archives. It read:

"9 May 1919

Three machines were shot down in air combat and forced to land here. One has received a severe chest wound; however, he will probably recover. One has nosed over on landing and has broken a leg and suffered a concussion. One was compelled to land with a jammed aileron by 5 of our single-seaters. His name is Jakubov. He is unharmed.

Thus to each of you who is allowed to venture over our territory or drop leaflets!

[Signed]
A German *Fliegerabteilung*" [64]
The pilots, Lt. Narciss Puskelis and Sergeants Janis Priedtis and Voldemars Jakubovs were all Latvians. Debriefed by VI. Reserve Corps intelligence, they revealed the strength of the Red air force in the area: the 1st *Aviaotriad* was based at Weissenhof, across the river from Riga. After the loss of the three Nieuports, it was left with seven aircraft, none of which was believed airworthy. 2. *Aviaotriad*, based 5 km northeast of Wenden, had four planes, and the third *Aviaotriad* was located at Rudeli, east of Dünaburg [65].

All three pilots subsequently joined the fledgling Latvian air force. The Germans' ruse held up almost a year -- Soviet authorities launched a search for surviving family members in order to compensate them for their loss [66], but on 21 February 1920, a report from the Commissar of Aviation and Aeronautics of the Soviet 16th Army reported the three pilots were alive and well in Riga [67]. Puskelis died accidentally 25 July 1920 [68], but the other two survived until 1940, when the Soviets, after regaining control of Latvia, settled accounts by murdering them.

Action on the Lithuanian Front

The political situation in Lithuania was considerably less involved than in Kurland. The region was not dominated by an ethnic German elite, so Germany had

Above: LVG C.VI 9092/18 of the Lithuanian Aviation Company, purchased from Germany February 1919. Lithuania's early aircraft insignia, the White Knight Vytis in a tri-colored dimond, has to be one of the handsomest ever to embellish an aircraft. (Moshe Bukhman)

no stake in local politics, aside from desiring a pliant, friendly state. The Lithuanians had also been quick to organize resistance to the Bolsheviks, who were clearly less aggressive than they were farther north.

After the Armistice, German *Ober Ost* troops remained in Lithuania because Soviet forces had not advanced as far into the area as they had farther north. Saxon units made up a significant portion of the German forces there, and they adapted themselves to meet their new enemy. One of the first German aviation detachments in the area was the *Sächsische Freiwillige Flieger-Geschwader Staffel* 327, part of the 46. *Sächsische Landwehr Division - Freiwillige*. It appears in the Division's *Kriegsgliederung* (Order of battle) on 28 February and that of its successor, the Brigade *Südlitauen*, on 1 April [69], but disappears from the records shortly thereafter. It is ignored, for some reason, by the official *Reichswehr* histories, and little else is known about it.

Alexsotas airfield, across the Niemen River from Kowno, was home of the first German aviation unit to be formed in the Baltics. There, on 18 December 1918, *Hauptmann d. R.* Hermann Gröckel assembled the surviving men and equipment of the former *FA(A)* 214

Above: Founding officers and flyers of the Lithuanian Aviation Company. Starred names were former *FA* 425 Personnel. L to R, Seated: Eugenijus Salkauskas (O), Ignas Adamkavicius (O), *General* Slobsevicius, *Overste* (*Hptm.*) Olaf Dahlbeck (CO), Boleslovas Martinaitis (O). Middle row: *Oblt.* Paul Dietrich* (O), Pranas Jancevicius (O), *Oblt.* Albrecht Weinschenk* (P), Adolf Schwesig* (O), *Ltn.* Hermann Leichner* (O), Sigizmundas (Zigmantas) Knystautas (O), Zivatovas [Viktor Zhivatov] (O), Pranas Hiksa (P), Juozas Grigiskis (O), Pantalionas Soltanas (O), *Off-stv.* Harry Rotter* (P), *Off-stv.* Fritz Schulz* (P) Back row: Erich Schulz* (O), *Off-stv.* Erich Kulikowski* (P), Stasys Jakstys. (Moshe Bukhman)

into a new unit, designated *Freiwillige Flieger Abteilung* 425.[70]

On 8 January, the Lithuanians were forced to abandon Schaulen by the advancing Soviet army. Schaulen sat on the rail line between Murajewo and Kowno, and its loss would have isolated the German forces in southern Lithuania, so on 11 January, the German high command sent a *Freikorps* unit, the *Detachement* von Randow, to Kielmy, southwest of Schaulen, as the first step in recapturing that city Shortly afterward, the Randow *Detachement* was followed by *Grenzschutz Flieger Abteilungen* 408 and 409, organized at the *Flieger Beobachter Schule* (Aviation Observer School) in Königsberg.[72]

Once Kielmy had been secured, *FA*s 408 & 409 moved there from their temporary base in Tauroggen. Headed by *Hpt.* Johannes Missfelder and *Hpt.* Julius Schulz, respectively, these small units had only seven aircraft between them at the beginning of March [73].

Other German aviation units continued to arrive in Lithuania. Farther south, at Mariampol, *FA* 424 took up residence. Its commander was *Oblt.* Erwin Martin. In addition to ten Fokker D.VIIs, eight C-types (including LVG C.VIs) and one Friedrichshafen G.IIIa, the *Abteilung* boasted a howitzer battery [74]. The battery was apparently attached to the *FA* as an expedient to get it into Lithuania -- it did not, as far as is known, remain with *Oblt.* Martin's unit.

FA 426, under *Oblt.* Hellmuth Weinschenck, would anchor the southern end of this line of airfields at Suwalki. At the end of February, *Fl. Abt.* 426 had 6 two-seaters and 4 single seat fighters [75].

Collectively, these units were placed under a newly formed tactical command, *General Kommando z.b.V.* 52.

On 5 February, the Lithuanian Aviation Company obtained its first aircraft by chance, a Sopwith 1A2 Strutter abandoned by its Red air force pilot, and on 14 February, the *Jagdkommando* of the Randow *Abteilung* successfully assaulted Radziwilischki, destroying

Above: Ginkunai airfield. It had been both a Russian and German base during World War I and would serve as the home of *FA*s 408 and 409 in 1919. Although this photo was taken in August 1916, the field had not changed much in 1919. A similar shot appeared in von Eberhardt's *Unsere Luftstreitkräfte* titled "*Deutsches Fliegerlager in Litauen*". (Moshe Bukhman)

railroad equipment and facilities. This marked the turning point in the Lithuanian campaign; German troops and their Lithuanian allies had stopped the Bolsheviks and began a counter-offensive to drive them out of the country.

Not wanting to depend solely on the German aviation units for their support, the Lithuanian Aviation Company purchased eight LVG C.VIs from Germany. These were taken on strength 28 February [76]. They were supplemented by the transfer of three Albatros B.II and one each Rumpler C.I and Halberstadt CL.IV from *Flieger Abteilung* 425, reducing that unit's strength to three aircraft[77]. Short of trained crews of its own, Lithuania contracted four German pilots, *Oblt.* Albrecht Weinschenk and *Off.-stv.* Fritz Schulz, Harry Rotter, & Erich Kulikowski, observers *Oblt.* Paul Dietrich, *Ltn.* Hermann Leichner and Adolf Schwesig, along with six mechanics, to fly combat missions and train Lithuanians to take over their roles. The Company was commanded by a contracted Swedish officer, Olaf Dahlbeck (Dahlbeckis to the Lithuanians).

On 1 March, the Lithuanian LVGs made their first test flights, and two days later, Schultz (identified as Sulcas in Lithuanian records) and Lithuanian officer Kostas Fugalevicius conducted the new air force's first mission, dropping leaflets and conducting reconnaissance over Kaisiadoriu and Vilnius.[78] Their flight report also contained a comment that would take on great significance in the final days of the West Russian Army:

"Note: the engine with benzene does not work well and therefore it is dangerous to make further flights; at Vilnius, the engine backfired a couple of times and lost revs."

Though much reduced in strength, *FA* 425 conducted reconnaissance in area of Schaulen, as did *FA* 433. When the area appeared to be secure, *FA*s 408 & 409 moved from Kielmy to the northeastern outskirts of Schaulen.

The airfield was located at Ginkunai, a country manor known to the Germans as Gut Ginkunen. It was a familiar site; a German *Flieger Abteilung* had been stationed there in 1916.

On 12 March 1919, the Lithuanian Aviation Company was upgraded to an Aviation Group, with Kostas Fugalevicius named to command it. This event is commemorated as the birthdate of Lithuanian military aviation.

In an advance begun 11 March, the troops of *Generalkommando Z. b. V.* 52 and its 45th Reserve

Right: Officers of *FA* 426. L to R: *Oberleutnant* Hellmuth Weinschenck (CO), *Ltn.* von Osterroht (*O.z.b.V.*), *Ltn.* Erich Kroll (*O.b.z.V.*, *Abt.* VIII), and *Ltn.* Hans Arimond, (CO, *Abt.* IV). (Moshe Bukhman)

Division drove the Bolsheviks out of Radziwilischki and Schadow. By the 16th, they had linked up with the troops of von der Goltz' VI. Reserve *Korps*. The Germans and their allies, now holding a unified front from the Gulf of Riga to the Polish Front, settled in to rest and resupply. The previous day, *FA* 429, *Hpt.* Wagner commanding, arrived in Wilkowischki, a few kilometers to the northwest of Mariampol, from Langfuhr airfield near Danzig.

From mid-March till early April, *FA* 426, operating out of Suwalki, was in the thick of the action. Per the *Darstellung*, on 18-19 March:

"... readily-available aerial reconnaissance made possible by the standing alert of *Fl.Abt.* 426 greatly simplified surveillance of the enemy. On 26 March the German fighter pilots successfully engaged a Russian truck column north of Orany."[79]

" On 5. April the Volunteer Battalion 26/45 under *Oblt.* Lohmann (Klüfer Regiment) accompanied the Border Guard Command Suwalki on a reconnaissance probe to Orany, which led to a violent combat continuing up till April 7, including artillery on the Russian flank, in which *Fl. Abt.* 426 cooperated."[80]

The Lithuanian air force came up to speed quickly, conducting reconnaissance flights in early April. On 2 May, it launched a strong bombing and strafing attack on Red forces at Ukmerge.

The Lithuanian situation seemed to be well in hand, and not a moment too soon.

With the approaching signing of the Versailles Treaty, tension was building on the Polish Front. If Germany refused to accept the treaty, fighting was sure to break out, and the forces facing the Poles would need all the resources they could muster. Accordingly, *Grenzschutz Ost* prepared to transfer some of its units from Lithuania under the cover name "Tannenberg".

On 25 April, FA 409 was ordered to Bavaria to participate in the assault on Red Munich. It apparently left its aircraft at Ginkunai, because it received new machines upon its arrival in Schleissheim. This was followed by plans to transfer additional *Flieger Abteilungen* to Germany on 3 May. *Oberkommando* Nord instructed various units, including *Flieger Abteilungen* 424, 426, and 429, to prepare for evacuation to Germany.[81] The operation began on the 10th, but only *FA* 429 returned home, departing Wilkowischki 18 May for Schloss Schönberg, Deutsch Eylau, West Prussia.

Transfer orders for the other two *Abteilungen* were canceled, *FA* 424 moving instead to Gut Weitenfeld, near Alt-Auz, on the 10th, and FA 426 remaining at

Above: *FA* 429 personnel pose with one of the unit's Halberstadt C.Vs at Wilkowischki. (Peter M. Grosz Collection/STDB)

Above: In the field at Wilkowischki. *FA* 429 Halberstadt C.V and tent hangars. (Peter M. Grosz Collection/STDB)

Above: *Oblt*. Weinschenck (3rd from left), NCOs and enlisted men of *FA* 426, plus a couple of unidentified ladies. The unit's 426 sleeve patch is apparent on several uniforms, including Weinschenck's. (Brian Lewis)

Pfalz D.IIIa 6014/17
***Fl.Abt* 410 / *Artillerie-Fliegerstaffel* 120**
Deuthen, 1919.

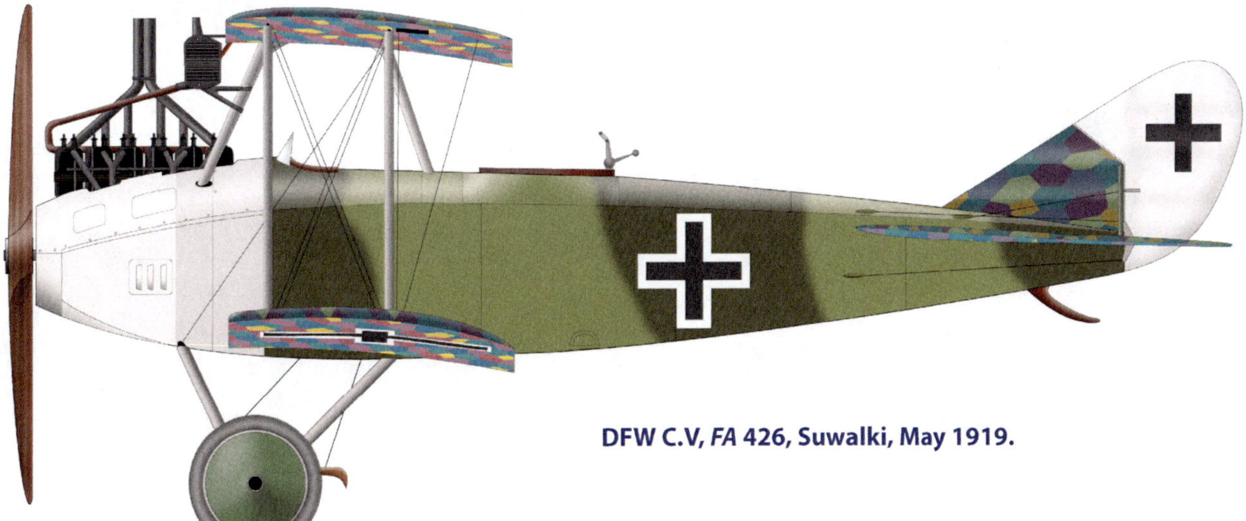

DFW C.V, *FA* 426, Suwalki, May 1919.

Above: On 13 April, *Feldwebel* Bachert wrecked this Halberstadt CL.IV within sight of *FA* 426's hangars and barracks. A Roland-built CL.IV, sharing the same white nose, stands in the distance. (Brian Lewis)

Above: This DFW C.V met a similar fate 24 May at the Suwalki airfield. The white nose was the unit marking. (Author's collection)

Above: *Oblt.* Weinschenck and three of his enlisted men posing with Halberstadt CL.IV (Rol) 8176/18 at Suwalki. This aircraft and the one to its left sport *FA* 426's white nose marking. (Greg Vanwygarden)

Halberstadt CL.IV (Rol) 8176/18, *FA* 426, Suwalki, 1919.

Suwalki. Command of *FA* 425, still based at Aleksotas, passed to *Hpt*. Ernst Dörffler on 3 May.

Assault on Riga

22 May 1919 was chosen as the day for "The Second Liberation of Riga", as it was termed in the *Reichswehr* official history [83]. Troops began moving forward shortly after midnight, and the flyers went into action at first light. *FA* 433's Kanisch and Hillmann described the overall plan:

"So arrives 22 May, the day of the assault on Riga. The plan of attack was also drawn up in detail for us *Fliegerabteilungen*. Each machine had its particular mission. One to two attack machines were always covered by one fighter machine. The troops were to be equipped with flare pistols and signaling apparatus so they could maintain contact with us. As the first *Kette*

Above: *Flieger* Max Thiemicken, Radziwilischki, 16 August 1919. In addition to his Iron Cross ribbon and black wound badge, he wears the sleeve patch of *FA* 426. (Brian Lewis)

[literally, "chain"; the term refers to a two- or three-plane tactical formation] the following were selected: *Vizefeldwebel* Kanisch - Observer *Ltn.* Deichmann. *Gefreiter* Wolf - Observer *Ltn.* Rauch. *Vzfw.* Jedwill flew the assigned fighter aircraft. First we were to follow the Kalnzen - Osilne road between the Babitsee and Tirul Swamp, on which the Baltic *Landeswehr* was to advance. We observed the advance in small groups. Enemy flyers didn't show themselves this day.[84]

"The morning of 22 May passed with steady thrusts against Red columns and transport. No serious resistance was offered. Only here and there did we come under enemy machine gun fire. Where the enemy was encountered, we blanketed him with bundled explosives and grenades. The pilot of our fighter plane, *Vzfw.* Jedwill, struck against one marching Bolshevik detachment that offered a tempting target. After expending his ammunition, Jedwill turned away, whereupon he received lively fire from the dug-in Bolsheviks. In spite of a shot that passed through his main tank and shattered his left thigh, causing a great loss of blood, Jedwill succeeded, by summoning up all his energy, in flying his machine back to the airfield. Twice he attempted landings, but each time the runway was occupied by machines taking off. Finally, the place was unoccupied. Cleanly, he set the machine on the ground and stopped the engine, as three ground crew hurried up to lift him, unconscious, from the machine. The fuselage of the aircraft was splattered with blood back to the tail surfaces. No one believed it possible that anyone in such condition could still fly and land."[85]

Hauptmann a. D. Erich Freund of *FA* 433 reminisced in *Reiter gen Osten*:

"22 May. Up 1:30 a.m., to the airfield at 2:30 a.m. This morning at 1:30 a.m. the advance on Riga began. Our first crates start at 2:30 a.m. My turn will be at 5:30 a.m. With *Vfw.* v. Bartdorff started on "DFW. 847" at a height of 100 meters, sometimes even lower (cloud) along the path. It's getting better at the Aa. Seen by the troops in the woods and dunes south of Binadnek. There is currently fierce fighting at Vw. Zenne. Then to the north, on the Binadnek - Osilne road, columns march east. I look for *Oberst* Ballod's white Baltic detachment and I go down to 20 meters in gliding flight. However, it's Bolsheviks. They and me surprised. Horses romp around, wagons drive into the ditch, fur-hatted guys look for cover in the ditch, sometimes they shoot up. But then I got there with a machine gun, put two mines on top and off over the Babitsee. Pulled up to 300 meters. Beyond again fire (even audible) to Assern [Essern?], then spiraled higher over the sea and back to the harbor.

Landed smoothly in Tuckum at 6:25 a.m. Got four hits in the crate. Had breakfast.

Started again at 10:45 a.m. Quite turbulent in the midday heat. Over Kalnzem to Olai. The advance of the Iron Division was noted. Main troop with tethered balloon 800 meters high (we are at almost 1000) at Olai, advance guard at Bonde. The Reds in a hurry, individual troops. Main part already in Riga-West. Bombs on it! Are really hit by flak (about 25 shrapnel), then via Pupe rail station to the mouth of the Aa. Steamers seen. Situation there unclear. Further back toward Riga. The Reds are just walking over the Düna bridge. A few more bombs on it, anti-aircraft fire again. Then to Osilne, the Chief of Staff discovered there and dropped the message.

Our advancing troops, moving 50 to 100 meters next to the country road, cheered up by waving. Then back to Tuckum. Landed at 12:07 p.m. Lunch, slept a bit. *Vfw.* Jedwill comes straight back to the field in a single-seater fighter, in spite of a gunshot wound to the thigh.

Above: Albatros C.XII 1144/17. Some sources have identified this plane as belonging to *FA* 425, but the lake-studded terrain below is more reminiscent of the area around Suwalki, home base of *FA* 426. (August Blume)

Albatros C.XII 1144/17
***FA* 426 / *Truppen-Fliegerstaffel* 38**
Suwalki, 1919.

Leutnant Hamster has to make an emergency landing.

Started again at 6:02. To Riga. Our troops are just moving into the city center with great cheers. Accompanied the entry at a height of 50 to 100 meters. Everything beckons. Flew halfway around the church steeple. Then south. To Getlingsmoor railway station. There are still Bolsheviks and two trains ready to go. Some bombs on it. Then again over the city and after Hoeden to the Weissenhof airfield. Landed there at 7:05 p.m. as the second plane. The Reds are about to leave for Bolderaa, about 1500 meters away. Talked to people, looked things over, but there were no infantry there, started again at 7:55. Again around the city and the surrounding area, more about its battles in a moment. Home to Tuckum, landed smoothly there at 8:32 in the evening. Busy but also successful day out! Home. Straight to bed 9:30 p.m."[86]

Flieger Abteilung 427 suffered a loss when *Uffz.* Grund and his observer, *Ltn.* B. Hermann, returning from the front to drop messages to the Iron Division northeast of Mitau, flew into a balloon cable, dying in the ensuing crash. The balloon probably belonged to *Freiwillige Ballonzug* (Balloon Company) 108. [87, 88, 89]

The Fokker D.VII of *Vzfw.* Kurt Stremple, reassigned

FA 424 – Versteh 'mer' richtig!

Some *Flieger Abteilungen* left a more complete record of their activities than others. Sometimes that had to do with their size, combat accomplishments and longevity, but a key factor seems to be whether they had a photo section. When it wasn't loading cameras and processing and interpreting film, the photo *Abteilung* had ample time to record activities around the base and snap candid shots of unit personnel. Copies of these photos were distributed to members of the unit, often in commemorative albums, as a token of their service. Surviving examples of photos from these albums have been indispensible resources for modern historians. The Sachsenberg *Geschwader* was one of the most thoroughly documented this way, as was Maercker's *Flieger Abteilung* 423. Conversely, *FA* 427, which originated as *Eiserne Flieger Abteilung* 8 a few days after the Armistice and survived to the bitter end of Bermondt's army a year later, seems to have been chronically camera-shy.

Flieger Abteilung 424 was one of the well-documented ones. It left a rich visual record of its activities in 1919, thanks to a photographic section that chronicled its day-to-day goings on. Although few written accounts have survived to provide details, these photos suggest a close-knit unit able to leaven business with pleasure. It also may have enjoyed more than the usual official sources of financing. *FA* 424's personnel sported a metal sleeve badge announcing their unit affiliation, a luxury not enjoyed by any other aviation detachment. These photographs suggest *FA* 424's personnel enjoyed many of the comforts and amenities available to their brethren stationed back in the Fatherland, although they were based far from home.

FA 424 was created by *Idflieg* 18 December 1918 and worked up to operational strength at *Flieger Beobachter Schule* (Aviation Observer School) Königsberg in the following weeks. Its commander was *Oblt.* Erwin Martin, a veteran observer who had served with *Feldflieger Abteilung* 19 in France. For part of his stint there, he flew with *Lt. d. R.* Willy Rosenstein, one of the better-known Jewish pilots in the *Luftstreitkräfte*. On 28 April 1916, the pair was attacked by a French Nieuport, both sustaining severe leg wounds.[P33/299-301 (82)] Rosenstein managed to regain their base, and both flyers were hospitalized. Unlike Rosenstein, who recuperated quickly and returned to flying, Martin seems to have been laid up for an extended period, and *FA* 424 appears to have been his first combat duty assignment thereafter.

Early in 1919, the *Abteilung* arrived in Mariampol, Lithuania, reporting to *Generalkommando der Zusammenges Reservekorps*, a collection of German volunteer units helping the new Lithuanian army anchor the southern portion of the line against the advancing Red army. Its aircraft inventory included Fokker D.VIIs, LVG C.VIs, and Halberstadt CL.IVs, and at some point, one or two SSW D.IVs. A single Friedrichshafen G.IIIa distinguished it as one of the few *Abteilungen* to boast a bomber. As far as is known, it was used only as a transport and liaison aircraft and did not perform any bombing sorties.

Above: *Oblt.* Erwin Martin, *Abteilungsführer* of *FA* 424 / *Truppen-Fliegerstaffel* 37 / 3. *Plastun* Aviation Division. (04AEF-B127-Album114-PB135, History of Aviation Collection, Special Collections and Archives Division, Eugene McDermott Library, The University of Texas at Dallas)

Above: *Ltn.* W. Hein Blöcker, Technical Officer of *FA* 424. (*Der Reiter gen Osten*)

Above: Officers and aircrew of *FA* 424. *Oblt*. Erwin Martin is in the front row, third from right. To his left, in a white cap, is *Oblt*. Hans-Helmut von Boddien, who served with Manfred von Richthofen at *Jasta* 11 before taking over command of *Jasta* 59. Boddien was assigned to the unit 21 August 1919 and appointed its acting CO the following day; by this time *FA* 424 was the 3rd *Plastun* Aviation Division of the West Russian Army. (Neal O'Connor)

Right: *Oblt*. Martin (second from left) with some of *FA* 424's pilots and ground crew by an LVG C.VI. The fuselage and serial number appear to have been repainted. (04AEF-B127-Album114-PB151, History of Aviation Collection, Special Collections and Archives Division, Eugene McDermott Library, The University of Texas at Dallas)

Above: *FA* 424's sports team. During and after the Great War, periodic athletic competitions called *Sportfests* drew teams from various military and *Freikorps* units as a morale booster. This is probably the *Abteilung*'s track and field squad, as evidenced by their jerseys with a winged propeller badge. In the 21 September *Sportfest* at the Iron Division's Replacement Depot, *FA* 424 acquitted itself well, with *Uffz.* Behrens placing second in the 1500 m run and *Gefreiter* Voigt second in the triathalon and third in the 100 m dash. In the 26 km pack march, pilot Haupt and *Flieger* Vilgis finished third and fourth. This photo is believed to have been taken at the conclusion of that meet.[82] (04AEF-B127-Album114-PB120, History of Aviation Collection, Special Collections and Archives Division, Eugene McDermott Library, The University of Texas at Dallas)

Below: "*Hauptquartier FA* 424 Photo-*Abt.*" A good deal of schnapps was probably consumed in the production of this photo. *"Versteh'mer' richtig!"* (Understand correctly!, or in English vernacular, "Are we clear?!"), chalked on the boxcar, appears in other photos and may have been an unofficial unit slogan. (04AEF-B128-Album 114-PB121, History of Aviation Collection, Special Collections and Archives Division, Eugene McDermott Library, The University of Texas at Dallas)

Right: NCOs of *FA* 424. *Vzfw*. Fritz Lebiedz is fourth from left. The two men on the right are sporting cutout versions of the unit's sleeve badge. (04AEF-B128-Album 114-PB121, History of Aviation Collection, Special Collections and Archives Division, Eugene McDermott Library, The University of Texas at Dallas)

Below: Relaxing with the ever-present cigars, swagger sticks and rustic birch railing. (04AEF-B127-Album114-PB115, History of Aviation Collection, Special Collections and Archives Division, Eugene McDermott Library, The University of Texas at Dallas)

Vzfw. Fritz Lebiedz and pup with a Fokker D.VII. Lebiedz had been a pilot with *FFA* 56 and *FA (A)* 212. (04AEF-B127-Album114-PB141, History of Aviation Collection, Special Collections and Archives Division, Eugene McDermott Library, The University of Texas at Dallas)

Above: *Ltn.* Karl August von Schoenebeck (8 victories, former CO of *Jasta* 33) poses with an SSW D.IV. The aircraft is most likely D.6177/18. The civilian, who appears in other photos of the unit, is unidentified. (Jack Eder)

Above: Intrepid noncoms and ground crew with the unit's SSW D.IV. Young man perched on the prop spinner sports a Russian cockade on his *Feldmütze,* likely dating this photo to the autumn of 1919, after *FA* 424 joined Bermondt's *Westrussischer Armee* to become its 3rd *Plastun* Aviation Division. This photo may also explain the dents in the D.IV's prop spinner in the von Schoenebeck photo above. (04AEF-B127-Album114-PB151, History of Aviation Collection, Special Collections and Archives Division, Eugene McDermott Library, The University of Texas at Dallas)

Above: *Flieger* Joseph Fürst served with *FA* 424 and its successors from at least 1 September till the end of Germany's Baltic adventure. In addition to his Baltic Maltese Cross, which was authorized on 1 November 1919, he wears the iron Cross 1st and 2nd class, a wound badge, the Silesian Eagle badge and a Prussian pilot's badge, in addition to a Russian cap cockade. (Joseph Fürst Jr.)

Facing Page, Above: Two of *FA* 424's flyers posing with Halberstadt CL.IV (Rol.) 8167/18. This photo shows up many of the Halberstadt's construction details to good advantage. (04AEF-B127-Album114-PB116, History of Aviation Collection, Special Collections and Archives Division, Eugene McDermott Library, The University of Texas at Dallas)

Facing Page, Below: *FA* 424 Staff Car. (04AEF-B127-Album114-PB153, History of Aviation Collection, Special Collections and Archives Division, Eugene McDermott Library, The University of Texas at Dallas)

Below: Coffin of an unidentified *FA* 424 member. The *Abteilung* gave him a good sendoff, complete with an abundance of evergreen boughs and an inscribed ribbon. (04AEF-B127-Album114-PB145, History of Aviation Collection, Special Collections and Archives Division, Eugene McDermott Library, The University of Texas at Dallas)

Halb.C.L.II Rol. 8167

FRW. FLIEG. ABTL. 424

Above: *"Versteh' mer' richtig"!* (04AEF-B127-Album114-PB117, History of Aviation Collection, Special Collections and Archives Division, Eugene McDermott Library, The University of Texas at Dallas)

Left: Another photo of the Fokker D.VII (OAW) of *FA* 418/ *Truppen-Fliegerstaffel* 27 shown on page 81. (Greg VanWyngarden)

from the Sachsenberg *Geschwader*, was downed by fire from what he described as British and French ships off the coast. He managed to land safely, but a hip wound kept him from leaving his plane. Stremple credited that wound for saving his life; the swamp he landed in would probably have swallowed him up. He sat in his plane overnight and was rescued the next day by a German patrol that bridged the marsh to rescue him.[90]

The swamp that almost claimed Stremple was the Tirul Marsh, an expanse of wetland between the Aa and Düna Rivers. In the summer it's normally dry, and in the winter, it's frozen over, but in the spring, thawing and seasonal rains can make it nearly impassable. One rail line and a few roads and trails were the only safe ways to cross it, simplifying the defense of Riga from the west. These were the routes the *Landeswehr* and Iron Division would have to follow to take Riga. It was in this area that one of the few documented friendly fire incidents of this war took place; the von Medem *Freikorps*, in the vanguard of the assault, was attacked by a plane of *FA* 427.[91] Regardless, the *Landeswehr* pressed on, and led by its *Stosstruppe* (Shock troops), forced the highway bridge into Riga early in the afternoon.

A key objective for the *Landeswehr* and German forces was the Reds' airfield, located near a cement plant in Weissenhof, now Spilve, across the Düna River from Riga. This had been the home of *Flieger Abteilung* 22 and the *Eiserne Flieger Abteilung* prior to the Germans' withdrawal in December. Obtaining this base would shorten the flyers' turnaround time as they continued to attack the Reds fleeing east. Once it was evacuated, the Germans lost no time occupying it, as Kanisch and Hillmann relate:

"About 4:00 PM, I landed with *Ltn.* Deichmann as the first machine at the airfield in Riga. The Red garrison had abandoned the field just a few minutes before in an orderly retreat. I had the existing Red aircraft pushed out of the hangars and onto the field to make room for our machines. We also took over the quarters of the Red flyers and crews. I was pleased that a short time afterward, five machines of our *Abteilung* were landing, even though there was still violent fighting around the field."

"On the day following the capture of Riga, the Riga airfield was already occupied by us. The Riga field lay in Thorensberg [sic], on this [west} side of the Düna. One of the workers' quarters in the nearby cement plant served as our barracks. The speed with which our troops had driven the Bolsheviks away had yielded a great deal of abandoned material, which was welcome booty to us. Thus we received a very elegant English car, which, with its electric lights and magneto, was a motorist's pride, and which later was in the service of the *Luftfahrt-Überwachungsabteilung Mitteldeutschland*. Ironic that the fate of this car, which had been taken away from the Bolsheviks in the Baltic, was to be employed later in combat in the homeland against the Bolsheviks in *Mitteldeutschland* – against Max Hölz[92] and his comrades. For the time being, however, we took a day's well-deserved rest there."[93]

Angelroth's account gives bragging rights to *Leutnant* Winter and *Vzfw.* Todheide as the first to arrive, in a Junkers CL.I, the evening after Riga's capture.[94]

Shortly after the victory at Riga, the *Landeswehr* and *FA* 433 were recipients of praise from their commander, Major Fletcher:

"... As you have received the thanks of the liberated Riga citizens, the entire Homeland, and indeed, the entire civilized world, so your leaders also thank you. You and your brave sister troops, the *Flieger Abteilung* 433, which has so faithfully accompanied us and vigorously supported us during the advance. Much was done. Yet much remains before us. We are still not allowed to stick the sword back into the sheath. ..."[95]

Between transfers to *FA* 433 and combat losses, *Geschwader* Sachsenberg had been reduced to 20 aircraft by 23 May[96]. At the time, *FA* 433 had 11 aircraft, and *FA* 427, 12, for a total of 43 among the three units. This provides some insight into the attrition that took place in the prior months; at the beginning of March 1919, these units boasted nearly 140 aircraft, and additional machines had continued to arrive between then and the assault on Riga.

To maintain the fiction that *Geschwader* Sachsenberg was still resting quietly at Peterfeld, awaiting transport back to Germany, von der Goltz created an ad hoc unit, *Baltische Staffel*, to which the few remaining personnel and machines of *FA*s 413 and 417 were transferred. The new *Staffel* joined *FA* 433 at Weissenhof, leaving *FA* 416, Sachsenberg's reconnaissance *Abteilung*, behind at Peterfeld.[97]

Rather than continue fighting for control of Riga, the Reds quickly withdrew so far to the east that by 25 May, German reconnaissance could find little trace of them, aside from a small concentration near Friedrichstadt, upriver from Riga, which they bombed a day or two later.[98]

Overreach

With Riga in his pocket, von der Goltz began laying the ground for his next move. The Reds had quickly withdrawn to the East; the Soviets had decided to write off the Baltics in favor of opposing Admiral Kolchak's Siberian forces and General Denikin's Volunteer Army in south Russia. With the Bolsheviks no longer in play, von der Goltz opted to strike through Estonia toward Petrograd. The Estonians, with a great deal of material

Above: The wreck of this FA 417 Ju. CL.I testifies to the intensity of the 22 May battle for Riga. Because of their all-metal construction, Junkers machines were expected to better resist combat damage and were preferred for the dirtiest low altitude work. However, they were not armored and probably fared little better than their wood and fabric counterparts. Records make no mention of deaths of any flyers serving with *FA* 417 or on loan to *FA* 433 that day, but it's difficult to believe anyone survived this crash. (SDTB)

Above: This is not another photo of Halberstadt C.V 6902/18, overturned at Wainoden. It's Halb. C.V (Av) 6911/18, based at Peterfeld. Trying to land the close-coupled C.V frequently produced this result. (Peter M. Grosz collection/STDB)

assistance from the British and Finns, had already cleared the Reds from most of their territory and struck a temporary alliance with White Russian General Rodzianko's Army of the North, which was planning an attack on Petrograd. This was another marriage of convenience -- the Estonians, with good reason, had no faith the Russians would respect their independence, but the Russians, needing a base of operations, were on their best behavior. Von der Goltz's ambitions dovetailed neatly with the Russians' plan -- if he could ally with Rodzianko for a combined drive on the Russian capitol, it might fall into their hands.

The *Landeswehr* and Iron Division had already swung north from Riga and headed toward Estonia. On 3 June, two of *FA* 433's aircraft crashed at Rodenpois station, injuring their crews and reducing the *Abteilung*'s strength to six C-type and three D-type aircraft, along with one 50 cm and two 25 cm cameras and two pairs of radio transmitters and receivers.

With his army already marching northward, von der Goltz needed to establish contact with Rodzianko. The most logical way was to send an emissary via air. To that end, on 8 June, *Oblt. z. See* Friedrich Patze of the Sachsenberg *Geschwader* (now, technically, the *Baltische Staffel)* departed Weissenhof for Narva in a DFW C.V; his passenger was Senator Dmitri Neidhardt, a member of Russia's pre-revolutionary State Council. Neidhardt was a respected figure in White Russian circles, and von der Goltz hoped he'd be able to obtain Rodzianko's agreement for a joint venture. Assuming the Estonians would not support his plans for an alliance, he chose to literally go over their heads. It was a mistake -- the Estonians didn't trust the Germans any more than the White Russians, and when they got wind of von der Goltz's plans to join up with Rodzianko, they feared an attempt to overthrow their new republic. They had already seen what von der Goltz had done in Latvia.

So when Patze and Neidhardt landed near Narva on 8 June, their reception was less than cordial -- they were arrested by Estonian troops and hauled off to prison. The following day, von der Goltz made another attempt, dispatching two more Sachsenberg aircraft, Halberstadts C.V 6898 and 6905/18 (Av), with Neidhardt's secretary, Herbert Buschmann [99].

On landing between Narva and Yamburg, C.6905 overturned, and the planes were immediately seized by the Estonians and impressed into their own air force. Their pilots, *Ltn.* Preis and *Flugmaa*t Horst Sawatski, joined Patze in captivity. They were to spend over a month in confinement under conditions that ranged from pleasant to downright frightening, as described in *Beilage zur Trommel*:

"After they were obliged to carry out a forced landing, to make clear that hostile intentions were far from their thoughts, they waved white cloths. About 50 poorly-dressed Estonian soldiers came running toward them, and even though the white flags must have been seen by everyone, shot at them and thus wounded the flyer Preis, who, with his fall, joined the injured observer [sic] *Leutnant* Patze. They might have been shot dead immediately, but only after they were beaten soundly. One tried to pull the ring from the finger of the flyer Sawatzki. It refused to budge, and the soldier wanted to cut off his finger. After that, Sawatzki pulled the ring off himself and turned it over to the Estonian soldiers. It appears that afterward, Estonian officers, who treated the prisoners amicably, offered them cigarettes and calmed their men. Their words should now be thanked, for otherwise they [Sawatski and Patze] might have been shot. On calming them, however, they bound their hands together behind their backs with their belts. The flyer Sawatzki's jacket was taken off. After that the captive flyers were taken to Narva under strong escort.

They were interrogated by the Staff, where they were treated very kindly.

"The injured *Leutnant* Patze and the injured flyer Preis were admitted to the hospital. On the way, they met a Russian officer of the North Army, who attempted to get the Estonian soldiers to unbind their hands. He failed to do it. Probably from fear that the North Army might still be able to liberate the prisoners, they took them to Narva on the same day. At the third station on the way to Narva, they were taken out, where they were presented to the railroad station commandant. This one cursed them profusely. The prisoners were not exposed to further harassment, because the Estonian officer in charge of the transport took them back to the auto. The seriously wounded flyer Preis, who was badly injured, had suffered great pain during the trip, and his condition worsened greatly. Arriving in Reval, the prisoners were taken to the command staff in an auto. Only the flyer Preis remained about one hour long in the waiting room, gazed at and admonished by the crowd. The behavior of the Estonian command staff was absolutely correct. They questioned the flyers and sent them to the prison. They remained here for one day and then were removed by the chief guard to solitary confinement, where they received relatively good food and treatment. On 2 July their confinement deteriorated when they were taken to the notorious Nargö prison camp. In this prison camp were imprisoned Bolsheviks and lawbreakers, who made a very deplorable impression. The conditions were absolutely pathetic; they had only ¼ pound of bread and one soup daily. There was a shortage of water (the prisoners could not wash themselves), the vermin were out of control, and spotted typhus was rampant. As a result of poor nutrition, numerous cases of hunger typhus and

Above: Halb. C. V 6905/18 (Av) overturned at Narva, Estonia, 9 June 1919. (Moshe Bukhman)

typhus appeared. The imprisoned flyers soon found themselves enclosed together with 10 Germans in the prison camp. Their quarters were a covered hole. They were required to lie on planks, with large holes, obtained from the British. Straw sacks and coverings, naturally, made no improvement. So it was, that they suffered chills and rheumatic pain from the humid air and strenuous trip. After a while, the food and treatment improved, as the Estonian *Feldwebel* among them admitted knowing a German who had helped him in a German prison camp. On July 24 the German prisoners went into quarantine. However, they were de-loused in swimming baths, freeing them from the vermin that the daily washings had failed to achieve. On July 27 the prisoners were taken to Rahumeli prison. This was a prisoncamp with barracks. Found here were also German prisoners confined together with Bolsheviks, among whom women were also found. Here, as well, no mattresses with covers were to be found. The weather there was very poor, and all the prisoners, inadequately dressed, were to continue suffering together.

"The flyers, freed from captivity through arrangements with the Danish Red Cross, were brought on board the Hamburg ship, the "Dania", sailing under the English flag. On August 13, they traveled to Riga. The English were very astonished at the appearance of the Germans and treated them exceptionally correctly and kindly. They arrived in Riga on the 15th, and on the 19th, they were allowed to travel on to Mitau.

"The pilot Preis had fared worse than the pilot Sawatzki. He was to be shot by Estonian soldiers immediately after being taken to the hospital. Only the intercession of the chauffeurs and a nurse managed to save him at the time and later, as well, to prevent a planned knife assassination and poisoning attempt. He was often harassed and cursed by Estonian flyers, sick soldiers. At his request, he was placed in another barrack, where the personnel were friendlier. He was very contented with the food and medical treatment. *Ltn.* Patze still remains in Estonia."[100]

Failing in his attempts to contact Rodzianko, von der Goltz decided to push on. The Reds had abandoned Livonia, between Estonia and Riga, and Estonian forces, reinforced by reconstituted Latvian units loyal to Ulmanis, were pushing southward. The Baltic *Landeswehr* and Iron Division were moving north on a collision course with them. On 7 June, the *Landeswehr* drove the Estonians out of Wenden. A hastily-called truce stopped fighting for a few days, but neither side would yield ground. Anticipating a resumption in combat, von der Goltz called on *FA* 433, now down to five serviceable aircraft, to reconnoiter Estonian troop positions from its new home in Weissenhof, and on 16 June, ordered it to set up an advance airfield near Wenden, which it did two days later [101]. On the 21st, one of its aircraft was lost near Gross Roop, when its fuel tank was punctured by ground fire. The pilot made an emergency landing in no man's land and escaped. During the night, enterprising Estonian soldiers pulled the plane to their lines with a rope. The same day, a second German plane fell near Auciems.

Meanwhile, on the 18th, aircraft from *FA* 427 moved from Paulsgnade to Bienenhof, on the west bank of the Düna across from Riga, to carry out attacks on Lemsal and Gross Roop, west of Wenden. The following day, it lost one of its observers, *Ltn. d. R.* Wittemayer, killed on a combat flight [102].

Above: A downcast Sawatzki looks on as his wounded compatriot, *Ltn*. Preis, rides a *panje* wagon into captivity. (Moshe Bukhman)

On 22 June, it moved to the southern Riga suburb of Wiedersholm, where it remained until 2 July, when it hopped to nearby Lutzausholm Island in the Düna. The same day, the Estonians and the allied North Latvian Brigade shocked the Germans by retaking Wenden and driving them back toward Riga. *FA* 433 lost a third aircraft to ground fire near Wever [103]. The remains of the *Abteilung* returned to Weissenhof. By 3 July, with the Germans in full retreat toward Riga, the Allies got both sides to agree to a ceasefire. With the fighting ended, *FA* 427 and *Artillerie-Fliegerstaffel* 101 probably didn't even bother to unpack on Lutzausholm before returning to their permanent base in Paulsgnade.

After intensive negotiations, the Allies announced all German forces would withdraw from Riga, Windau and Libau and soon afterward, from the Baltics entirely. Ulmanis and his cabinet would establish their government in Riga, and the Baltic *Landeswehr* would be handed over to Latvian control.

Realignment & Consolidation

In the Baltic, the summer would be a period of relative peace as both sides took up new positions and reorganized their forces. Within days of the ceasefire, von der Goltz had moved his headquarters back to Mitau.

He no longer had any excuse to delay the transfer of the 1. *Garde Reserve* Division back to Germany. In preparation for that move, command of the Sachsenberg *Geschwader*'s *FA* 417 passed to *Ltn.* Ernst Kredel on 5 June. The *Abteilung* was back at Peterfeld by the 18th. On the 25th, it was merged with *FA* 416 *to form Artillerie-Fliegerstaffel* 108 and placed under the command of *Oblt.* Adolf Averdam. Together with *FA* 413, the fighter squadron, it accompanied the 1. *G.R.D.* to Stettin, Pomerania. The division was redesignated *Reichswehr* Brigade 34, reporting to *Grenzschutz Ost's A.O.K. Nord,* and *FA* 413 became *Truppen-Fliegerstaffel* 34. *Artillerie-Fliegerstaffel* 108 would continue on to Graudenz, where it was unofficially dubbed *Polenstaffel* Graudenz. It remained there until it was ordered to Jüterbog for demobilization per *Armee-Tagsbefehl* (Army daily order) 145 of 14 August 1919 [104].

Above: Ju. CL.I "M" met its end at Peterfeld in June. As noted earlier, this might have been Majewski's personal aircraft, but details of the markings suggest it was not the same machine that flew out of Wainoden in the Spring. (04AEF-B308-F23-PB005, History of Aviation Collection, Special Collections and Archives Division, Eugene McDermott Library, The University of Texas at Dallas)

On 30 June, *Hauptmann* Alt relinquished command of *FA* 433. Four days later, the *Abteilung* was detached from the *Landeswehr* and sent back to Tuckum. On 19 July, a new aviation unit, *Baltische Fliegerabteilung* 1, was created and assigned to the *Landeswehr*. The order [105] establishing it stated it was was FA 433, renamed. At best, it was a spinoff, because FA 433 was returned to the Reichswehr. *Baltische Fliegerabteilung* 1 was placed under the temporary command of *Feldmeister* Skribanowitz on 26 July [106]. It remained idle at Tuckum until 27 July, when it moved to Weitenfeld, near Alt-Auz, and merged with *FA* 425, which had arrived from Kaunas on the 7th. *FA* 425 was dissolved 14 August, and the surviving combined unit was renamed *Flugpark* Kurland or *Armee Flugpark* 1 (*AFP* 1). *Hpt.* Dörffler, who had led *FA* 425, was named *Führer* of the *Flugpark*.

In the meantime, *FA* 427, still part of the Iron Division, took up residence in Gut Paulsgnade by Mitau with its 10 aircraft, along with the single captive balloon of *Ballonzug* 108. The Iron Division had also picked up a newly formed aviation unit, *Artillerie-Fliegerstaffel* 101, with 6 aircraft [107]. This created one of those unit identity anomalies that would characterize this period -- under *Idflieg*'s master plan, *Artillerie-Fliegerstaffel* 101 was to be the new designation of the former *FA* 427. Instead, it remained an independent *Staffel*, serving alongside 427.

Lithuanian Operations

On 17 May, to support a strong offensive in the northeastern part of their country, the Lithuanians moved four of their aircraft to an advanced base at Kedainius, north of Kaunas. The next three days, they flew over Panevezys, Kupiskio and Karsakiski, strafing Red forces and a supply train. On the 28th, they carried out reconnaissance over Red forces as far as Utena and suffered their first aircraft loss, when an LVG C.VI crewed by Kulikowski & Petronis caught fire after its engine seized. The plane was a total loss, burning out after crash landing, but the crew escaped unharmed.

The Lithuanian offensive continued. On 13 June, Utena had been secured, and 3 or 4 LVGs and their crews, including all the Germans, were moved to a temporary airbase there. By this time, the Soviets had

German Aviation Units in the Baltic, 24 May - 30 June 1919

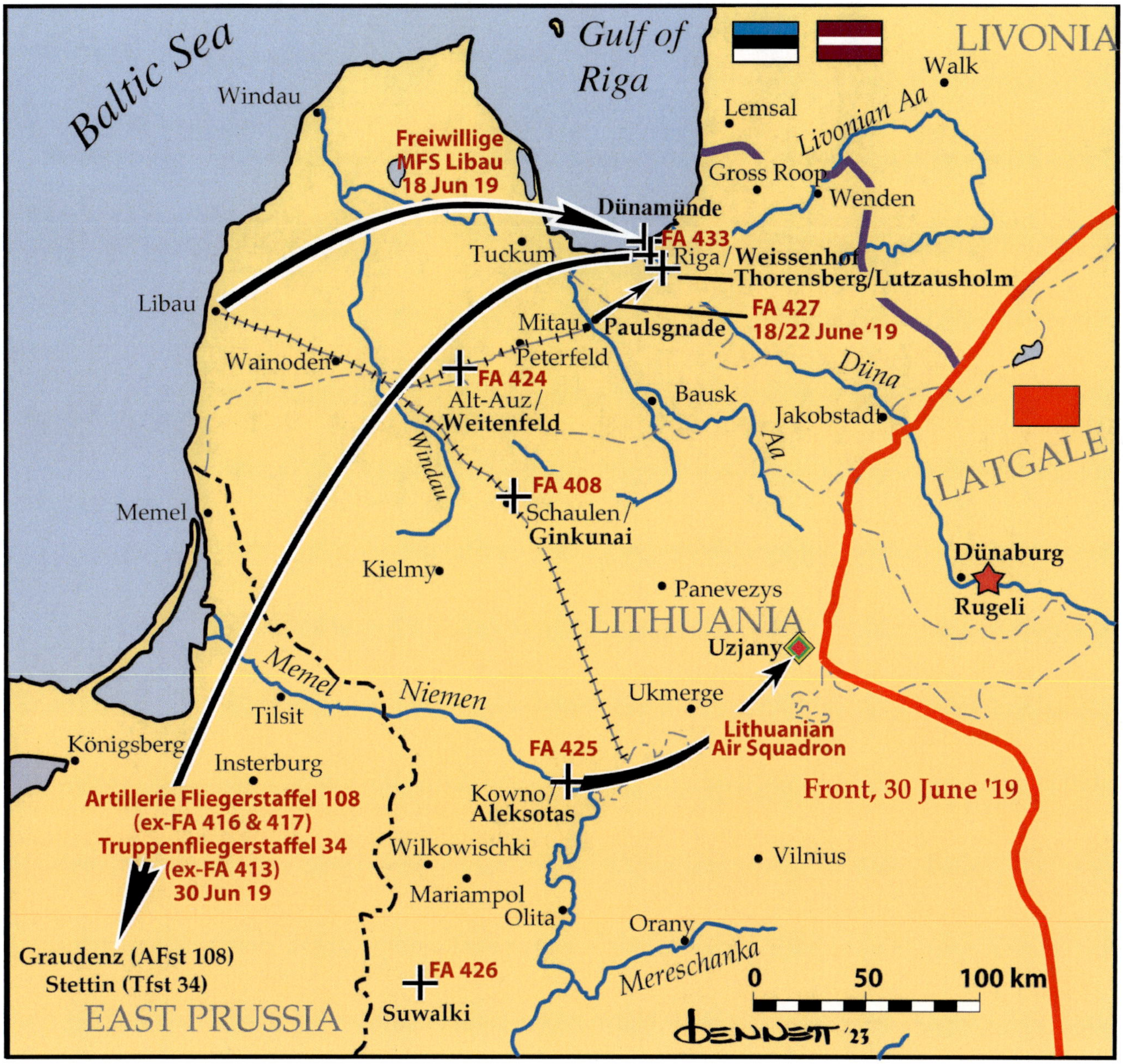

been cornered in northeastern Lithuania. Fighting subsided, and both sides dug in. During this time *FA* 425 began shedding its inventory, selling three Albatros B.II and one Ru. C.I to the Lithuanians on 10 June, followed on the 28th by one Albatros B.II, 1 Rumpler C.I, 1 Halberstadt CL.IV[108] and 2 LVG C.VI aircraft. All this was a prelude to the *Abteilung*'s 7 July transfer to Weitenfeld by Alt-Auz. In a move that essentially wedded German forces in Lithuania to the West Russian Army, Brigade Schaulen was subordinated to the VI. *Reserve Korps* 28 June and then dissolved and replaced by the Diebitsch *Freikorps* 13 August.[109] *FA* 408 was to be transferred to Munsterlager, Germany, 18 August. That order was rescinded, however, and the *Abteilung* would remain at Ginkunai, near Schaulen.[110]

Since the Armistice, the Suwalki region had been disputed between Poland and Lithuania. By virtue of its alliance with Germany, which occupied the area, Lithuania stood to inherit it when the Germans left. This expectation was dashed in July, when the Peace

Above: Suwalki, Sunday, 25 May 1919. Parishoners gather for Mass at the Church of St. Alexander in this aerial photo taken by *Vizefeldwebel* Gödeke & *Ltn.* Johannesson of *FA* 426. (Author's collection)

Commission declared the so-called "Foch Line" [111] as the provisional boundary between Poland and Lithuania, placing Suwalki on the Polish side and compelling Lithuanian and German forces to leave. It was generally understood that the displaced *FA* 426 would return home to Graudenz, likely for demobilization. By 12 August, however, its withdrawal had stalled at Radziwilischki, 240 km to the north [112].

The transfer to Graudenz appears to have been a cover story to conceal the fact that *FA* 426 was never intended to leave the Baltics. The Versailles Treaty had made it redundant on the Polish Front, so it made more sense to leave it in the Baltic to support whatever future schemes von der Goltz might cook up. A little over two weeks later, Major Bischoff's speech to his troops triggered the mutiny that led the German volunteers to join the West Russian army. *FA* 426's sleight-of-hand may have been a preliminary move in a larger scheme to keep German troops in the Baltics under the Russian flag, and more specifically, equip them with another experienced, combat-ready aviation wing.

Becoming Russians

Von der Goltz may have been under orders to remove his depleted forces from the Baltic, but getting him to obey would be another matter. The general was determined to convert the Baltic territories into German satellites, even if Berlin's support blew hot and cold.

During the summer months, a wholesale restructuring of forces took place. In early July, Colonel Fletcher relinquished command of the *Landeswehr* to a British officer, Lt. Col. Harold Alexander, as its interim commander, and its *Reichs* Germans would be replaced by Latvian troops. It was also agreed the *Landeswehr* would be transferred to Latvia's Eastern Front to confront the Soviets, rather than its former brethren in Bermondt's part-German army. By this time, the *Landeswehr* had already been stripped of *FA* 433, to be replaced by a new Latvian-staffed aviation unit commanded by *Ltn.* Richard Schaefer, a Baltic German.

On the 23rd of July, Berlin ordered von der Goltz to remove his forces from Libau and Windau. He was now confined to a landlocked, roughly rectangular area straddling Latvia and Lithuania, bounded by the East Prussian border in the west, Tuckum in the north, the west bank of the Düna to Friedrichstadt in the east, and Radziwilischki and Tauroggen in the south. Its skeleton was the network of rail lines from Murajewo to Mitau, Mitau to Schaulen and Radziwilischki and from there to Tilsit. The Iron Division settled sullenly in Mitau, along with several *Freikorps* units no longer needed or welcome in Lithuania.

Von der Goltz' repeated stalling and defiance of orders to return with his troops to Germany had finally driven the Allies to exasperation, and on 5 August, they threatened Berlin with economic sanctions if German troops weren't brought home. The German government responded by telling von der Goltz, in no uncertain terms, to return his men to Germany by the end of the month. The general, always one jump ahead, hurried to Berlin, where he persuaded the government to agree it would have no authority over any Germans who volunteered to join a Russian army. There's ample evidence the government knew what was about to happen and was quite content to play innocent bystander. Nevertheless, the General Command of the VI. Reserve *Korps* had cut orders for the return of the aviation units to Germany. This was to begin after 20 August.

On the 24th, troops of the Iron Division, assembled at the Mitau rail station to debark for Germany, were elated when their commander, Major Bischoff, arrived and declared, "I hereby forbid the withdrawal of the Iron Division!" [113] This set in motion a wholesale defection of other *Freikorps* units, which then banded together as the German Legion. It would soon join the West Russian Army to justify its continued presence in the area.

Hpt. Rudolf Berthold, former *Führer* of *Jagdgeschwader II*, congenitally unable to stand by while fighting was going on anywhere, decided to join the mutiny. In mid-September, with nothing to do in Germany, he led his *Eiserne Schar*, one thousand strong, to Mitau and reported to the Iron Division.

The unlikely inspiration for this mutiny had appeared in Mitau in the Spring. Colonel Pavel Bermondt, an out-of-work Tsarist officer, arrived to recruit an anti-Bolshevik army from Russians recently released from German POW camps. Bermondt was a classic example of those people who, in normal times, would have been dismissed as interesting oddballs, but these were not normal times. He was an unapologetic monarchist; his goal was to see the Romanovs back on Russia's throne, and he believed an alliance with von der Goltz's army could bring that to pass. To gain the credence he thought necessary to lead the force he named the West Russian Volunteer Army, he assumed the title of Prince and the surname Avaloff from his adoptive father, a Georgian noble. His overpowering ego was matched only by his incompetence as a military leader. The refusal of other White Russian generals and Colonel Fletcher to take him seriously didn't dent his self-confidence in the slightest. Parading around in his Cossack uniform, Bermondt issued reams of orders to his new army, readying it to return the Baltics to the arms of Mother Russia.

Von der Goltz quickly took Bermondt under his wing, offering the German Legion and Iron Division, along with the *Flieger Abteilungen* remaining in the Baltic, to the West Russian Army. He also equipped Bermondt and his penniless mob with everything they needed to play army. With his German troops making up better than half of Bermondt's force, he also expected to co-opt Bermondt's operation and use it to his own ends, with Bermondt playing figurehead. He seriously underestimated the Russian's obstinate drive to do things his own way.

Above: Colonel Prince Pavel Bermondt-Avaloff in all his Cossack glory. He had served with a Cossack unit in the Tsarist army, so he could legitimately claim the title. (Author's collection)

Contraction & Consolidation of German Aviation Units in the Baltic, 1 July - 31 August 1919

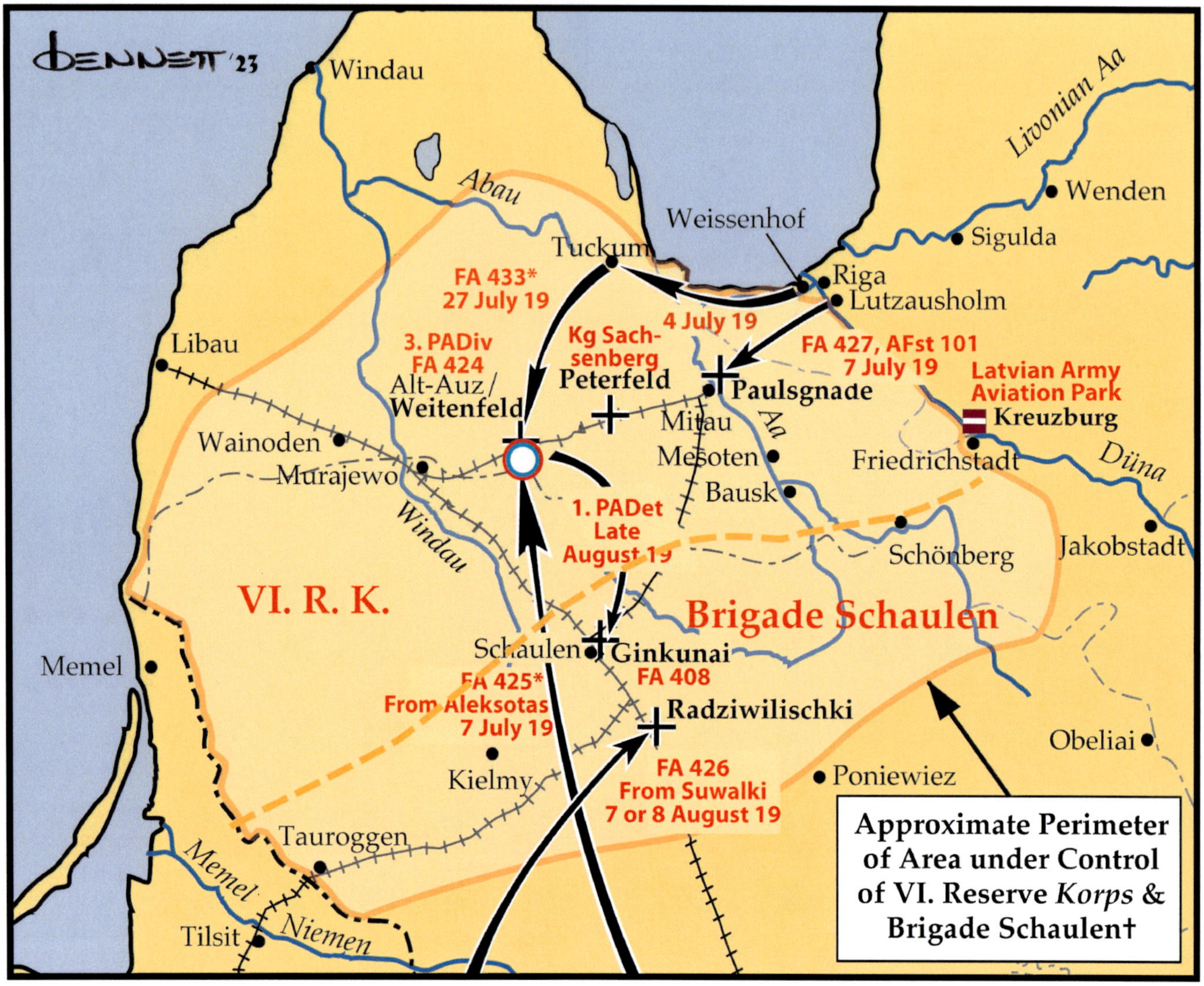

* On 14 August 1919, *FA* 425 was dissolved and merged with *FA* 433 to form *Armee Flugpark* (*AFP*) 1, also known as *Flugpark* Kurland, at Weitenfeld.

† Brigade Schaulen was disbanded and replaced by the Diebitsch *Freikorps* on 13 August 1919. *FA* 408 remained in Ginkunai under Diebitsch.

During August, *FA* 424, *FA* 426 and the Sachsenberg *Geschwader* joined the West Russian Volunteer Army of Colonel Bermondt-Awaloff, where they were redesignated as *Plastun* Aviation Divisions, each consisting of 2 to 4 *Plastun* Aviation Detachments, as follows:.

2. *Plastun* Aviation Division - *FA* 426/*Truppen-Fliegerstaffel* 38
3. *Plastun* Aviation Division - *FA* 424/*Truppen-Fliegerstaffel* 37
4. *Plastun* Aviation Division - *Kampfgeschwader* Sachsenberg

In addition, the 1. *Plastun* Aviation Division was formed from scratch, staffed by Russian personnel.

Map Abbreviations:
PADiv - *Plastun* Aviation Division
PADet - *Plastun* Aviation Detachment

The West Russian Air Force

Bermondt spent the summer months organizing his army, and by September, it had taken shape. It consisted of four main units, two Russian and two German, arrayed along the west bank of the Aa River:

1. The *Graf* Keller Corps
2. The Wirgolitsch Corps
3. The German Legion and
4. Bischoff's Iron Division

On 14 May, Bermondt had appointed a Russian aviation officer, Lt. Col. Boris Firsov, as commander of his incipient air force [114]. In the following weeks, an aviation detachment, reporting to the *Graf* Keller *Freikorps*, would be staffed, initially with pilots and crews from Russia's old imperial air force. As the orphaned German Baltic *Flieger Abteilungen* joined his ranks in August and September, the Colonel could now boast the makings of a potent air force. On 31 July, Firsov was named commander of the corps aviation detachment, overseeing all of Bermondt's aviation resources. His former position of leader of the (now First) *Plastun* Aviation Division, part of the *Graf* Keller Corps, passed to Lt. Col. Livotov [115]. As the German flyers signed up with Bermondt, he created three more *Plastun* Aviation Divisions. The 2nd *Plastun* Aviation Division, formerly *FA* 426 / *Truppen-Fliegerstaffel* 38, led by *Ltn.* Hellmuth Weinschenck, was assigned to the other major Russian force, the Wirgolitsch Corps. The German Legion received *Oblt.* Martin's *Flieger Abteilung* 424 / *Truppen-Fliegerstaffel* 37, now known as the 3rd *Plastun* Aviation Division and based at Weitenfeld, along with the reconstituted Sachsenberg *Geschwader*, dubbed by the Russians the 4th *Plastun* Aviation Division, at Peterfeld. Rounding out the lineup were the Iron Division's *FA* 427 and *Artillerie-Fliegerstaffel* 101. The Diebitsch *Freikorps* and *FA* 408 in Lituania, by virtue of their subordination to the VI. Reserve *Korps*, were now an adjunct to the West Russian Army, although they played little part in the coming events.

In the summer, *FA* 424 / *Truppen-Fliegerstaffel* 37 / 3rd *Plastun* Aviation Division was split into two *Plastun* Aviation Detachments -- the 1st, populated mostly with Russian personnel, was formed under the leadership of Russian Lieutenant Colonel Livotov 31 July[116] and departed to Ginkunai the following day. There, it was divided into two sub-units, each with six aircraft. The 2nd Detachment, which retained most of the German personnel, remained at Alt-Auz / Gut Weitenfeld under the command of *Oblt.* Martin.

VI. Reserve Corps General Command, issued 24 August 1919, specified Russian markings to be applied to German aircraft flying with the *Graf* Keller Corps: "5.) Ic / *Grufl.* The Russian aircraft (Det. Keller) carry the following identification marks:
a) On the left side (seen in the direction of flight) of the upper and lower wings, as well as on the right and left side of the fuselage, the polygonal Russian Baltic Cross [this was the black Maltese cross used on Bermondt's flag] in the white field.
b) On the right side of the upper and lower wings, as well as on both sides of the rudder, cockades in the Russian national colors (white-blue-red)."[117]

These markings were to be painted over the German wing, fuselage and rudder crosses.

On 12 September, the West Russian Army's command got around to specifying aircraft markings for Bermondt's other Russian unit, the Wirgolitsch Corps. They were to be essentially identical to those of the Keller Corps, except for the crosses, which were to be white on a blue field [118]. Cross style was not defined. Some modern sources have interpreted this as a straight-armed cross on a circular field, but in the absence of photographic evidence, this can only be considered conjectural.

Photos of Bermondt's aircraft are scarce, so it's difficult to determine how thoroughly (or, indeed, if) these orders were carried out. At least one surviving photo shows the wingless, gutted fuselage of Albatros C.XV 7818 / 18. The narrow-bordered *Balkenkreuz* on its fuselage appears to have been reworked into a three-barred Russian Orthodox cross on a white ellipse, and the vertical stabilizer and rudder carry a remnant of the Russian tricolor cockade.

In a faded photo of another Keller machine, Albatros C.XV 7815 / 18, close examination suggests the *Balkenkreuz* on the underside of the left wing has been converted into the Orthodox cross by adding two black crossbars and painting out its narrow white border. Unfortunately, the camera angle obscures the fuselage and tail insignia.

Aircraft of *FA* 426 / *Truppen-Fliegerstaffel* 38, although part of the Russian Wirgolitsch Corps, appear not to have been altered -- all known photos of machines that survived to fall into Lithuanian and Latvian hands after fighting ended show them in German markings.

Each *Plastun* Aviation Division was subdivided into two to four *Plastun* Aviation Detachments (eleven in all), each with a nominal complement of six aircraft, essentially the equivalent of an old German two-seater *Flieger Abteilung*. In addition to these combat units, Bermondt's air force was supported by two *Flugparks*, functioning like the *Luftstreitkräfte*'s old *Flieger Ersatz Abteilungen*, depots holding the air force's inventory of everything from spare aircraft to uniforms and the source of replacement personnel to the combat units. With the exception of cap and uniform insignia, all the material, whether used by the Germans or Russians, was standard German issue. *AFP* 1 (*Flugpark*

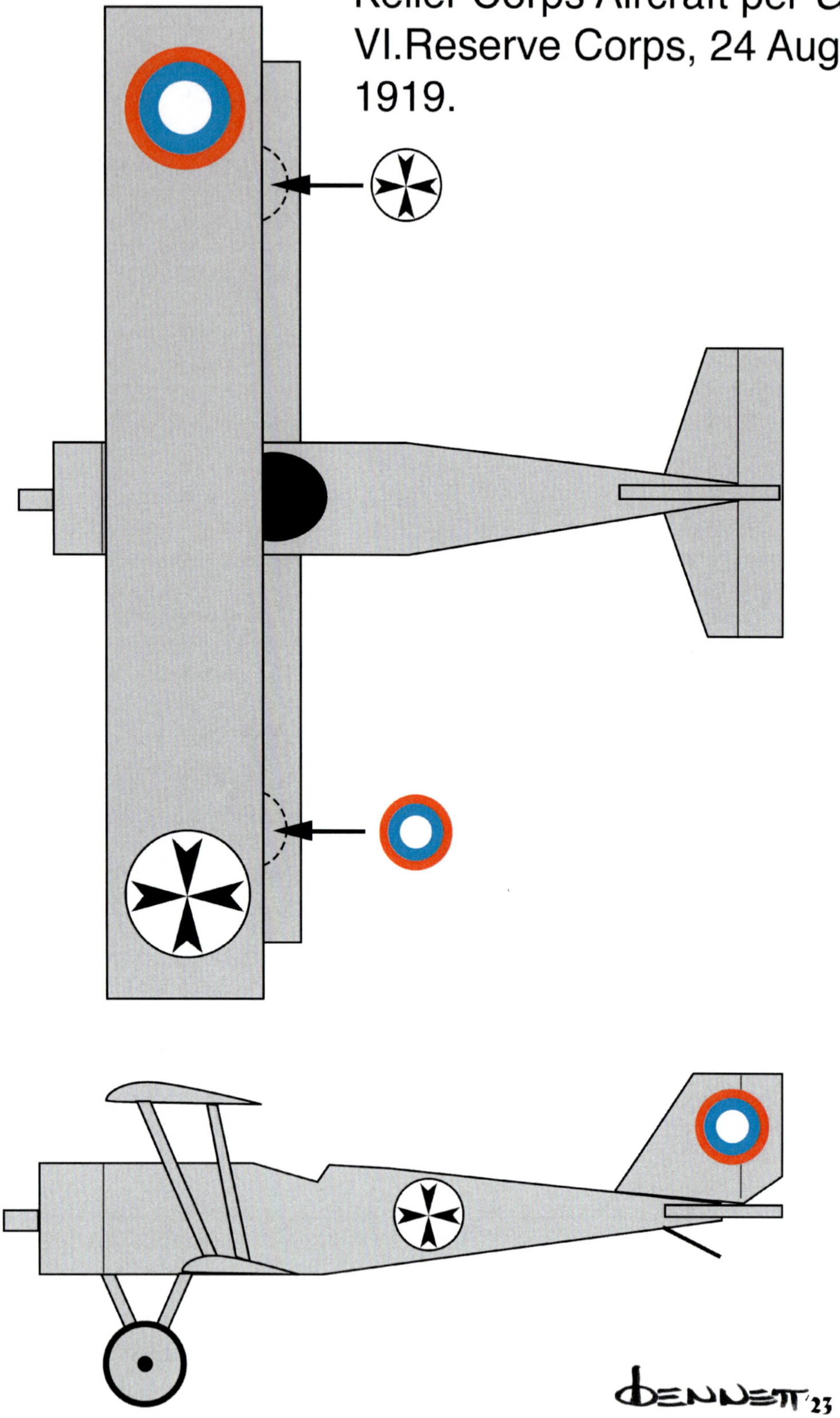

Proposed markings for *Graf* Keller Corps Aircraft per *Grufl.*, VI.Reserve Corps, 24 August 1919.

Albatros C.XV 7818/18, *Graf* Keller Corps, West Russian Army, Autumn 1919.

Above: Albatros C.XV 7815/18, with what appears to be a Russian Orthodox cross beneath the left lower wing. In its long career, it served five masters – Germany, the West Russian Army, Lithuania, Central Lithuania and Poland. (Moshe Bukhman)

Kurland), offspring of the *FA* 433 - *FA* 425 merger, remained under *Hpt.* Dörffler's leadership and was primarily responsible for resupplying the German units. As it was the principal conduit through which German equipment reached the Baltic, it also supplied the Russians' *Graf* Keller *Flugpark* 2. The Russians' allocation was established by an agreement between the two groups.

Allied sources credited Bermondt with having scraped together a total of 64 machines [119] from former *Freikorps* units in the Baltic, including a fair number of totally obsolete types like Albatros C.IIIs. Bermondt, however, claimed 120 aircraft under his command [120]. The disparity between his claims and Allied intelligence estimates may have reflected the sudden arrival of a large number of German aircraft in the Baltic in August. Throughout the month, agents in Kurland reported numerous trains carrying dismantled aircraft eastward. These machines were made surplus by the drastic contraction of the German air force following signing of the Versailles Treaty. Many of these shipments were clandestine operations, but others boldly exposed the German government's continued support of von der Goltz' Baltic schemes. *Die Freiheit*, newspaper of the *USPD*, was especially zealous in unearthing Germany's under-the-table support of Bermondt, reporting that pilots from *FA* 426 had arrived at Adlershof in September to collect three aircraft and fly them back to the Baltic, all with the full approval and cooperation of the German

Freiwillige Westarmee
General Prince Pavel Bermondt-Avaloff

Kommandeur der Flugwesens (Kofl.) der Westrussischer Korps Armee
1.) Lt. Colonel Boris Firsov (31.7.19), 2.) Lt. Col. Livotov (23.10.19), 3.) *Oberstltn.* Marschall von Bieberstein (Deputy to Firsov & Livotov from 2.9.19)

Verwaltung der Flugwesens

- Flugparks
  Lt. Col. Mikhail Vilkovuisky (1.9.19)
  - *Flugpark* Kurland [1. *Flugpark*] (ex-*FA* 425 & 433)
    Hpt. Ernst Dörffler (18.8. - 30.9.1919)
  - *Flugpark Graf* Keller (2. *Flugpark*)
- *Korps Graf* Keller
  Col. Potocki
  - 1. *Plastun* Aviation Division
    Lt. Col. Livotov (1?.10.19)
    Shtabskaptain Golikov (18.9.19)
    - 3. *Plastun* Aviation Detachment (Forming 9.19)
    - 4. *Plastun* Aviation Detachment (Forming 9.19)
      Ltn. Hans Arimond (7.19 - 11.19)
    - 5. *Plastun* Aviation Detachment (Forming 8.19)
      Ltn. Hellmuth Weinschenk (7.19), *Ltn.*Ebel (30.8.19)
      [18.9.19]
- *Korps* Wirgolitsch
  Col. Wirgolitsch
  - 2. *Plastun* Aviation Division
    (*FA* 426/*TFst* 38)
    Ltn. Hellmuth Weinschenk - stv. (2.9.19),
    Lt. Col. Livotov (18.9.19)
    - 6. *Plastun* Aviation Detachment
      Ltn. Ebel (29.8.19), *Ltn.* Johannsohn - *stv. (1.9.19)*
      Ltn. Hans Arimond (11.19 >)
    - 7. *Plastun* Aviation Detachment
      Ltn. Stephan - stv. (2.9.19 - ?)
    - 8. *Plastun* Aviation Detachment
      Ltn. Stephan, Oblt. Kratz - stv. (3.9.19), Ltn. Stephan (4.10.19>)
- *Deutsche* Legion
  Kapt. z. S. Paul Siewert, *Hpt.* Otto Wagener (17.9.19)
  - 3. *Plastun* Aviation Division
    Ltn. Erwin Martin
    - 1. *Plastun* Aviation Detachment
      (Russian & *FA* 424 Personnel)
      Lt. Col. Livotov (31.7.19), *Shtabskaptain* Golikov (18.9.19)
    - 2. *Plastun* Aviation Detachment
      (*FA* 424/TFst 37)
      Ltn. Erwin Martin, Oblt. Hans-Helmut Boddien - stv (22.8.19 - ?), *Oblt. Frhr.* von Münch (10.19)
  - 4. *Plastun* Aviation Division
    Lt. z. S. Gotthard Sachsenberg
    - 9. *Plastun* Aviation Detachment
      Ltn. Bruno Majewski (>13.8.19)
    - 10. *Plastun* Aviation Detachment
      (ex-*Afst* 108 Personnel)
    - 11. *Plastun* Aviation Detachment
      (ex-*Afst* 108 Personnel)
- *Eiserne* Division
  Maj. Bischoff
  - *Flieger Abteilung* 427
    Oblt. Walter Kirschstein (20.7.19 - 2.1.20)
  - *Artillerie Fliegerstaffel* 101
    Hpt. Friedrich Erdmann (13.8 - 9.9.19), *Oblt.* Walter Kirschstein (10.9.19)
- *Freikorps* Diebitsch
  Oberstltn. von Diebitsch
  - *FA 408 (Truppen-Fliegerstaffel 30)*
    Hpt. Johannes Missfelder, *Hpt.* Friedrich Köckler (19.7.19)

Freikorps Rossbach
Oblt. Gerhard Rossbach
- *Fliegerstaffel der FK* Rossbach
  Ltn. Ernst Krull

authorities [121, 122]. Actually, *Die Freiheit* was late to the game -- surviving documents show Adlershof released three Albatros C.XV, three Fokker D.VII, and three sets of Halberstadt CL.II wings, rudders, and elevators to *FA* 426's *Ltn.* Bracht on 21 August. Instructions from the Reichs Utilization Office even authorized Bracht to test-fly the six aircraft before they were packed for shipment to Radziwilischki [123].

Even more revealing is Corps Daily Order *Nr.* 18699, dated 11 September 1919, issued by *Generalkommando* VI. *Res. Korps Abteilung* IIa [124]. It stipulated *FA* 424, now serving under Bermondt as the 3rd *Plastun* Aviation Division and identified in German documents as *Truppen-Fliegerstaffel* 37, was to report to *Grufl.* (*Gruppenflieger* -- Flying Group) 37, headed since 13 August by *Hauptmann* Ascheberg, formerly *Kofl.* of the VI. *Reserve Korps*.

The very existence of *Grufl.* 37 in the Baltic is confusing. In the *Kriegsgliederung* (Order of battle) of the VI. Reserve *Korps*, it appears on the corps staff as early as 20 May, ten days after *FA* 424 moved from Mariampol to Alt Auz / Gut Weitenfeld. Based on standard *Reichswehr* conventions for designating units, *Truppen-Fliegerstaffel* 37 and *Grufl.* 37 should have been part of the 37th Brigade. At that time, 37. Brigade didn't even exist. It was created on paper in June 1919, to be based at Schwetz, West Prussia. On 12 July, that plan was cancelled and in August, a second incarnation was created at Gross Plowenz, in the Marienwerder district of West Prussia, over 500 km from the Baltic [125]. The logical explanation was that *FA* 424 was originally intended to be transferred to 37. Brigade when the latter unit was established. By that time though, Germany had inked the Versailles Treaty and was winding down much of *Grenzschutz Ost*. *FA* 424 was about to become an orphan, but was saved from demobilization by the creation of the West Russian Army. Rather than generate new *Fliegerstaffel* and *Grufl.* numbers to replace the ones already assigned, the German command apparently retained it in the Baltic to make the numbers match up.

Another theory is that these *Grufl.* numbers no longer had any relationship to *Reichswehr* brigades. *Idflieg*'s instructions for the demobilization of the *Reichswehr* aviation forces [126] list all the surviving air force units and their projected demobilization sites. It calls for *Grufl.* 31, 36 and 37 to report to Insterburg, 32 to Stolp, 33 to Neuruppin, 34 to Brieg, and 35 to Schwerin for disbandment. Many of these bases were far removed from the brigades corresponding to those numbers, and there was no logical reason to require the *Grufls*. to travel that far to shut down. Additional support for this interpretation is the inclusion of *Kofl.* 21 and 22 on the list. *Kofl.* 21 is known to be part of *Grenzschutz Ost*'s *AOK Nord*, and *Kofl.* 22 is believed to be part of *AOK Sud*'s command structure. Neither of these numbers correspond to any field commands in *Grenzchutz Ost*. In any event, the 37th Brigade had no involvement with its namesake aviation detachment.

The same order subordinated *Truppen-Fliegerstaffel* 1 (formerly *FA* 433) to *Grufl.* 37, ignoring the fact that *FA* 433 had merged with *FA* 425 to create *Flugpark* Kurland a month earlier. So despite the German government's claim to have washed its hands of the Baltic operation, it was still intimately involved with it. The VI. Reserve *Korps* claimed authority over aviation detachments that had supposedly deserted to the West Russian Army and of the repossessed FA 433, which the German government had gone to great lengths to identify as an arm of the Latvian Baltic *Landeswehr*, not the *Reichswehr*. German documents referring to the former *FA* 426 continued to call it *Truppen-Fliegerstaffel* 38, in spite of its official absorption into Bermondt's army as the Second *Plastun* Aviation Division. Its component flying units, called *Plastun* Aviation Detachments by the Russians, were still identified in German documents as *Abteilungen*. Even some Russian daily orders slipped up, referring to air units by their German identities. The Russians could call them whatever they wanted; functionally, they still were part of the *Reichswehr*. *Die Trommel*, a newspaper published in Mitau for the German troops in Bermondt's army, mentioned *FA* 424 by this supposedly outdated designation in its 15 October issue [127].

Clearly, Germany's hands-off claim was a sham -- Bermondt's West Russian Army was a veneer concealing von der Goltz's, and Germany's, continued attempt to control the Baltics.

Regardless of how many aircraft reached him, Bermondt lacked aircrews to fly all of them, and many remained unassembled. This correlates with the scarcity of information about several of Bermondt's *Plastun* Aviation Detachments. Although they appear on organization documents as "in formation" in September 1919, and personnel assignments to them were being made early that month, no subsequent documents have surfaced that would verify their existence as fighting units. Bermondt launched his first assault on Riga early in October, and any personnel and equipment replacements would most likely have gone to make up losses in existing units, rather than build new ones from scratch.

Although Bermondt may have been able to boast an abundance of aircraft, getting fuel to keep them flying was a different matter. Records of orientation flights for pilots and observers carried out 20 and 24 August [128, 129] showed them to be very brief -- none lasting more than 35 minutes, and some as short as 5 minutes, barely enough for a circuit around the airfield. On 25 August, *Oberkommando Nord*, to which the VI. *Reservekorps*

Disposition of Russian & German Aviation Units in the Baltic, 1 September - 8 October 1919

Map Abbreviations:

PADiv - *Plastun* Aviation Division
PADet - *Plastun* Aviation Detachment

Former German Identities of West Russian Aviation Units:

2. *Plastun* Aviation Division - *FA* 426/*Truppen-Fliegerstaffel* 38
3. *Plastun* Aviation Division - *FA* 424/*Truppen-Fliegerstaffel* 37
4. *Plastun* Aviation Division - *Kampfgeschwader* Sachsenberg

AFP 1 *(Armee Flugpark* 1*)* - Merger of former *FA* 425 and *FA* 433

Above: Junkers D.I, Fokker D.VIIs, Halberstadt C.V, and LVG C.VI (?) at Swinemünde. This gathering of aircraft is said to belong to the Sachsenberg *Geschwader*. If so, the summer foliage indicates this photo was probably taken in late August or September, during a stopover in the unit's return to its old leader in Kurland.

reported, issued these stringent instructions, suggesting a desperate shortage of fuel:

"3.) Ic / *Grufl.* Aviation restriction. As a result of the very poor fuel situation, the *O.K. Nord* again issued regulations restricting all flight operations. For the Corps area, in addition to orders already issued, the following orders are summarized:

6.) All flight operations are to be suspended except for urgent tactical and essential courier flights.

7.) Tactical flights are hardly necessary in the current situation. If necessary, they will be ordered by the General Command.

8.) Authorization for courier flights is only granted by the chief of the General Staff with a written ID. The aircraft are only provided [fuel] by *Grufl.* 37 after showing this ID.

9.) Absolutely necessary training flights conducted in machines and aircrew schools are dependent on permission of *Grufl.* 37. They should not exceed 10 minutes in duration and should not be carried out at low altitudes over cities and towns, as they often cause accidents.

10.) *Grufl.* 37 submits a security briefing of the flights that took place during the week and an inventory record of the existing and flown operating equipment every Saturday." [130]

A little over a month later, Bermondt launched his offensive against the Latvians, confident he had enough fuel to last through a campaign he expected to be a walkover. However, this might have come at the expense of fuel quality. In the final weeks of the German-Russian campaign, forced landings due to engine failures were disturbingly frequent. They put a serious crimp in the effectiveness of Bermondt's aerial operations, and they would cost several German fliers their lives.

What caused the engine reliability problems isn't known for sure, but mechanical issues aren't likely -- the engines were proven designs, and mechanics should have had the knowledge and skills to keep them in good working order, as long as spare parts and lubricants were available. Fuel, on the other hand, was problematic, judging from the instructions quoted above and incidents in prior months. Contamination with water and solid debris could have been the source of trouble, but aircraft had long been fitted with fuel

Above: Aviators of the reconstituted Sachsenberg *Geschwader*, Peterfeld, Autumn 1919. Left to Right: *Flugmeister* Horst Sawatsky and *Vizeflugmeisters* Hans Antonious, Mayer, Alexandre Zenses, Karl Sharon, Hans Goerth (wearing what appears to be an RNAS jacket), and Engelhardt. Some sources date this photo to April 1919, but Fokker D.VIIIs were not known to be in the *Geschwader's* inventory in the spring. More likely, they were brought to Kurland with the returning personnel. (Moshe Bukhman)

Below: Erich Schaefer (third from right) and Ju. D.I 3188/18 at Peterfeld, 9 October 1919. A Fokker D.VII is in the background.

Above: Lineup of Junkers D.Is at Peterfeld, 9 October 1919, the day after Bermondt launched his ill-fated assault on Riga.

Above: Ju. CL.I 12921/18 "V" on its nose at Peterfeld. After the Germans' departure, the machine was taken over by the Latvian air force.

Junkers CL.I 12921/18 "V", *Staffel* Majewski, *Kampfgeschwader* Sachsenberg (4. *Plastun* Aviation Division), Peterfeld, Autumn 1919.

filters to strain out solids and drain taps to remove water from their fuel tanks. In addition, filtering through chamois before fueling aircraft was a standard field practice. More likely, the problems stemmed from adulteration of the gasoline with some other substance, likely benzine or kerosene, to stretch its supply. In limited amounts, these chemicals will burn in a gasoline engine, but because they are less volatile than gasoline, they are more difficult to ignite, and their combustion may be less complete. Used to excess, they can deposit carbon in the engine cylinders, foul spark plugs and cause engine failure [131]. This jibes well with reports that many of the engine failures took place some time after the aircraft took off.

Despite their lip service to Bermondt's cause, the Germans were operating in support of von der Goltz' agenda, and their Russian brothers in arms were just a convenient cover. The Russians weren't naïve; they were aware of their junior status, and friction inevitably arose between these allies. On 18 September, Lt. Col. Firsov, head of the West Russian Army's air force, to which the *Flugparks* were subordinate, submitted a lengthy memo to Bermondt's high command, complaining bitterly about Dörffler's treatment of the Russian units, accusing him of withholding agreed-upon equipment and supplies, substituting worn out items for new, and diverting equipment back to Germany, where it was certainly no longer needed. Firsov added that even some of the German *Abteilungen* were receiving the same backhanded treatment [132]. There was apparently some substance to his claims, because Dörffler was relieved of his command 12 days later. He was replaced by Russian Lt. Col. Mikhail Vilkovuisky. *Flugpark* 1 was shut down, and on 22 October, its aircraft, equipment and supplies were allocated equally to the twelve aviation detachments [133]. If we are to believe the documents, this included a stunning twelve aircraft per detachment, plus 24 spares. Bermondt might have been able to provide each unit 12 machines, but it's highly doubtful his promised 288 additional aircraft were on hand. Combat losses, returns to Germany and booty taken by the Baltic states in the next two months don't come close to matching this total.

The Flugpark incident didn't appear to have put a permanent stain on Dörffler's record; on his return to Germany, he was placed in command of the *Mitteldeutschland Polizei Fliegerstaffel* in Gotha, and when he later joined the Luftwaffe, he rose to the rank of *Generalmajor*.

Geschwader Sachsenberg Flies Again

The 4. *Plastun* Aviation Division bears a closer look. Although Sachsenberg's *Abteilungen* (*Artillerie-Fliegerstaffel* 34 and *Truppen-Fliegerstaffel* 108) returned to Germany for eventual demobilization, Sachsenberg and a remnant of his *Geschwader* remained in the Baltic, retaining the name *Kampfgeschwader* Sachsenberg, and it committed to Bermondt's West Russian Army in the summer. Its reconstituted *Schlachtstaffel* was named *Staffel* Majewski after the former *FA* 417's leader, *Ltn.* Bruno Majewski. Meanwhile, many of the personnel of the demobilized units promptly returned to Kurland to rejoin their former leader. Surviving records suggest Sachsenberg was expecting their return; his *Kampfgeschwader* and *Staffel* Majewski were already in operation by 9 August 1919 [134], five days before *Truppen-Fliegerstaffel* 108 was ordered to demoblize in Jüterbog. *Staffel* Majewski would become the 9th *Plastun* Aviation Detachment. The other two components of the *Jagdgeschwader*, the 10th and 11th Detachments, drew heavily on returning personnel of the now-disbanded *Artillerie- Fliegerstaffel* 108. By the beginning of October, the *Geschwader* had rebuilt at Peterfeld [135] with a strength of 32 aircraft. Given that it was down to 20 well-worn machines after May's Riga

Above: Wreckage of Halb. CL.IV (Rol) 9432/18, in which Fritz Schulz lost his life 28 September. The officer seated in the rear of the car is an American, apparently a representative of the Inter-Allied Armistice Commission. (Moshe Bukhman)

Halb. CL.IV (Rol) 9432/18
***Off.-stv.* Fritz Schulz**
Lithuanian Air Service
Aleksotas, 1919.

assault, the returnees must have brought new aircraft with them. As they say about rigged prizefights, the fix was in.

Winding Down in Lithuania

By virtue of their successes in the field, the Lithuanians were already anticipating the day they could dispense with German assistance. Bermondt, fixated on returning all the Baltic territories to Russia, began trampling clumsily over what was left of the Germans' rapport with Lithuania, treating it as one of his theaters of operations. His Order #20, issued 2 September, commanded the Count Keller Corps, with its 1. *Plastun* Aviation Division, to depart for Obeliai, Lithuania, for a planned September 5 attack on Illuxt, Latvia. The Division, comprising the 3. And 4. *Plastun* Aviation Detachments, had a paper strength of 12 planes, but the order failed to take into account that the 3rd Detachment was still forming. Consequently, only the six machines of the 4th Detachment took part in the operation. It was also a classic example of bad timing, coming a little over a week after the General Command of the VI. Reserve *Korps* issued stringent limitations on the use of fuel for aviation activities [136].

On 28 September, Fritz Schulz, one of the Germans contracted to fly for Lithuania, stalled his Halberstadt CL.IV (Rol) 9432/18, crashing fatally on the Aleksotas airfield [137]. This Halberstadt had formerly belonged to *FA* 425.

The same morning, Albrecht Weinschenk and his Lithuanian observer, Živatovas, returned across the lines. Forced down by engine failure at Zarasai following a sortie to Dünaburg, they sheltered near their plane overnight and recrossed the lines the following morning. During their escape, Weinschenk was wounded in the arm by Red fire [138].

Six days later, the 8th *Plastun* Aviation Detachment of *FA* 426, led by acting commander *Ltn.* Stephan, departed Radziwilischki with eight aircraft and their crews [139] to set up a base in Mesoten, along with an advance field in the vicinity of Bausk. Its assignment was to reconnoiter along the Düna River between Riga and Jakobstadt. This was to be the *Deutsche* Legion's zone of advance when Bermondt attacked Riga. The rest of the *Abteilung* remained in Radziwilischki [140].

On 19 October, *FA* 408 departed Ginkunai, stopping over at its old base of Kielmy before returning to Königsberg, where it demobilized 4 November.

Bermondt Goes on the Offensive

Skirmishes between the West Russian Army and the Latvians began early in October and intensified quickly. On the morning of the 8th, three of Bermondt's planes dropped a few bombs on Latvian positions[141] as his army kicked off a five-pronged advance on Riga. Throughout the day, German and Russian planes roamed the area, conducting reconnaissance and bombing and strafing Latvian positions.

Eiserne Schar Berthold participated in this operation. After two days of see-saw battles, two of Bermondt's columns had fought their way into the southwestern suburb of Thorensberg, on the west side of the Düna. Berthold's band was among them, and they received a hot reception:

> "8 Oct. Upon arriving in the streets of Riga, the Bavarian *Eiserne Schar* of aviation *Hauptmann* Berthold, along with portions of the 2nd Infantry Regiment, came under persistent frontal and flank attacks supported by armored cars and were forced to fall back on Thüringshof during the night. A number of wounded left behind at that time were found the following day, cruelly mutilated by the Letts."[142]

The Iron Division and Berthold's *Eiserne Schar* were ultimately extracted by Gerhard Rossbach's *Sturmabteilung*, which made an epic forced march from its base in Culmsee, West Prussia, to Riga. The rescue force had left Germany accompanied by four LVG C.VI, two of which had been converted to type P.I limousines. These were civilian conversions of military two-seaters, where the rear cockpit was modified to provide seating for two or four passengers. Lack of fuel forced Rossbach to leave the aircraft behind at Alt-Auz, [143] while his *Sturmabteilung* continued on to Riga.

The Latvians pulled back to the east bank of the Düna, leaving Bermondt's forces in possession of much of the west bank and its suburbs. Unable to force the bridges across the river, they began spreading laterally along its west bank.

The war in the air was lopsided; by virtue of numbers, Russian and German aircraft had clear air superiority, making nearly daily bombing attacks on Latvian troops, emplacements, and armored trains without risking serious retaliation. Between 8 and 20 October, they lost only two aircraft on the Latvian Front, one downed by ground fire over Riga, the other forced down while trying to distribute copies of the troop newspaper *"Die Trommel"*.

A *Beilage* (extra edition) of newspaper, published 10 October 1919, detailed the mishap that kept its readers in the dark:

> **"*Die Trommel* at the Front.**
>
> The pennant-decorated aircraft of *Leutnant* von Frantzius[144], the *"Trommel"* pilot, became a popular, welcome bird with the advancing troops in a few days. Wherever troops are marching or resting in the woods and swamps, the white drummer beats down on them once a day. This is how the *"Trommel"* best greets its readers in the field and, for its part, communicates most quickly, from the observations of the airmen, how the picture is shaping up at the front.
>
> Over Riga the plane of the *" Trommel "* was not given a particularly warm welcome. When *Leutnant* von Frantzius dropped a few hundred *"Trommels"* over Riga's Alexander Boulevard on 9 October, he was shot at heavily. He paid little attention to it, knowing that the countless people waiting in Riga for liberation from the Ulmanis carnival did not like to take leaves and will soon know the pennant of his plane and greet him with joy.
>
> Between five and six o'clock in the afternoon the plane returns daily to Mitau from its flights to Riga and to the front. Also in Mitau one already knows how to recognize the pennants, and one pays attention to them.
>
> In the afternoon of 10 October, the flight with the *"Trommel"* to the combat troops on the Riga front had to be canceled because the plane had to make an emergency landing due to an engine defect near Mesothen.[145]"

Bezugspreis: monatlich 10 Mk. Einzelnummer 40 Pf
Annahme von Inseraten und Abonnements:
Mitau Kannengießer-Straße Nr. 22, 1 Treppe rechts.
Preis der sechsteiligen Kleinzeile 60 Pf.

Post-Adresse der Schriftleitung der „Trommel"
Deutsche Feldpost Nr. 3086.
Sprechstunden des Hauptschriftleiters von 11–12
Mitau, Kannengießer-Straße Nr. 22, 1 Tr. r.

Die Trommel

Freie Deutsche Soldaten-Zeitung

unter Mitwirkung aller Freiwilligen

und

Kurländisches Nachrichtenblatt.

Erscheint täglich mit Ausnahme der Sonn- und Feiertage.

Nr. 16 — Mitau, Donnerstag, den 20. November 1919 — 1. Jahrgang

Above: Masthead of "*Die Trommel*", a free newspaper distributed to German troops in the West Russian Army. It was air-dropped to front line troops. (Moshe Bukhman)

Two other aircraft, probably German, were lost in apparent accidents over Lithuania, one crashing near Radziwilischki, the other at Luokei.

The Latvians had a small air arm, formed late July or early August, [146], flying reconnaissance and bombing missions out of Jumpravmuiza, mostly against the Bolsheviks to the east. Sergeants Jakubovs and Priedītis, who had defected to the Sachsenberg *Geschwader* 9 May, were its principal pilots. In addition to a Sopwith 1A2, the detachment operated N4300, one of the two Nieuports surviving from the May defection at Peterfeld. On 17 September, the unit moved from its base at Kreutzberg to Weissenhof. Within a few hours of the onset of Bermondt's 8 October attack, it fled across the Düna River to an emergency field at Jungfernhof. A few hours later, it had to move farther east to Segewald as troops of the German Legion approached. Finally, on 13 October, it settled at an advance field at Kaiserwald, northeast of Riga [147]. On 18 October, the Latvian high command requested the assistance of the Baltic *Landeswehr*'s two aircraft, but *Ltn.* Richard Schaefer, the new commander of its *Flugstaffel*, had to refuse -- per prior agreements, the *Landeswehr* was restricted to fighting the Bolsheviks in the east [148]. The following day, however, the Latvians gained another Sopwith, courtesy of a Latvian Soviet pilot who faked engine trouble to cover his defection [149]. The gain was short-lived, though, when the unit's original Sopwith was written off in a crash 23 October.

Shorthanded as it was, the Latvian unit carried out 47 sorties totaling nearly 51 hours flying time between 8 October and 19 November, despite losing its Sopwith to engine failure 10 November [150]. Although most of these sorties were strictly reconnaissance, the flyers also strafed German troops and dropped 413 kg of bombs on trains, rail stations and warehouses [151].

The battle for Riga degenerated into a stalemate, with both sides swapping artillery and machine gun fire across the river, with regular bombing and strafing sorties by Bermondt's aircraft. This might have gone on indefinitely but for the presence of French and British warships in the Düna River. They were under orders to protect the Allied Diplomatic and Military Missions and insure the safety of commercial traffic on the river, while remaining strictly neutral in the fighting. They took their charge with great forbearance, frequently moving up and down river to avoid active firing zones and the Bermondtist's habit of dropping rounds near their ships. Eventually, the *Westarmee* pushed its luck too far, hitting some of the ships and provoking retaliatory fire. This turned the tide; Allied naval gunfire now supported the Latvians, and Bermondt's hopes of crossing the Düna evaporated.

The other provocation that doomed Bermondt was his army's occupation of Fort Dünamünde, a 17th

Above: Ex-Sachsenberg aircraft at Peterfeld -- a Ju D. I, two SSW D.IVs, an LVG C.VI, and two others, one of which appears to be an Albatros D.V. The LVG is marked with the letter "M", similar to the one applied to one of the *Geschwader's* Junkers CL.Is.

Century fortress on the west bank, near the mouth of the Düna river. The Allies judged it a threat to shipping and demanded Bermont withdraw. After a fruitless exchange of messages, Allied naval guns settled the issue [152]. Bermondt's forces pulled back from the river, and Latvian troops regained a toehold on the west bank. This marked the end of Bermondt's assault on Riga. Although he had control of the air, it was no help in capturing the bridges, the keys to the city. Fighting slacked as both sides rested and reorganized for their next push.

By this time, Berlin had finally begun tightening the screws on the whole Baltic operation. *Generalleutnant* Walter Eberhardt, former head of *Idflieg* and later, editor of *Unsere Luftstreitkräfte*, had arrived in Mitau 12 October to replace von der Goltz, who was ordered home to Germany. Eberhardt's orders were to extract the German forces from the Baltic as quickly and cleanly as possible. Financing for the German troops, which had gone on under the table despite Berlin's denials, was suspended, and a week later, the deadline for them to return home expired. The border was officially closed to reinforcements and military equipment heading to the Baltic, and this time, Germany meant it. Previous border closures had been a joke; German border guards had looked the other way as men and materiel flowed eastward, while von der Goltz continued conducting his war.

Breakup in Lithuania

Farther south, the Germans had exhausted most of their good will in Lithuania, and it was clear their stay there was about to end. One of the final nails in the coffin was the Enver Pasha affair. Enver Pasha had been the Ottoman Minister of War during World War I, but had been sacked and tried for war crimes following Turkey's defeat in 1918. He fled to Germany and then involved himself in a succession of intrigues involving Germany and Soviet Russia. On 16 October, a German Junkers F.13 transport, "Annelise", carrying him & Djemal Pasha, another former Ottoman leader, was forced to land in Lithuania en route from Russia to Germany. The passengers & crew were taken to Kaunas, but a few days later, Harry Rotter, one of the German pilots in the Lithuanian air force, commandeered a Lithuanian plane and flew the two Turks to Germany. The Lithuanians fired their German mercenary aviators, and when Bermondt ordered the Wirgolitsch Corps south to Radziwilischki to reinforce the German units in the area, they began preparing for military operations against Bermondt and his allies.

Although the Lithuanian government had taken an informal hands-off policy toward the Germans, expecting them to leave soon, bands of irregular fighters ramped up their attacks on the Germans. In response, General von Eberhardt demanded the end of hostile actions against his troops. The Lithuanians ignored him until 30 October, when a ceasefire put a temporary end to the fighting.

Bermondt's Last Stand

On 22 October, a two-seater of the 8th *Plastun* Aviation Detachment, described as a radio-equipped Albatros[153], departed Mesoten to bomb a bridge over the Ogre River. The attack failed to destroy the bridge, but killed a woman and injured two workers and a national guardsman. It was forced by ground fire to land in a marsh near Rembata. The crew survived, the plane was extracted from the swampy ground, and both were transported to a nearby farm, protected by a cavalry detachment until the plane could be repaired and flown back to its base [154]. Learning of the machine's existence, the Latvians sent a raiding party to the farm the night of 27 - 28 October. Trying to avoid capture, the pilot, Christian (Christel) Mesch, a 13-victory ace with *Jasta* 26, tried to make a run for it and was killed. His observer, *Ltn.* Klöwer, surrendered and was taken POW. Unable to remove the disabled plane, the raiders stripped it of its instruments and equipment and blew it up with grenades.[155]

In a weeklong campaign beginning 3 November, Colonel Janis Balodis' Latvian forces, aided by shelling from British & French ships, forced the Bermondtists back from Riga's northwest suburbs and then outflanked them as they withdrew toward Mitau. The Germans' much-anticipated Third Liberation of Riga was not to be.

Despite the dire situation on his eastern front, Bermondt ordered attacks to the west against Libau and Windau on 4 November. Late that afternoon, one of his planes sortied over Libau, dropping bombs and propaganda leaflets. Continuing on to make a treetop strafing pass on Latvian troops holding the defensive perimeter, it was brought down by their ground fire, crashing in no-man's land and killing its crew. Both sides tried to recover the wreck, but the Germans eventually won the game of Steal the Bacon, repossessing their machine. The operation against the seaports was assigned to the *Graf* Keller Corps, so the downed plane probably belonged to *Ltn.* Arimond's 4th *Plastun* Aviation Detachment or the 5th Detachment, formerly part of *Ltn.* Weinschenck's 2nd *Plastun* Aviation Division (*Truppen-Fliegerstaffel* 38). That evening, after bombing Libau's military harbor, another of Bermondt's aircraft was brought down by ground fire as it passed low over the town's defensive perimeter.[156]

Supported by gunfire from Allied ships in the harbor, the Latvians held Libau. The West Russian Army continued its futile attacks on the city through the 14th. Repulsed every time, they finally gave up and withdrew east on the 15th and 16th. A day later, Bermondt's force threatening Windau also abandoned its assault and began to withdraw.

In difficult times, sustaining morale usually requires stretching the truth, and these two items from *"Die Trommel"* bear that out:

> **"From Our Flyers.**
>
> On November 18, several German aircraft made attack flights. The enemy was attacked with missile mines and machine guns on his hasty retreat to Thorensberg and shot out of farmsteads west of the Mitau - Olai road. A short distance before Olai, our pilots shot up 4 Latvian baggage wagons. All aircraft received strong ground fire, but landed safely at our airfield. You can imagine how horribly the attacks of our low-flying machines affected the raw instincts of the Latvian guys.
>
> **The Latvian Farmsteads on the Eckau,**
>
> where the Latvians had taken shelter during the advance towards Mitau, were set on fire by our artillery and our fire squads in the night of November 18/19. About 40 homesteads went up in flames. The fires offered a scary and beautiful spectacle at night. The residents of these farmsteads in Olai who had fled were found by planes.[157]

While these operations may have taken place as described, they were not going to alter the course of the war. Bermondt's position was crumbling.

Two days later, Latvian troops shot down one of Bermondt's aircraft as it was attacking them in Eckau.[158] Farther south, Latvian troops overran Mesoten, base of *Ltn.* Stephan's 8th *Plastun* Aviation Detachment, capturing the unit's seven surviving aircraft. Six of them had been burned in their hangar by the fleeing Germans [159]. They had seen little use; after two of them had to abort orientation flights, apparently due to substandard fuel, the Detachment's pilots had refused to take them up any longer.

By then, Bermondt had essentially thrown in the towel. On 15 November, he sent a telegram to von Eberhardt, claiming, falsely, that the Latvians had allied with the Bolsheviks [160] to overrun both his flanks. His troops, sick, unpaid and demoralized, were in retreat, and his wounded, along with German women and children residing in Mitau under his protection, were at risk of falling into Latvian hands unless they were evacuated immediately. With that, he dumped the entire mess into von Eberhardt's lap [161].

Four days later, there occurred what might have been the only aerial combat during the entire Baltic adventure -- one of the *Artillerie-Fliegerstaffel* 101's aircraft disengaged after exchanging inconclusive shots with a Latvian 1½ Strutter over Thorensberg [162].

On the 21st, Mitau was evacuated by the collapsing West Russian Army. The airfield at Paulsgnade was abandoned, [163] and the Iron Division retreated west along the rail line to Alt-Auz. The chaotic situation was aptly summed up in this article in *Rigasche Rundschau*:

> "*Latwijas Sargs*" [Latvian Sentinel, a newspaper published in Libau] reports from Mitau. According to an order from Bermondt, Mitau should have been evacuated by Saturday [22 November]. All the provisions of food there have been consumed, the last pieces of uniform clothing distributed to the soldiers, most of the horses have been sold or shot. The wagons at the station are marked Schaulen [It was common practice at the time to chalk the next destination on railway cars]. As a result of constant desertion, the stations will be garrisoned with machine guns. The Germans lack fuel for their planes. Most of the planes were hauled away from Mitau by horses, and now there aren't enough horses either. The activity of the enemy's artillery has therefore become weak, although there are still many guns. 15 deserters were shot dead on Sunday." [164]

What was left of Bermondt's army crowded into north-central Lithuania, concentrating around Schaulen and Radziwilischki. *FA* 426/*Truppen-Fliegerstaffel* 38/2. *Plastun* Aviation Division, at Radziwilischki since early August, was immersed in this crush, and from this time forward, became primarily a ground unit, as evidenced by this passage in *Von der Heimat geächtet*:

> [20 November] "… *Oberleutnant* v. Maltzan [Commander of Batallion von Maltzan of Regiment von Brandis] gathered everyone who could carry a gun. In addition to its field company, the Weinschenck *Fliegergeschwader* had set up a detachment of about 100 men with M.-G. and revolver cannons." [165]

The revolver cannons referred to were the 3.7 cm anti-aircraft cannons, whose tracer rounds were the "flaming onions" dreaded by American fliers in the Great War.

With the Lithuanians closing in, the aviators were no longer above and away from the ground fighting, and on 21 and 22 November, 15 members of *FA* 426 fell into Lithuanian hands at Radziwilischki [166]. Among them was *Ltn.* Heinrich Rauch, formerly an observer in *FA* 433 and later a *Luftwaffe Generalleutnant*. He was repatriated to Germany 16 December [167].

In the midst of all this, *Generalleutnant* von Eberhardt was trying to coordinate the evacuation of the remnants of Bermondt's army to Germany. He had to deal with a shortage of rail transport, Latvian and Lithuanian attacks on his units, and ever-changing demands by the Inter-Allied Baltic Commission, which seemed calculated to frustrate his task as much as possible. In the end, the evacuation was a hybrid activity. Most units able to march home did so, while those unable to, like the aviation detachments, boarded the few available trains.

Leaving the Baltic

On 22 November, the *Deutsche* Legion left Schaulen to return to *Wehrkreis Kommando* (Military District Command) I, in Königsberg via Kielmy, Skaudwile and Tauroggen. It was slow going; the Legion didn't cross into Germany until 13 December.

One day later, the Lithuanians overran Radziwilischki, seizing large quantities of aviation equipment, including about 30 aircraft, along with spare engines, propellers & bombs. They moved the material to Kaunas to secure it from recapture. What was left of the West Russian Army was now penned up in the vicinity of Schaulen, with the remains of its air force based at Zokniai. Given the tactical and fuel situation, it's doubtful Bermondt's air force was functional any longer.

That night, the last contingent of the Sachsenberg *Geschwader* evacuated Peterfeld and headed west. The train carrying the bulk of the *Geschwader*, which had already left on the 16th with much of its equipment, pulled into Tilsit the same night. The following day, Latvian troops took possession of Peterfeld, claiming several aircraft and substantial stores of aviation equipment, including 38 aero engines still in their packing crates [168]. According to one Latvian newspaper report [169], four aircraft fell into Latvian hands, but a photo of the booty in the Peterfeld hangar shows six.

The *Abteilung*'s arrival in Tilsit was not without incident. *Ltn.* Ernst Kredel, officer in charge, reminisced:

> "I led the last detachment, a transport train with aircraft, weapons and supplies. Our destination was the airship hangars at Seerappen near

Above: Unk., unk., *Ltn.* Carl-August von Schoenebeck (*Js* 33), *Oblt.* Erwin Martin (*FA* 19), unk., *Rittm.* Wilhelm *Freiherr* Marschall von Bieberstein (*Bogohl* I/2), unk. Bieberstein's presence dates this photo to sometime before 2 September 1919, when he joined the West Russian Army, eventually rising to deputy *Kofl.* of Bermondt's air force. (*Das Buch des deutschen Freikorpskämpfer*)

Koenigsberg. As we waited for the rest of the journey to begin at the gloomy train station at Tilsit, I heard that wewould be supervised by an officer of the *Entente* powers. I was determined to resist that! An hour or so later, a British officer appeared on the platform accompanied by a German railway officer. The Briton, Cpt. Geoffrey E. Lloyd Wilson, demanded in English that we surrender our weapons and submit to his supervision. I resolutely refused his demands and had a machine gun placed in position on the train platform! The Captain and his German companion withdrew and did not come back... Late in the night, with a whistle from the locomotive, we steamed unhindered toward Koenigsberg." [170]

In a report to France's General Vincent of the Commission for the Evacuation of the Baltic Provinces, Wilson agrees substantially on the facts, but suggests the encounter was tenser than Kredel admitted:

"Sir, I have the honour to inform you that on the night of the 23rd November about one in the morning, a train stopped at the station of Tilsit. Amongst the passengers on this train was a detachment of German soldiers of the 34th Squadron with their aeroplanes, motors and material complete, commanded by the Lieut. Knedel [sic].

Having been informed that the German Liaison Officer would be waiting for me on the platform, and as I did not know him, I addressed myself to an officer who was at that moment talking to some German soldiers, asking him if he was a Liaison Officer. This Officer immediately answered me in an extremely insulting manner and having roundly cursed me, finally ordered me to clear off the platform. I went to the other end of the train hoping to find the Liaison Officer. Lieut. Knedel followed me with his men and grasping my shoulder ordered me again to clear off the platform immediately, at the same time he spat on me. He seemed quite mad. As he had with him his soldiers and I had only a French orderly I retired to the Commission's Office where I waited for the German Liaison Officer. I told the orderly who accompanied me to remain on the platform and to see that the train did not leave the station without my having examined it. About twenty minutes later the orderly returned saying that Lieut. Knedel had ordered the platform to be cleared and that the same Lieutenant was placing machine guns on the line. Later Lieut. Knedel refused to leave the machine guns at Tilsit and again spoke and acted in an insupportable manner." [171]

In reading this vivid description of events, it's easy to overlook a significant fact -- Wilson describes the German detachment as the "34th Squadron", i.e.; *Truppen-Fliegerstaffel* 34, the heir to *Flieger Abteilung* 413, which, according to German records, no longer existed, having demobilized at Jüterbog three months earlier.

Farther west, the Alt Auz / Weitenfeld aerodrome remained in the hands of *FA* 424 / *Truppen-Fliegerstaffel* 37, but it became increasingly isolated as the West Russian Army retreated farther south into Lithuania. During this time, *Ltn.* Carl-August von Schoenebeck had been wounded by ground fire while conducting a low level attack. Because he hadn't returned to Germany before its repatriation deadline, the government wouldn't pay for his transportation or medical treatment. Friends stepped up and covered the cost of his hospitalization [172]. Although von Schoenebeck isn't mentioned by name, this passage from Marschall von Bieberstein's biography, in its overblown Nazi-era prose, appears to describe his downing and rescue:

"Mitau cannot be held any longer. On November 19, 1919, the retreat begins... Accompanied by a bunch of refugees, he [Marschall] arrives at his old detachment in Weitenfeld. His duty in the staff of the Russian Western Army is over.... An old comrade and fellow Baden man [173] who served with the Richthofen [sic] fighter squadron during the World War is shot down over the enemy positions 60 kilometers away. Weakened by severe blood loss, he is no longer able to reach the German airfield and has to go down hard behind the lines. Marschall is looking for the comrade who fell and actually finds him. With difficulty he coaxes him to consciousness. A dangerous night is spent between the lines. After the condition of the wounded man had improved towards morning, he brought him back with great effort." [174]

By this time, it was obvious there was insufficient rail transport to evacuate the remains of the West Russian Army and its equipment back to Germany, so *FA 424*'s commander, *Oblt.* Erwin Martin, authorized his flyers to ferry their aircraft from Weitenfeld directly back to the homeland. The rest of the *Abteilung* would pack up its equipment and bring home as much as possible by rail. The plan did not work out well. On 29 November, less than 30 km into his flight, *Oblt.* Hans-Helmut von Boddien was forced down at Okmiany either by Lithuanian ground fire or engine failure. He was shot and killed after landing.

Adulterated fuel, the undoing of the 8th *Plastun* Aviation Detachment at Mesoten, also appeared to add to *FA* 424's toll. On 2 December, *Ltn.* Alexander von Winterfeldt, *Ltn.* von Sallerau & *Vzfw.* Weiss, attempting to fly two-seaters back to Germany, force landed 12 km north of Kielmy and were captured by Lithuanian troops. They were still listed missing on 6 December, when Vice Admiral Hopman, German Commissioner for the Baltic, inquired about their status to Lithuanian General Yukowski [175]. An 8 January 1920 memo from *Oblt.* Erwin Martin to the staff of the *Westrussische Armee* stated von Gallera (probably a misspelling of Sallerau), Weiss and fitters Reich and Kretschmar had been killed by the Lithuanians near Kielmy [176]. Winterfeldt survived to serve with the *Luftwaffe* in WWII. Two other pilots, Neetzel and Jankowiak, were wounded under unknown circumstances, but were repatriated to Germany to recover.

The evening of 2 December also saw Ginkunai manor fall into the hands of the Lithuanians, and with it, 12 aircraft and five automobiles [177]. German aviation activities in the Baltic had come to an end.

By 6 December, the VI. Reserve *Korps* and Russian-German forces were scrambling to get out of the Schaulen area to Germany via the rail line to Tilsit. Despite their difficult situation and in defiance of Allied transportation and demobilization instructions, they devoted an inordinate amount of energy to bringing as much equipment as possible back to Germany and destroying whatever they were forced to leave behind. Spite, rather than military necessity, seemed to be their motivation. Ignoring orders of the Armistice Commission to leave their aircraft behind for the Lithuanians, the Germans loaded as many as possible aboard flatcars and hauled them back across the frontier. One news report stated these trains had originally been authorized by the Allies to repatriate refugees to Germany:

"German weapons trains seized.
At the German request, the Allied Commission allocated some trains with refugees from Šiauliai [Schaulen] to Germany. At the German border, controlling these trains on Sunday [19/7.XII], as

Above: The Inter-Allied Baltic Commission intercepted a shipment of German aircraft on its arrival in East Prussia and returned it to Lithuania. This train carried LVG C.VI 12017/18 (ex-FA 433), Halb. C.V (Av) 6859/18, Alb. D.III (OAW) D.5160/17, LVG C.V 9652/17, one DFW C.V, and three Alb. J.IIs. (Moshe Bukhman)

LVG C.VI 12017/18,
***FA* 433, Baltic *Landeswehr*,**
Kurland, 1919.

"L.S." reports, instead of refugees, they found them full of the most valuable war material. The trains were returned to Šiauliai, where they are now being unloaded. After this case, border control was strengthened. The representatives of Latvia and Lithuania also participate in the control, which, hopefully, will also prevent the transfer of stolen goods across the border." [178]

The representatives of the Inter-Allied Baltic Commission ordered the booty to be returned to Lithuania.

The night before the train was to return, the Germans did their best to destroy the aircraft, attacking them with axes, setting them afire and throwing grenades into cockpits. An inventory of returned material conducted by the Lithuanian military on 13 December [179] found none of the aircraft in flyable condition; some would fly again after extensive repairs; others would be no more than a source of spare parts.

By mid-December, all German and Russian troops had returned to Germany, ending the Baltic operation, and on 18th of the month, the *Reichswehr* assumed control of the border with Lithuania.

War Booty

Of all the aircraft Germany committed to von der Goltz' and Bermondt's adventures, relatively few eventually found their way into the air forces of the new Baltic republics. An exhaustive study by researcher Moshe Bukhman concludes:

Above: Inter-Allied Baltic Commission personnel inspect aircraft after their return from East Prussia to Lithuania. German sabotage is evident – Albatros D.III (OAW) 5160/17 has what appears to be two axe cuts in its fuselage. Its oversize military serial indicates it had served as a training machine before finding its way to the Baltic. (Moshe Bukhman)

Estonia took over five German machines, including the DFW C.V and two Halberstadt C.Vs lost in von der Goltz' abortive effort to contact General Rodzianko.

Latvia profited by 13 aircraft, including at least six left behind by the Sachsenberg *Geschwader* at Peterfeld 24 November.

Lithuania was the big winner, largely because the remains of the West Russian Army's air force was concentrated around Schaulen at the end of the war. Its haul of 70–74 aircraft included:

30 from *FA* 426/*Truppen-Fliegerstaffel* 38, captured at Radziwilischki 23 November

Twelve seized at Ginkunai 2 December

Thirteen sent back from Königsberg on a train, 13 December

Remains of eight on a train returned from East Prussia 28 February 1920

Three found at various locations around the country.

It may have also included the surviving aircraft from the purchases made from *FA* 425 and the German government in February 1919.

The Lithuanians were diligent in putting these captured aircraft into their nation's service. Their task was made easier, thanks to an airship hangar the Germans had built at Aleksotas in 1915–16. It provided ample room to house all the damaged aircraft, including those sabotaged by the Germans in December, along with stores of engines, propellers, instruments, wings and control surfaces, till they could be rebuilt into functioning machines for the Lithuanian air force. One such project was the restoration of a damaged Friedrichshafen G.IIIa bomber. The rebuilt machine crashed, unfortunately, on its maiden flight and was written off.

Endnotes - Chapter 5

1. Located adjacent to a cement manufacturing plant, Weissenhof was also referred to as the Cementfabrika airfield. Between the wars, it became Spilve, Riga's commercial airport.
2. *Reichskriegsministerium, Die Rückführung des Ostheeres, Darstellungen aus den Nachkriegskämpfen deutscher Truppen und Freikorps, Erster Band*, E. S. Mittler & *Sohn*, Berlin, 1936, pp. 180–4.
3. "*Die Tätigkeit der Flieger bei der 8. Armee vom December*

Right: Lithuania's treasure trove of German aircraft found a new home in the Aleksotas airship shed, itself formerly German property. At the bottom center is the ill-fated Friedrichshafen G.IIIa bomber under reconstruction. Other identifiable aircraft are Fokker D.VII, Halberstadt CL.II and IV, DFW C.V, LVG C.V, Albatros C.III, C.XV, and J.II. Fifth aircraft to the right of the Friedrichshafen is the Halberstadt CL.IV shown in the accompanying profile. (Moshe Buchman)

1916 bis Januar 1919". Dresden *Archiv*, pp. 84-5. (via Moshe Bukhman)

4. "Terms of the Armistice with Germany", Catalog Reference WO 106/412, nationalarchives.gov.uk. https://www.nationalarchives.gov.uk/pathways/firstworldwar/transcripts/after math/armistice_terms.htm
5. *Reichskriegsministerium, Die Rückführung des Ostheeres*, p. 19.
6. *"Die Tätigkeit der Flieger.*
7. *Inspektion der Fliegertruppen, Ehrentafel, der in Flugdienst während des Weltkrieges gefallenen Offiziere der deutsche Fliegerverbände*, E. S. Mittler *und Sohn, März* 1920, Berlin, p. 97. (via Moshe Bukhman)
8. *"Die Tätigkeit der Flieger"*, p. 87.
9. In late June - early July, the seaplanes would be based in the Düna River under the direct command of the VI. Reserve *Korps* and are believed to have participated in an unsuccessful attack on an Estonian gunboat in the Gulf of Riga. (Moshe Bukhman personal correspondence, 7 July 2020).
10. 1st Latvian Aviation Detachment Flight logs, 16.02.1919. (via Moshe Bukhman)
11. *"Absturz eines Fliegers.", Libauische Zeitung*, No. 34, 10 *Februar* 1919, p.3. (via Moshe Bukhman)
12. *"Frontbericht vom 14. Februar. Windau von der Bolschewiki besetzt.", Libauische Zeitung*, No. 38, p. 2. (via Moshe Bukhman)
13. *Die Rückführung des Ostheeres*, p. 154.
14. *"Die Tätigkeit der Flieger"*. p. 87.
15. *"Flieger in den Nachkriegskämpfen", Reiter gen Osten, Jahrgang* 7, *Folge* 10, *Seite* 14 (October 1936). (via Moshe Bukhman)
16. Some accounts identify Loerzer as *FA* 427's commanding officer. Surviving documents do not support this. Erdmann, an experienced administrator, continued in command and may have delegated responsibility for flying activities to Loerzer, a gifted combat pilot.
17. von Raumer, Kurt, Wittram, Reinhard, Fletcher, Alfred, Sielmann, Burchar, Dopkewitsch, Helene & Frantz, Gunther, *Der Bolschewismus und die Baltische*

Front, Verlag von S. Hirzel, Leipzig, 1939.

18. *Korpstagsbefehl Nr.* 1380, *Generalkommando* VI. Res. *Korps*, Libau, 28 February 1919. (via Moshe Bukhman)
19. Ludwig, *Ltn.*, *"Bericht über die Dienstreise des Beobachter Leutnant* Ludwig *und des Flugzeugführer* E. Hoffmann *nach Deutschland zwecks Beschaffung von Flugzeugen für die Baltische Landeswehr."* Wainoden, 11 March 1919. (via Moshe Bukhman)
20. Franke, *Generalmajor* Hermann, *Handbuch der neuzeitlichen Wehrwissenschaften, Dritter Band,* 2. *Die Luftwaffe, Verlag von* Walter de Gruyter & Co., Berlin, 1939, p. 185.
21. *Tagsbefehl Nr.* 72, *Oberstab der Baltischen Landeswehr,* 24 April 1919. (via Moshe Bukhman)
22. Angelroth, *Oblt. a. D.* Herbert-Armin, *"Fliegerkampfgeschwader Sachsenberg im Baltikum* 1919*"*, *Unsere Luftstreitkräfte* 1914–1918, *Generalleutnant* Walter von Eberhardt (ed.), Berlin, 1930. p. 445.
23. *"Die Ebert-Regierung als Seelenverkäufer", Die Rote Fahne, Nr.* 2, Berlin, 2 January 1919.
24. Osterkamp, Theo, *Du oder ich, deutsche Jagdflieger in Höhen und Tiefen,* Albert Nauck & Co., Berlin-Charlottenburg, 1938, pp. 225-6.
25. Angelroth, *"Fliegerkampfgeschwader Sachsenberg im Baltikum* 1919*"*, p. 446.
26. Kredel, Christoph-Ernst, "Backseat Fighter. The Reminiscences of Dr. Ernst Kredel", *Cross & Cockade Journal*, Vol. 25, No.1, 1984, pp. 43–50.
27. *Berliner Volks-Zeitung,* No. 45, *Erstes Beiblatt, Morgen-Ausgabe, Sonnabend,* 1. *Februar* 1919, *S.* 2. (via Moshe Bukhman)
28. *Телеграмма, Наштарм Латвии, Разведывательная. РГВА* (Telegram, Latvian *Nashtarm* Intelligence. RGVAz). (via Moshe Bukhman)
29. The sheds were razed between the wars; their roof trusses survive in Riga's Central Market.
30. von Salomon, Ernst, ed., *Das Buch des deutschen Freikorpskämpfer,* Wilhelm Limpert, Berlin, 1938, p.171.
31. Angelroth, *"Fliegerkampfgeschwader Sachsenberg im Baltikum* 1919*"*, p. 446.
32. The word "Gut" preceding a placename identifies it as a country estate or manor.
33. Many of them survive as community museums and cultural centers – they would make an interesting basis for a historical tour.
34. *Reichskriegsministerium, Der Feldzug im Baltikum bis zur zweiten Einnahme von* Riga. *Januar bis Mai* 1919. *Darstellungen aus den Nachkriegskämpfen deutscher Truppen und Freikorps, Zweiter Band,* E. S. Mittler & *Sohn*, Berlin, 1937, p.11.
35. *"Frontbericht vom 14. Februar. Windau von der Bolschewiki besetzt.", Libausche Zeitung,* No. 38, p.2. (via Moshe Bukhman)
36. *"Absturz eines Flugzeug, Frontbericht vom 24. Februar,", Libauische Zeitung* No. 47, p. 3, 25 February 1919. (via Moshe Bukhman)
37. *Divisionsbefehl* Ia / *Nr.* 83 *geh., Eiserne* Division, 2 March 1919. (via Moshe Bukhman)
38. *Reichskriegsministerium, Der Feldzug im Baltikum bis zur zweiten Einnahme von Riga,* p. 65.
39. *Korpstagsbefehl Nr.* 3110 2. Ic), *Generalkommando* VI *Reservekorps,* Libau, 19.3.19. (via Moshe Bukhman)
40. *Korpstagesbefehl, Generalkommando* VI. *Res. Korps, Abt.* IIa *Nr.* 5684, 15 April 1919. (via Moshe Bukhman)
41. *Korpstagebefehl, Generalkommando VI. Res. Korps, Abt.* IIa 4 April 1919. (via Moshe Bukhman)
42. *Korpstagesbefehl, Generalkommando* VI. *Res. Korps, Abt.* IIa *Nr.* 5684, 15 April 1919. (via Moshe Bukhman)
43. *Korpstagsbefehl Nr.* 5565, *Generalkommando* VI. Res. *Korps*, Libau, 14 April 1919. (via Moshe Bukhman)
44. *Tageblatt* N. 720, Report of Shock Troop Section Intelligence officer. (via Moshe Bukhman)
45. *Korpstagsbefehl, Abendmeldung vom* 5.5.19., *Generalkommando* VI. Res. *Korps, Abt.* IIa *Nr.* 7698, 6 May 1919. (via Moshe Bukhman)
46. 9. *Nachrichtenblatt, Generalkommando* VI. Res. *Korps, K. H. Qu., Abt.* Ic *Nr.* 4243, 1.4.19. (via Moshe Bukhman)
47. ibid.
48. Bennett, Geoffrey, *Freeing the Baltic,* Birlin Limited, Edinburgh, 2002. p. 98.
49. Grimm, Claus, *Vor dem Toren Europas 1918 –1920, Geschichte des Baltischen Landeswehr,* August Friedrich *Velmede Verlag,* Hamburg, 1963, p. 191.
50. ibid., p. 192
51. *Korpstagesbefehl, Generalkommando* VI. *Res. Korps, Abt.* IIa *Nr.* 7116, 30 April 1919. (via Moshe Bukhman)
52. *Korpstagesbefehl, Generalkommando* VI. *Res. Korps, Abt.* IIa *Nr.* 7381, 3 Mai 1919. (via Moshe Bukhman)
53. Angelroth, *"Fliegerkampfgeschwader Sachsenberg im Baltikum* 1919*"*, *Unsere Luftstreitkräfte 1914–1918, Generalleutnant* Walter von Eberhardt (ed.), Berlin, 1930, p. 448.
54. ibid., p. 448.
55. von Salomon, *Das Buch des deutschen Freikorpskämpfer,* p. 172.
56. Krüger, Hans-Eberhard, *"Jagdstaffeln über Kurland und Oberschlesien", Jägerblatt Nr.* 2 / XXXVIII, pp. 18–27, Köln, April / Mai 1989, p.23.
57. Kanisch, *Vizefeldwebel & Unteroffizier* Hillmann,*"Flieger über Kurland, Als Infanterieflieger gegen Bolschewisten", Reiter gen Osten, Folge* 10, 35 / XII, *Seite* 10 (1939). (via Moshe Bukhman)
58. *Flieger Abteilung* 433 *Frontbericht Tagsbefehl Nr.* 87, 9.5.1919 and *Tagsbefehl Nr.* 87, *Oberstab der Baltischen*

Landeswehr, 9 May 1919, state the forced landing was due to an engine defect, not hostile fire. This may be another instance where the truth was embellished in this Nazi-era retelling of events.
59. Krüger, *"Jagdstaffeln über Kurland und Oberschlesien",* p. 23.
60. Angelroth, *"Fliegerkampfgeschwader Sachsenberg im Baltikum* 1919", p. 447.
61. von Salomon, *Das Buch des deutschen Freikorpskämpfer,* p. 174.
62. Angelroth, *"Fliegerkampfgeschwader Sachsenberg im Baltikum* 1919", p. 447.
63. Zuerl, Walter, *Pour le Mérite-Flieger,* Curt Pechstein *Verlag,* München, 1938, pp. 402–4.
64. Transcription of 9 May 1919 Sachsenberg *Geschwader* note to Soviet authorities in Riga. Russian State Archives. (via Moshe Bukhman)
65. 15. *Nachrichtenblatt, Generalkommando* VI. *Res. Korps, K. H. Qu., Abt.* Ic *Nr.* 8559, 16.5.19. (via Moshe Bukhman)
66. 1st Latvian Aviation Detachment, Air Division of the Army of Soviet Latvia at the 1st Latvian Rifle Division of Soviet Latvia, May 9, 1919. (via Moshe Bukhman)
67. *Raport #69,* Commissar to the Military Commissar of Aviation and Aeronautics of the 16th Army 44 a/o RKKVVF, February 21, 1920. (via Moshe Bukhman)
68. Narciss-Puskelis .https://timenote.info/lv/Narciss-Puskelis.
69. Schroeder, *Generalmajor a. D., Die Sächsischen Freiwilligen-Truppen in Litauen* 1919, *Buchdruckerei der* Wilhelm *und* Bertha von Baensch *Stiftung* 1933, Leipzig, pp. 121-3. (via Moshe Bukhman)
70. *Reichskriegsministerium, Der Feldzug im Baltikum bis zur zweiten Einnahme von* Riga. *Januar bis Mai* 1919, p.3.
71. Kowno *Bl.* 11. (XII-15-F), *Fest. Verm. Personl* Kowno 1916. (via Moshe Bukhman)
72. *Bauaufstellungen in Januar* 1919, *Inspektion der Fliegertruppen Nr.* 82/1&19 *Abt.*Ia.", Charlottenburg, 9.1.19. *Deutsche Sozialiste Republik Zentralrat* Archives 48. (via Moshe Bukhman.
73. *Reichskriegsministerium, Der Feldzug im Baltikum bis zur zweiten Einnahme von* Riga. *Januar bis Mai* 1919, p. 147.
74. *"Fliegerabteilung* 424 in *Baltikum", Reiter gen Osten, Jahrgang 7, Folge* 10, *Seite* 13 (October 1936). (via Moshe Bukhman)
75. *Reichskriegsministerium, Die Kämpfe im Baltikum nach der zweiten Einnahme von Riga,* p. 203.
76. *Aviacijos Dalies Įsakymas* N 112, 15 *liepos* 1919 *m., Ūkio Skyriaus,* Kaunas. (via Moshe Bukhman)
77. http://www.elisanet.fi/pekka.hiitola/PekkaHiitola/Sotilasilmailu/Lithuania/lithuania.htm
78. *Kpt.* J. Pyragius, *Mūsu Sparnai. Aviacijos Leidinys,* Kaunas, 1929, p. 24.
79. *Reichskriegsministerium, Der Feldzug im Baltikum bis zur zweiten Einnahme von* Riga, p. 163.
80. ibid., p. 168.
81. *Die Kämpfe im Baltikum nach der zweiten Einnahme von Riga. Juni bis Dezember 1919. Darstellungen aus den Nachkriegskämpfen deutscher Truppen und Freikorps, Dritter Band,* E. S. Mittler & *Sohn,* Berlin, 1938, pp. 171–2.
82. *"Sportfest des Ersatzdepots der Eiserne Division", Die Trommel,* 6. *Extrablatt,* Libau, 15 *Oktober* 1919, pp. 2–3. (via Moshe Bukhman)
83. *Reichskriegsministerium, Der Feldzug im Baltikum bis zur zweiten Einnahme von* Riga, p. 111
84. von Salomon, *Das Buch des deutschen Freikorpskämpfer,* p. 173.
85. ibid., p. 174.
86. Freund, *Hauptmann a. D.* Erich, *"Aufklärungsflieger über Riga. Auszug aus dem Kriegstagebuch des Hauptmanns a. D.* Erich Freund", *Reiter gen Osten, Folge* 2, p. 21 (n.d.). (via Moshe Bukhman)
87. *Ehrentafel,* p. 97.
88. Bischoff, Major Josef, *Die letzte Front: Geschichte der Eisernen Division im Baltikum,* 1919, *Buch- und Teifdruck G. m. b. H. Abteilung Buchverlag,* Berlin, 1935, p.104.
89. Krüger, *"Jagdstaffeln über Kurland und Oberschlesien",* p. 26.
90. Eder, Jack R., " With the Baltic Flying Squadron", Cross & Cockade Journal, Vol. 14, No.4, 1973, p. 357.
91. *"Sturm auf* Riga", *Reiter gen Osten,* ? *Jahrgang, Folge* 11, *Seite* 7, Berlin, November? (via Moshe Bukhman)
92. Max Hölz (1889–1933) was a German Communist activist who led an anti-government guerilla force in the *Vogtland* in 1920–21. – https://en.wikipedia.org/wiki/Max_Hoelz.
93. Kanisch, *Vizefeldwebel* & *Unteroffizier* Hillmann,*"Flieger über Kurland, Als Infanterieflieger gegen Bolschewisten", Reiter gen Osten, Folge* 10, 35/XII, *Seite* 10 (1939).
94. Angelroth, *"Fliegerkampfgeschwader Sachsenberg im Baltikum* 1919", p. 449.
95. Grimm, *Vor dem Toren Europas* 1918–1920, p. 220.
96. Angelroth, *"Fliegerkampfgeschwader Sachsenberg im Baltikum* 1919", p. 449.
97. Moshe Bukhman personal correspondence, 2021.
98. Bischoff, *Die Letzte Front,* pp. 122, 125.
99. Moshe Bukhman personal correspondence, 25 October 1988.
100. *"In estnischer Gefangenschaft", Beilage zur Trommel, Nr.* 7, *Folge* 1, *Montag, der* 1 September 1919. (via Moshe Bukhman)
101. *Lagen-Karte Gen. Kdo. VI.R.K. Abt.* Ia *Nr.* 619. *Stand vom* 18.6.19. (via Moshe Bukhman)

102. *Ehrentafel, p.97.*
103. Kitvel, Toivo, & Johannes Tik, *Eesti lennukroonika: tekste ja pilte aastani* 1940, Aviopol, 2003, p. 80.
104. *Armee-Tagsbefehl* 145. Ia, *Kofl., Ober-Kommando Grenzschutz Nord*, Bartenstein, 14 August 1919. (via Moshe Bukhman)
105. *Tagesbefehl Nr.* 146, *Oberstab der Baltischen Landeswehr*, 19 July 1919. (via Moshe Bukhman)
106. DBFP 1919–1939, 1st Series, Vol. III 1919, London, HM Stationery Office, 1949.
107. Bischoff, *Die letzte Front*, pp. 263–4.
108. This may have been C. 9432/18.
109. Although the entire southern zone of the German/Russian occupied area was placed under the control of the Diebitsch *Freikorps*, *Oberstleutnant* von Diebitsch apparently refused to accept this assignment and limited his force's responsibility to protecting the Schaulen -Tauroggen rail line, the West Russian Army's lifeline to Germany. The official *Reichswehr* history chose to ignore this act of insubordination and perpetuated the myth that Diebitsch was in command of the entire area. In fact, von der Goltz, through his proxy, Bermondt, exercised true authority. (Moshe Bukhman personal correspondence, 8–9 July 2023).
110. *Reichskriegsministerium, Die Kämpfe im Baltikum nach der zweiten Einnahme von Riga*, p. 43.
111. https://en.wikipedia.org/wiki/Foch_Line.
112. *"Was Noske nicht weiss.", Die Freiheit, Nr.* 417, *Seite* 3, Berlin, 30 June 1919. (via Moshe Bukhman)
113. Waite, Robert G. L., *Vanguard of Nazism. The Free Corps Movement in Postwar Germany 1918–1923,* Harvard University Press, Cambridge, Massachusetts, 1952, Reissued 1970, p. 125.
114. Documents of the Western Volunteer Army. *Gr.* Keller (detachment of General Bermont-Avalov), Order of the General's Detachment from the Cavalry of Count Keller, No. 120, § 8, 15.VI 1919. (via Moshe Bukhman)
115. Order of the *Graf* Keller Western Volunteer Corps. No. 15, § 9, 31.VII 1919. (via Moshe Bukhman)
116. ibid.
117. *Generalkommando VI. Res. Korps Tagsbefehl, K. H. Qu.*, 24.8. 1919, Dept. IIa No. 17486. (via Moshe Bukhman)
118. ibid.
119. Waite, Robert G. L., *Vanguard of Nazism*, p.125.
120. Awaloff, General *Fürst, Im Kampf gegen Bolschevismus, Druck und Verlag von* J. J. Augustin, Glückstadt & Hamburg, 1925, pp. 217–9.
121. *"Auch ein Demenki!", Die Freiheit, Nr.* 444, Berlin, 14 September 1919. (via Moshe Bukhman)
122. *"Wie gelogen wird", Die Freiheit, Nr.* 483, Berlin, 15 October 1919. (via Moshe Bukhman)
123. *Reichsschatzministerium, Reichsverwertungsamt Aktenzeichen J.Nr.* 90065, 21 August 1919, Berlin. *Bundesarchiv.* (via Moshe Bukhman).
124. *Generalkommando* VI. *Res. Korps Abteilung* IIa Nr. 18699 *Korps* Daily Order, *K. H. Qu.*, 11.9 1919. (via Moshe Bukhman)
125. Tessin, Georg, *Deutsche Verbände und Truppen* 1918 – 1939, Biblio *Verlag*, Osnabrück, 1974, pp. 137–8.
126. *Inspektion der Fliegertruppen Tgb. Nr.* 1283/8.19 *Abt.* I. II. *Ang.*, Berlin, 28.8.1919. (via Moshe Bukhman)
127. *"Sportfest des Ersatzdepots der Eiserne Division"* pp. 2–3. (via Moshe Bukhman)
128. Order of the *Flieger Abt.* of the *Westl. Freiwl. Korps Graf* Keller No. 6, § 6. 26.VIII.1919. (via Moshe Bukhman)
129. Order of the *Flieger Abt.* of the *Westl. Freiwl. Korps Graf* Keller No. 7 § 6. 27.VIII.1919. (via Moshe Bukhman)
130. *Armee-Tagsbefehl* 154.3 *Kofl., Ober-Kommando Grenzschutz Nord*, Bartenstein, 25 August 1919. (via Moshe Bukhman)
131. Obodeh, O. & N. C. Akhere, "Experimental study on the effects of kerosene-doped gasoline on gasoline-powered engine performance characteristics", Journal of Petroleum and Gas Engineering Vol. 1(2), pp. 37–40, June 2010.http://www.academicjournals.org/JPGE.
132. Secret Report No. 100, Western Volunteer Army, 18 September 1919. (via Moshe Bukhman)
133. There were only 11 *Plastun* Aviation Detachments, so the identity of the twelfth is unclear. It might have been the Iron Division's *Artillerie-Fliegerstaffel* 101.
134. 9 August 1919 note from the *Flieger Abteilung* of the *Baltische Landeswehr* to *Kampfgeschwader* Sachsenberg *Schlasta* requesting dismissal documents for personnel transferred to the *Landeswehr*. Document bears *Staffel* Majewski receipt stamp.
135. Wagener, *Hauptmann a. D., Von der Heimat geächtet, Chr.* Belsersche *Verlagsbuchhandlung*, 1920, Stuttgart, p. 60.
136. *Armee-Tagsbefehl* 154.3 *Kofl., Ober-Kommando Grenzschutz Nord*, Bartenstein, 25 August 1919. (via Moshe Bukhman)
137. *"Amžina atmintis pilotui Šulcui Pričkui", Kariškių Žodis, Nr.* 22, 19/17.X, p. 168. (via Moshe Bukhman)
138. Pyragius, *Mūsu Sparnai*, p. 26.
139. *Deutsche* Legion *Besondere Anordnungen Nr.* 24, Item No. 20, *St. Qu.* 29.10.1919. (via Moshe Bukhman)
140. Wagener, *Von der Heimat geächtet*, p. 60.
141. Bennett, *Freeing the Baltic*, p. 172.
142. *Reichskriegsministerium, Die Kämpfe im Baltikum nach der zweiten Einnahme von Riga*, pp. 91-2.
143. von Salomon, *Das Buch des deutschen Freikorpskämpfer*, p. 205.
144. *Lt. d. R.* Gerhard Frantzius, *Flivo* of *FA* 426.

145. *Die Trommel, 4 Extrablatt*, Mitau, 10 *Oktober* 1919, p. 2. (via Moshe Bukhman)
146. *Kurzemes Vārds*, # 46, Liepāja, Friday, 19.8.VIII. (via Moshe Bukhman)
147. Moshe Bukhman personal correspondence, 18 June 2023.
148. Grimm, *Vor dem Toren Europas 1918–1920*, p. 281.
149. Moshe Bukhman personal correspondence, 20 June 2023.
150. *Raports. Arm. Aviac. Parka I Nodaļas Komandieris, Adm. Nozare*, 19/11.XI. (via Moshe Bukhman)
151. *Telefonogramma* # 0433, 19. nov. 1919.*g. plk.* 11.50, *Vidzemes divīzijas komandierim Man uzticētā pulka I. Bataljona komandieris ziņo*. (via Moshe Bukhman)
152. Bennett, *Freeing the Baltic*, pp. 172–9.
153. Probably an Albatros C. XV.
154. *Kurzemes Vārds*, #113, Liepāja, Saturday, 19/25.X. (via Moshe Bukhman)
155. *Latvijas Sargs* #177, Thursday, 19/30.X. (via Moshe Bukhman)
156. 6. *Rīgas Kājnieku Pulks, Rīgas Kājnieku Pulka Vēsture* 1919–1929, 1929, Rīgā, p. 33. (via Moshe Bukhman)
157. *Die Trommel*, Mitau, 19 November 1919. (via Moshe Bukhman)
158. *Telefonogramma* # 0433, 19. nov. 1919.*g. plk.* 11.50, *Vidzemes divīzijas komandierim Man uzticētā pulka I. Bataljona komandieris ziņo*. (via Moshe Bukhman)
159. *Kurzemes Vārds* # 135, Liepāja, Friday, 19/21.XI. (via Moshe Bukhman)
160. The Bolsheviks were repeatedly invoked as the bogeymen by Bermondt and Nazi-era histories of this period. In fact, the Reds had abandoned this theater of operations after Riga fell in May.
161. *Reichskriegsministerium, Die Kämpfe im Baltikum nach der zweiten Einnahme von Riga*, p. 118.
162. Moshe Bukhman personal correspondence, 18 June 2023.
163. ibid.
164. *Rigasche Rundschau*, # 101, Friday, November 21, 1919, p. 2. (via Moshe Bukhman)
165. Wagener, *Von der Heimat geächtet*, p. 146.
166. (Untitled) List of West Russian Army personnel taken captive by the Lithuanian army, November 1919, (via Moshe Bukhman)
167. Hildebrand, K. F., *Die Generale der deutschen Luftwaffe* 1935-1945, III. *Band, Biblio Verlag*, Osnabrück, 1990, pp. 88–9.
168. *Dobeles Balss*, # 47, Thursday, November 22, 1934. (via Moshe Bukhman)
169. ibid.
170. Kredel, "Backseat Fighter", p. 49.
171. DBFP 1919–1939, 1st Series, Vol. III 1919, London, HM Stationery Office, 1949.
172. Krüger, *"Jagdstaffeln über Kurland und Oberschlesien"*, p. 20.
173. Marschall was a native Badener. Schoenebeck was from Bernstadt, Silesia, but his first army posting was with the Baden *Leib-Grenadier* Regiment.
174. von Pölnitz, *Emir - Das tapfere Leben des Freiherrn* Marschall von Bieberstein, p. 79.
175. Letter from Vice Admiral Hopman to General Yukowski, December 6, 1919, Tilsit. (via Moshe Bukhman)
176. Memo, 3. *Flieger Division der freiw. russischen Armee, Flughafen, den* 8. *Januar* 1920. (via Moshe Bukhman)
177. Matius, Dr. J., Military Archive no, XI, 1939 (Dp 2080). (via Moshe Bukhman)
178. *Kurzemes Vārds*, #52, Liepāja, Thursday, 19.11.XII. (via Moshe Bukhman)
179. Transcript of report of aviation assets by Lithuanian reception commission, 13 December 1919. (via Moshe Bukhman)

Polizei Fliegerstaffeln

Above: *Flugplatz* Paderborn, home of the *Polizei-Fliegerstaffel* Paderborn, 17 January 1920. (Paul Rempe)

The Versailles Treaty, accepted by Germany 28 June 1919, made it clear the nation was not to have any military aviation at all. On 10 January 1920, the treaty was finally ratified, giving the *Reichswehr* until 10 March to shut down aviation operations, and the AIACC was now empowered to see that it did. The experience of the previous fourteen months had convinced the new republic's leadership that an air force was a key resource in maintaining domestic order and securing the integrity of its borders. If a military force would no longer be permitted to carry out those tasks, why not a police force?

Actually, the *Luftpolizei* had been established well before the ratification of the treaty, performing mundane duties like monitoring commercial air traffic, including keeping records of flights between cities. This met no objection from the Armistice Commission. The more contentious issue was the formation of a paramilitary air force to help restore order in the event of uprisings or strikes, and this stirred Allied unease.

An 18 September 1919 memo from the Prussian *Truppenamt* cited, at great length, occasions where aviation units supported government ground forces in establishing and maintaining domestic order, concluding:

"The joyful appreciation of the flyers by the fighting troops is the best evidence for these assertions. They wave enthusiastically to our

Facing Page: Officers of the *Polizei-Fliegerstaffeln* had their own uniforms, identifiable by winged propellers embroidered on black collar tabs and black braided epaulettes with rank pips -- two for a *Hauptmann*, one for an *Oberleutnant*, and none for a *Leutnant*, same as in the *Reichswehr*. *Luftstreitkräfte* veterans were permitted to continue wearing their wartime badges and decorations. (04AEF-B116-Album 069-PB071, 056, 063 & 050, History of Aviation Collection, Special Collections and Archives Division, Eugene McDermott Library, The University of Texas at Dallas)

Above: *Polizei-Fliegerstaffel* Paderborn took this aerial shot of the sprawling Krupp Works in Essen, 8 February 1920. (Paul Rempe)

flyers; missing them, the troops feel uneasy, which paralyzes their energetic engagement. Finally, the flyer is tied to no railroad, where strikes [occur] and no streets, which can be blocked. A few minutes after the beginning of a riot, he is there.

From all the foregoing, it suffices to say, that for the suppression of unrest, the flyer cannot be dispensed with." [1]

Once the terms of the Versailles Treaty became known, the government had planned to establish several police aviation units around the country. As part of its demobilization plans of August 1919, it had selected several *Fliegerstaffeln* to be the nuclei of new police flying squadrons. *Hpt.* Eberstein of *Truppen-Fliegerstaffel* 29 was informed that his might be one such unit. Accordingly, he departed Hundsfeld on 16 August with "some first-class people", *Oblt.* Hartmann and *Leutnante* Heidepriem, Steppuhn, Liebig, Martin, Käubler, Reichert, Meyer and Maulshagen, who accompanied him to the *Staffel*'s demobilization base at Paderborn, projected home of the new police *Staffel* [2].

In addition to *Truppen-Fliegerstaffel* 29, *Truppen-Fliegerstaffel* 15, formerly *FA* 422, commanded by *Hpt.* Joachim von Schroeder, *Artillerie-Fliegerstaffel* 116 (*FA* 423), *Hpt.* Victor Krocker, and *Artillerie-Fliegerstaffel* 103, ex-*FA* 432, commanded by *Hpt.* Egloff *Frhr.* von Freyberg-Eisenberg-Allmendingen, were selected to be the foundation for new *Polizei-Fliegerstaffeln* at Hamburg, Weimar and Döberitz, respectively. Additional *staffeln* were planned for Karlshorst and Potsdam, but by 24 August, all but Hamburg and Paderborn had been dropped from the list. *Artillerie-Fliegerstaffel* 116 departed Weimar 27 September, reporting to Paderborn, where it demoblilzed. Some of

Above: Halberstadt CL.IV of *Polizei-Fliegerstaffel* Hamburg at *Flugplatz* Blankensee bei Lübeck (Greg VanWyngarden)

Halberstadt CL.IV
***Polizei-Fliegerstaffel* Hamburg, 1919.**

its personnel may have joined ex-*Truppen-Fliegerstaffel* 29 fliers as part of *Polizei-Fliegerstaffel* Paderborn.

With Weimar no longer an active air base, protection of Thüringia appears to have passed to a new police detachment, *Mitteldeutschland Polizei Fliegerstaffel* in Gotha. Following the dissolution of *Armee Flugpark 1 in Kurland, Hpt.* Ernst Dörffler returned to Germany to assume command of this unit on 1 October. Dörffler held this position until 3 June 1920 [3].

Various locations around Berlin had been proposed as homes for a police *Staffel* serving the capitol. In addition to the originally planned Döberitz, Biesdorf and Karlshorst had been considered. However, Potsdam appears to have been the final choice.

Polizei-Fliegerstaffel Hamburg operated out of two airfields -- Fuhlsbüttel on the city's north side, and at nearby Blankensee bei Lübeck.

The disbandment of Bavaria's surviving air units

Above: Halb. CL. IV (Rol) 8197/18 of *Polizei-Fliegerstaffel* Hamburg. (Peter M. Grosz collection/STDB)

at Schleissheim and Fürth was met with the proposal to convert them into *Polizei Fliegerstaffeln*. The Allies were said to have quashed this proposal before it got off the ground, figuratively and literally. Yet, surviving records confirm the existence of *Polizei Fliegerstaffel* 1 at Schleissheim, whose personnel included *Leutnant* Rudolf Stark [4] and *Lt. d. R.* Rudolf Hess (yes, that Rudolf Hess) [5]. From 2 December 1919 till 14 February 1920, *Hauptmann* Eduard *Ritter* von Schleich served with the *Fliegerstaffel* of the *Polizeiwehr Bayern* [6] -- this may have been another name for *Polizei Fliegerstaffel* 1.

Hpt. Franz Josef Walz served with *Polizei Fliegerstaffel* Kitzingen from 1 February till 3 June 1920 [7], and *Lüftpolizei* Königsberg, based at Seerappen, counted *Hpt.* Erhard Milch [8] and *Oblt.* Karl Angerstein as commanding officers. Angerstein's tenure ended 31 May 1920 [9].

In Saxony, a memo proposed "reanimation" of its three aviation units, as follows:

> *Truppen-Fliegerstaffel* Bautzen would become *Flieger-Polizeistaffel Nr.* 1, *Artillerie-Fliegerstaffel* Grossenhain would become *Flieger-Polizeistaffel Nr.* 2, and *Fliegerkommando* Dresden-Kaditz would become *Fliegerkurierstaffel* Dresden-Kaditz.

The Armistice Commission was said to have choked off this plan before it was implemented, yet the service records of *Uffz.* Paul Rothe, a pilot, state he served with *Luftpolizei* Dresden [10].

Polizei Fliegerstaffeln also operated out of Allenstein / Deuthen, Böblingen, Breslau / Gandau, Schwerin, Münster and Stettin. Other cities believed to have hosted police units were Brieg and Gleiwitz.

The Allies repeatedly rejected the idea of a police air force, and Germany responded to every rejection with a modified proposal that might pry a concession out of them. In the meantime, new police units were quietly set up and did their best not to draw Allied attention to themselves.

The police units that survived into 1920 appear to have disbanded between the end of May and July 1920. *Polizei-Fliegerstaffel* Paderborn, one of the major *Polizei* units, did not relinquish its aircraft (1 LVG C.V, 5 LVG C.VI, 3 Fokker D.VII, 1 Halberstadt CL.II, 3 CL.IV and 1 C.V and one Albatros D.Va) until July 1920.[11]

The final police unit to disband appears to have been the Danzig *Polizei-Fliegerstaffel:*

> "In November 1919 a Police *Fliegerstaffel* was established, as in other large German cities, also in Danzig. The unit belonged to the Danzig Security Police, was stationed in the Langfuhr district and consisted initially of a complement of about 25 personnel. The flying personnel consisted of four pilots and two machine gunners.

Above: Family outing of *Fliegerstaffel* Allenstein of the *Sicherheitspolizei.* (04AEF-B116--Album069-PB038, History of Aviation Collection, Special Collections and Archives Division, Eugene McDermott Library, The University of Texas at Dallas)

The airfield of the Danzig Police flyers had 20 aircraft of various types, among them, in the spring of 1920, three Albatros C.XII and one Fokker D.VII, ready and equipped for flight. With these machines, diverse training and courier flights were undertaken in the Danzig region in the following months.

In May 1920, according to the framework of the provisions of the peace treaty of Versailles, the Allied high command placed an aviation ban on all Danzig Police aircraft and officially confiscated the aircraft park. On 21.11.1920 came the directive of the Allies, to immediately transfer all aircraft to Poland and to disband the Police *Fliegerstaffel*." [12]

The *Staffel*'s inventory also included a Friedrichshafen G.IIIa bomber. Although Danzig was now a Free State, no longer part of Germany, it was not exempt from the aviation restrictions imposed on its former homeland.

Finally, more than two years after the Armistice had been declared, the German air force was grounded.

Endnotes - Chapter 6

1. "*Auszug eines Truppenberichtes über die Tätigkeit der Militär Flieger im Kampf um die innere Ordnung*", *Reichswehrbefelungstelle Preussen Truppen-Departement. Nr.* 124II.19. A.7.L. IV.C., *BayHStA Abt.* IV, *Kriegsarchiv* document *Fl. Horst* Gersthfn. *Bd.* 13, p. 6.
2. "*Kriegstagebuch*" unpublished manuscript, *Hpt.* Ernst Otto Eberstein, p. 46.
3. Hildebrand, K. F., *Die Generale der deutschen Luftwaffe* 1935–1945, Biblio *Verlag*, Osnabrück, 1990, *Band* I, p. 201.
4. Browne, O'Brien, "Winged Artist - Rudolf Stark", *Cross & Cockade International*, Vol. 33, No. 2, p. 67.
5. feldgrau.net / forum / viewtopic.php?f=26&t=11104
6. Hildebrand, K. F., *Die Generale der deutschen Luftwaffe* 1935–1945, Biblio *Verlag*, Osnabrück, 1990, *Band* III, p. 178.
7. Hildebrand, K. F., *Die Generale der deutschen Luftwaffe* 1935–1945, Biblio *Verlag*, Osnabrück, 1990, *Band* III, pp. 470–1.
8. Hildebrand, K. F., *Die Generale der deutschen Luftwaffe 1935–1945*, Biblio *Verlag*, Osnabrück, 1990, *Band* II, pp. 394–395.
9. Hildebrand, K. F., *Die Generale der deutschen Luftwaffe 1935–1945*, Biblio *Verlag*, Osnabrück, 1990, *Band* I, p. 16–7.
10. Hannes Taeger personal correspondence 14 August

2022.
11. Ott, Günther, "*Zulassung und Kennzeichen der deutschen Zivilflugzeuge* 1914-1934, 3. *Die deutsche Luftfahrzeugrolle* 1919-1920 (LFR A)", pp. 9-10, *Arbeitsgemeinschaft der Luftfahrthistorik*, https://adl-luftfahrthistorik.de/dok/zulassung-kennzeichnung-teil-3-luftfahrzeugrolle-a.pdf
12. Frost, Günther, Zulassung und Kennzeichnung von Flugzeugen der Freien Stadt Danzig 1920–1939, republished as *Auszüge aus der Luftfahrgeschichte der Freien Stadt Danzig 1920–1939*, *Jet & Prop*, Nr.5/2006 - 4/2007.

Above: Another view of the scene shown on page 45. *Lt. d. R.* Otto Weisshaar of *FA* Haas with his Fokker D.VII. The oak branch on the fuselage is patterned after the collar insignia of the parent *Freikorps*, württembergische *Freiwillige Abteilung* Haas. (Tobias Weber via Hannes Taeger)

Eagles without Wings

Not all of the World War I aviators who volunteered to serve with the *Freikorps* or Provisional *Reichswehr* remained in aviation, choosing instead to join ground-based units. Unfortunately, this return to the relative anonymity of infantry or artillery service makes them more difficult to track down, but a few prominent examples stand out.

Hpt. Richard Bentevigni, wartime commander of giant airplane squadron *Riesenfleugzeug Abteilung* 501 (*Rfa* 501), returned to his hometown to establish and command the Düsseldorf *Freikorps* [1].

Lt. d. R. Carl Degelow, 30 victory *Kanone* and *Führer* of *Jasta* 40, returned to his home in Hamburg and organized the Hamburg *Zeitfreiwilligenkorps* (temporary volunteer corps) to help fight the Reds in February 1919. He was wounded while storming Hamburg's city hall [2].

Lt. d. R. August Raben, final *Staffelführer* of *Jasta* 18, "The Ravens", originally joined the *Schutztruppe* (protection troops) of the Division Lettow-Vorbeck before enlisting in Bermondt-Awaloff's West Russian Army, where he became a company commander and was wounded serving with the *Graf* Keller *Freikorps*. He may have returned to aviation in his final days with the West Russian Army -- an order dated 17 October announced the transfer of a *Ltn.* Raben to the First Aviation Detachment [3]. This was the unit spun off *FA* 424 and installed in Ginkunai.

In 1921, Raben led a volunteer detachment in Germany's intrusion into Upper Silesia. In this photo, probably taken after his return to Germany at the end of 1919, he is wearing a Russian cap cockade, the Order of St. Stanislaus 3rd Class, the Iron Cross, *Baltenkreuz*, and Maltese Cross of West Russian Army, in addition to his pilot's badge and silver Wound Badge.

A later photo published in Bermondt's memoir [4] adds the Silesian eagle, the Lettow *Schutztruppen* lion's head sleeve badge and a ribbon bar that spans two thirds of his chest.

Lt. d. R. Willy Rosenstein, *Oblt.* Erwin Martin's Emil (pilot) with *FA* 19 in France, returned home and joined a *Freikorps* ground unit fighting the Communists [5].

Lt. d. R. Josef "Sepp" Veltjens, 35 victories, commanding officer of *Jasta* 15, and Rudolf Berthold's protégé, joined an armored detachment of *Freikorps* Gerstenberg, part of the Lüttwitz *Freikorps*, formed in Berlin and sent to northwest Germany to seize the port cities from the revolutionaries. He sustained three wounds in the recapture of Bremen. Charles Woolley's book, *German Uniforms, Insignia & Equipment 1918–1923* [6], contains a wonderful picture of Veltjens, looking every bit the movie idol, exiting an armored car.

Above: *Lt. d. R.* August Raben sporting an eclectic mixture of German and Russian awards and medals. (Neal O'Connor)

Lt. d. R. Richard Wenzl, eleven-victory pilot with Jastas 31, 11, and 6, joined the von Epp *Freikorps* as a unit commander. He participated in the liberation of Munich [7].

The best known of the German flyers to abandon the skies for the earth was *Oblt.* Rudolf Berthold. The fiery commander of *Jagdgeschwader* II, with 44 victories on his balance sheet, was forced into a desk job by *Idflieg* late in 1918, when it became obvious his injuries made him a liability in the cockpit of an airplane. After the Armistice, he returned to his native Franconia and recruited volunteers to his *Eiserne Schar* (Iron Host, or Band) Berthold. Despite the *Reichswehr*'s attempts to absorb most of the *Freikorps* formations, Berthold remained stubbornly independent.

He and his *Eiserne Schar* failed to emulate his wartime success. In October 1919, after joining the Iron Division in Kurland, they were trapped and badly

Above: A gaunt Hpt. Rudolf Berthold sporting the wreathed "B" sleeve insignia of his *Eiserne Schar* Berthold. (Greg VanWyngarden)

Above: Sleeve badge of the *Eiserne Schar* Berthold.

beaten up by the Latvians in the battle for Riga.[8] After returning to Germany, they traveled to Hamburg to play a part in putting down the March 1920 Red revolt. In a story now well known, they made it as far as the suburb of Harburg, where they were besieged by the revolutionaries and forced to surrender, Berthold suffering a gruesome death at the hands of the Red mob. Peter Kilduff's recent biography[9] has finally put to rest the Hitler era myth that he was strangled with the ribbon of his *Pour le Mérite* order.

Endnotes - Chapter 7

1. von Salomon, Ernst, ed., *Das Buch des deutschen Freikorpskämpfer*, Wilhelm Limpert, Berlin, 1938, p. 391.
2. Kilduff , Peter (ed.), *Germany's Last Knight of the Air: The Memoirs of Major Carl Degelow*, William Kimber, London, 1979.
3. *Tagsbefehl Nr.* 3467, *Flieger Abteilung 2, Westl. Freikorps Graf Keller*, 23 October 1919. (via Moshe Bukhman).
4. Awaloff, General *Fürst*, *Im Kampf gegen Bolschevismus, Druck und Verlag von* J. J. Augustin, Glückstadt & Hamburg, 1925.
5. Gill, Robert B., "The Albums of Willy Rosenstein", Cross & Cockade Journal, Vol. 25, No. 4, 1984, p. 301.
6. Woolley, Charles, *German Uniforms, Insignia & Equipment 1918–1923*, Schiffer Military History, Atglen, 2002, p. 127.
7. Wenzl, Richard, *Richthofen-Flieger, Badische Zeitung*, Karlsruhe, 1930, p.89.
8. *Reichskriegsministerium, Die Kämpfe im Baltikum nach der zweiten Einnahme von Riga. Juni bis Dezember 1919. Darstellungen aus den Nachkriegskämpfen deutscher Truppen und Freikorps, Dritter Band*, E. S. Mittler & *Sohn*, Berlin, 1938, pp. 91–2.
9. Kilduff, Peter, *Iron Man: Rudolf Berthold: Germany's Indomitable Fighter Ace of World War I*, Grub Street, London, 2012.

Conclusion

Contrary to what has been written in most World War I histories, Germany's air force didn't disappear with the signing of the 11 November 1918 Armistice. Although the *Reich* did surrender most of the 1700 aircraft demanded by the Allies as war booty, it continued to sit on an inventory of over 15,000 machines, many the latest designs, to equip a reconstituted volunteer air force, eventually comprising over 60 tactical units and staffed by experienced pilots, observers and gunners. This air force, like the army it served, went through three incarnations in little more than a year -- first as a *Freikorps* (volunteer) force integrated with the remnants of the old Imperial army, second, as components of the Provisional *Reichswehr*, and finally, as transitional units leading up to the shutdown of military aviation in 1920. It also spun off numerous police aviation units that Germany hoped the Allies would tolerate in lieu of a military air force. With most of the Great War's combatants slashing their military forces to skeleton strength in 1919, it can be argued that defeated Germany boasted the largest and most potent air force in the world.

This air force was engaged in three principal theaters: within Germany itself, aiding in the suppression of leftist uprisings; along the Polish and Czecho-Slovakian borders to prevent territorial encroachments; and in the Baltic states of Latvia and Lithuania, first to aid them in repelling the Red army, and later, in support of a quixotic White Russian army formed to return those territories to a restored imperial Russia. In reality, all of it was an attempt to turn the new republics into German satellites.

The fliers of this new air force found themselves carrying out duties considerably different from those they faced in 1914-18. Domestically, they were largely constrained from offensive operations against the left-wing revolts and strikes that broke out throughout Germany. For fear of turning loyal civilians against the government, they had to settle for performing "demonstration" or "intimidation" flights, stunting and zooming over the heads of the strikers, firing harmless shots into the sky. Their principal duties, though, were reconnaissance of enemy positions and transporting mail, military communications and key personnel from one point to another to bypass struck rail and postal facilities.

On the Polish and Czecho-Slovak frontiers as well in the Baltics, the fliers were largely engaged in classic tactical duties -- short-range reconnaissance, ground assault and communication with ground units. On the Polish Front, there were also a few air-to-air engagements. Although Germans' strategic objectives were thwarted by the Versailles Treaty, their short-term success in 1919 may have encouraged Third Reich planners to construct the *Luftwaffe* primarily as a tactical air force.

Like the *Freikorps* and *Reichswehr* ground forces it supported, this air force carried out its assignments effectively and efficiently, but in a cause doomed to have no influence on the final outcome of the Great War -- its brief moment of glory came during a period of negotiations that would end in Germany having to relinquish military aviation, which it did with quiet resignation.

LVG C.VI 'M'
***Kampfgeschwader* Sachsenberg**
November 1919

Fokker D.VIII
Artillerie-Fliegerstaffel 103
Doberitz, 1919.

Fokker D.VII (OAW)
Ltn. Richard Kraut
Fl.Abt 434 / *Truppenfliegerstaffel* 9
Graudenz, 1919.

Halberstadt CL.IV (Rol) 9455/18 '2'
3. Marine Brigade

Halberstadt CL.IV, possibly ex-*FA* 426/*Truppen-Fliegerstaffel* 38/2. *Plastun* Aviation Division, Russian *Westarmee*. Remains among Lithuanian war booty at Aleksotas airship shed, 1920.

Nieuport Ni.24 *bis* #4300
Latvian Army Aviation Group
October 1919

Acknowledgments

What began as a solitary effort over 30 years ago brought me into contact with other historians and individuals, some of whom I already knew, and others who have become fast friends.

August Blume
Moshe Bukhman
Rick Duiven
Jack Eder
Joseph Fürst Jr.
Dr. Dieter H. M. Gröschel
Peter M. Grosz
Jack Herris
Mateusz Kabatek
Hans-Eberhard Krüger
Stephen Lawson
Brian Lewis
Piotr Mrozowski
Neal O'Connor
Colin Owers
Douglas Pardee
Terry "Taz" Phillips
Mary Ann Presman
Paul Rempe
Michael Seidel
Gunnar Söderbaum
Hannes Taeger
Greg VanWyngarden
Dr. Robert G. L. Waite
Charley Woolley
Justin Young
Reinhard Zankl
History of Aviation Collection, University of Texas at Dallas
  Patrizia Nava
  Larry Sall
  Paul Oelkrug
Balazekis Museum of Lithuanian Culture, Chicago, IL
Rockford Public Library, Rockford, IL
Hoover Institution On War, Revolution And Peace, Stanford University, Palo Alto, CA

My friends, I hope you'll find this account worthy of the contributions you made to it.

I want to single out a few for special attention, because they have defined generosity:

Peter Grosz, who became my mentor when I was delving into the Siemens-Schuckert D.III and D.IV fighters, was a continuing source of encouragement for years. Time after time, a postcard from some far-off location would arrive, inscribed with an arcane contribution to my knowledge base. If not that, it would be a photocopy of a page from some book or archival document with his trademark yellow Post-it note. And I know I wasn't alone -- Pete seeded all our research gardens with his discoveries.

Rick Duiven, Dr. Dieter Gröschel and Greg VanWyngarden, my *Over the Front* compatriots. Whenever they came across some tidbit or photo pertaining to the *Freikorps* flyers, they would forward it. More pieces of the jigsaw puzzle. Rick and Dieter are gone, but Greg continues to enlighten me on aircraft markings and share hard-to-find photos. You're the beneficiary of his wisdom in the color profiles that appear in this book.

Paul Rempe, son of *Lt. d. R.* Leonhard Rempe, pilot in *FA(A)* 261 and *FA* 423. Bless his heart, *Herr* Rempe was a packrat, and his collection of photos and memorabilia was the basis of Paul's excellent book and a valuable contribution to this one.

Reinhard Zankl, my expert on all things Bavarian. He was the source of numerous archival records that formed the bones of my Bavarian account and then helped flesh it out it in our on-line discussions.

Dr. Hannes Taeger. All too often, Saxony's contributions are an afterthought in German military histories. I'm grateful for Hannes' generosity in sharing information and comments about the Saxon flyers' role during the *Freikorps* and Provisional *Reichswehr* era.

Patrizia Nava, Senior Curator at the History of Aviation Collection at UT-Dallas. Thanks for all your help and patience as I stumbled through the Collection's archives in search of photos. It all paid off handsomely.

Finally, a very special thanks to Moshe Bukhman. Moshe's persistence and energy in rooting out archival information about aviation activities in the Baltic in 1919 and the early 1920s should set an example for us all. He has unearthed material, most of it from primary sources, the rest of us gave up for lost. I am not exaggerating when I say this book's section on Germany's Baltic adventure would have been only a fragment of what you have read here without his dogged research and generosity in sharing it. I hope this book's Baltic overview will set the stage for what should be his definitive work on the subject of early aviation in the Baltics.

Sadly, many of these friends have gone West while I continued to grind away on this account. God, I miss them. I hope the historians' Valhalla has a library with a copy of this book waiting for them to sign out. Like those long-gone flyers, they deserve to know their contributions haven't been forgotten.

Appendix 1

Roster of personnel reported to have served in *Freikorps* and Provisional *Reichswehr* flying units and aviation personnel who served in non-flying units

No comprehensive database of *Freikorps*/Provisional *Reichswehr* aviation personnel is known to exist, and it's likely none ever did. Based on the number of aviation units that existed during this period and assumptions about their average size, we estimate 4000 - 6000 men were members of these units at one time or another. Over 1000 have been identified and listed here. Personnel who served with *Freiwillige FA* Haas are the most numerous; so far this is the only *FA* whose entire roster is known to have survived.

The names in this roster have been gathered from numerous primary and secondary sources, and their listing is only as accurate as those sources. In many cases, no more than a family name is known. Some references contain details about an individual's previous assignments and dates of service; others simply say something like, "fought against the Communists in 1919."

Included in this list are:

1. Flying and non-flying personnel who served in the *Luftstreitkräfte* in 1914 - 1918 and who later served in a *Freikorps* or Provisional *Reichswehr* military unit, whether it was an aviation detachment or not.
2. Flying and non-flying personnel who served with a *Freikorps* or Provisional *Reichswehr* aviation unit, whether or not they are known to have served in the *Luftstreitkräfte* in 1914 - 1918.

West Russian Army personnel whose names appear to be Russian have been omitted; only German personnel are included.

Fields included in this database:

Surname. Personnel are listed alphabetically by their surnames; for the purpose of sorting, "von" is ignored if it's part of a person's surname. If a vowel is accented with an umlaut, it is treated as being followed by the letter e. Some names of personnel who joined the West Russian Army are probably phonetic misspellings; they were obtained from Russian documents into which they had already been transliterated into the Russian language.

First Name.

Rank. Highest rank the individual is known to have held during his service in the *Freikorps* or Provisional *Reichswehr*. Some personnel who enlisted in the West Russian Army were given a boost in rank, but this was not recognized in *Reichswehr* personnel records.

***Freikorps/Reichswehr/Polizei Fliegerstaffel* Unit and Function.** Many personnel served through the transition from *Freikorps* to *Reichswehr*, and in some cases to the West Russian Army, although the sources may refer to only one of these unit designations. For example, *FA* 424 was renamed *Truppen-Fliegerstaffel* 37, and finally, the 3. *Plastun* Aviation Division when it joined the West Russian Army. Someone who served with the unit from beginning to end could claim membership in all three incarnations of this unit, although only one of them might appear in the source. On the other hand, someone joining the unit in September 1919 would technically have belonged to the 3. *Plastun* Aviation Division, or *Truppen-Fliegerstaffel* 37, as the Germans continued to call it, but not *FA* 424, even though that unit might have been listed in his records. Only knowledge of that person's dates of service would positively establish the designations of the units he belonged to.

Unit designations are abbreviated as follows:

AFst - *Artillerie-Fliegerstaffel*
FA - *Flieger Abteilung*
FLA - *Flieger Luftschiffer Abteilung*
PADet - *Plastun* Aviation Detachment (West Russian Army)
PADiv - *Plastun* Aviation Division (West Russian Army)
TFst- *Truppen-Fliegerstaffel*

Numerous non-aviation *Freikorps* detachments appear in this database. In these instances, the individual was known to be a member of the *Luftstreitkräfte* during the 1914-18 War, but joined a non-flying unit postwar.

Where known, individuals' functions within their units are presented in parentheses after the unit designation:

CO - Flying Detachment Commanding Officer (*Flieger Abteilungsführer* or *Fliegerstaffelführer*)
OzbV. - Adjutant (*Offizier zur besonderer Verwendung*)
P - Pilot (*Flugzeugführer*)
O - Observer (*Beobachter*)
G - Aerial gunner (*Bordschützen*)
Tech. Off. - Technical Officer (*Technische Offizier*)
Flivo -- Aviation Liaison Officer (*Fliegerverbindungs Offizier*)
Artzt - Medical Officer

Other functions, such as mechanic, *Werkmeister* or paymaster, are specified if they are known.

1914 - 18 Units. Combat units the individual served with during World War I. Space doesn't permit a complete listing, so only the principal ones are noted.

German Flyers In *Freikorps* & Other Postwar Activities				
Last Name	***First Name***	**Rank**	***Freikorps* Unit**	***1914-18 Combat Units***
Aberglaub	Hermann		FA 426	
Abernetty	Otto	Ltn.	FA 409 (O)	FA 47, FA 6, Schlasta 13 (CO)
Achilles	Paul	Lt. z. S.	KG Sachsenberg	Seefrosta II, Marine LFJ V (CO)
Adomeit		Lt. d. R.	FA 426 (Staff Abt V)	
Ahling			FA 426 (P - Abt V)	
Albrecht	Karl	Vzfw.	FA 426	
Alfred		Flg.	FA 424	
Allgaier	Calestin	Frw.	FA Haas	
Alt	Erich	Hpt.	FA 433 (CO); FA 1 des Baltischen Landeswehr (CO); Fliegerkorps Kurland (CO)	FFA 27 (O), FA(A)253(CO), FA 37 (CO); FFA 27, FA(A) 266, FA(A)253 (CO-stv.), FA 37 (CO)
Altenburg		Ltn.	FA 426	
Angelroth	Herbert-Armin	Oblt.	FA 417 (O), AFSt 108	Fl. Abt. (A) 258
Angerstein	Karl	Oblt.	FA 420 (P), Fliegerstaffel des AOK Nord (CO), Luftpolizei-Abt. Königsberg (CO)	Fl. Halbabt. "W" & "K", FFA 52, FFA 23, KEST 1, KEK West (CO), BOGOHL 17/3, BOGOHL 18 (CO)
Anschütz	Oskar	Ltn.	Schlachtstaffel Anschütz (CO)	
Anton	Werner	Ltn.	Grenzjäger FA 2 (O)	FA(A) 212 (O)
Antonious	Hans	Vzflugm.	KG Sachsenberg	
Apell, von		Hpt.	FA 423 (CO - Abt Ia))	
Arimond	Hans	Ltn.	FA 426, TFst 38, 2. PADiv	
Armbruster	Wilhelm	Frw.	FA Haas	
Arndt	Dietrich?	Hpt.	FA 430 (CO)	
Arndt	Emil Hugo	Frw.	FA Haas	Rfa 500, 501
Ascheberg, von	Percy	Hpt.	Kofl XI, Grufl. 37	
Aschenbach			Flugpark Kurland (?)	
Aschoff	Walther		FA 410 (CO)	BG 3?
Assenheimer	Ernst	Feldw.-Ltn.	FA Haas	
Aue	Paul	Offst.	FA 423, TFSt 116 (P)	Kg 5, Js 10
Auerbach		Vzfw.	FA 412 (P)	
Averdam	Adolf	Oblt.	FA 417 (O), Afst 108 (CO)	

German Flyers In *Freikorps* & Other Postwar Activities				
Last Name	***First Name***	**Rank**	***Freikorps* Unit**	***1914-18 Combat Units***
Averes	Dietrich	Vzfw.	Grenzschutz Ost (P)	FA(A) 218, FA(A) 242, Js 74, Js Ob. Ost (81)
Baas	Lothar	Hpt.	FA 415/TRfst 2 (CO)	FFA 26
Bach		Feldw.	FA 427	
Bachert		Fw.	FA 426 (P?)	
Backhaus	Paul	Hpt.	FA 400 (CO)	
Bader	Wilhelm Friedrich	Frw.	FA Haas	
Bading		Uffz.	FA 423	
Bäder	Alfred	Vzfw	FA Haas	Js 65
Bahl		Ltn.	FA 426 (P - Abt VII)	
Bakse	Erwin	Frw.	FA Haas	FA(A) 243, Js 71
Bandele	Heinrich	Frw.	FA Haas	FA(A) 224
Banholzer	Franz	Vzfw.	FA Haas	
Bannier	Wilhelm	Frw.	FA Haas	Schusta 17
Banse	Gustav	Ltn.	Polizei Fliegerstaffel Mecklenburg (O.z.b.V.)	FFA 6
Barlen	Karl	Oblt.	FA 419 (O.z.b.V.), FA 410/AFst 120 (P)	FFA 8
Barnow	Karl		KG Sachsenberg	
Bartdorff, von		Vzfw.	FA 433 (P)	
Barthel	Josef	Flieger	FA 430	
Bassenge	Gerhard	Ltn.	FA 424, 3. PADiv (P)	
Bauer	Hans	Vzfw.	FA Krug, TFst 21a	
Bauerle	Willy	Gefr.	FA Haas	
Baumgartner			FA 426 (O - Abt. V)	
Baumgartner	Friedrich	Frw.	FA Haas	
Baur	Kurt	Oblt.	AFst 103 (O)	
Bautz	Werner	Oblt.	FA 432 O)	
Bazlen	Eduard	Gefr.	FA Haas	
Becht	Phillip	Ltn z. S.	KG Sachsenberg	Seefrosta Zeebrügge, Seefrosta I, MFJ I (CO)
Beck		Obergefr.	AFst 121 (P)	
Beck	Johann	Gefr.	FA Haas	
Beck	Paul Robert	Gefr.	FA Haas	
Beck	Peter	Frw.	FA Haas	
Becker	Hermann	Ltn.	FA 418 (P)	FA 2,57,BG V/7,SS11,Js12(CO)

German Flyers In *Freikorps* & Other Postwar Activities

Last Name	*First Name*	Rank	*Freikorps* Unit	*1914-18 Combat Units*
Becker	Hermann	Oblt. Z. S.		Seefliegerstationen Norderny, Borkum, List & Zeebrügge
Becker	Karl Hermann	Frw.	FA Haas (Korporalsschaftsführer I)	
Beer	Hans	Flg.	TFst 37, 2. PADet.	
Behrends		Hpt.	Kofl VII	
Behrendt	Johannes (Hans)	Oblt.	FA 423/AFst 116 (O, Flivo),	FFA 45, KAGOHL 41/7,
Behrens		Uffz.	FA 424	
Beitat			FA 426 (P)	
Belz	Ludwig	Gefr.	FA Haas	
Bene		Gefr.	FA 424	
Benke		Ltn.	FA 424	
Bentivegni	Richard	Hpt.	Freikorps Düsseldorf (CO)	Rfa 501 (CO)
Berengrün	Otto Max		FA 433	
Berger	Heinrich	Gefr.	FA Haas (Korporalsschaftsführer I)	Js B, Bg IV/Ks 19, Bg IV/Ks18
Bernhard	Joseph	Sergt.	FA Haas	Kg I/ Ks 5
Bernhardt		Uffz.	FA Haas (Korporalsschaftsführer III)	
Bernhardt	Friedrich	Offst.	FA 418 (P)	FA(A) 277
Berthold	Friedrich	Hpt.	TFst 21b	FFA 5b, FA 300, Schlasta 25 (CO), FA 304 (CO)
Berthold	Rudolf	Oblt.	Eiserne Schar Berthold (CO)	JG II (CO),Js18 (CO),14 (CO),4
Bertrab, von	Joachim	Ltn.		FA 71, Js 30
Berwaldt	Karl	Off -stv	FA 426	
Bickel	Rudolf	Flg.	TFst 37, 2. PADet	
Bidlingmayer	Friedrich Wilhelm	Frw.	FA Haas	
Bihlmayer	Max Paul	Frw.	FA Haas	
Bihnenkamp	Karl	Vzfw.	FA 423	
Bilke		Oblt.	FA 418	Js 41(OzbV)
Bindner		Ltn.	FA 423 (CO - Abt VII)	

German Flyers In *Freikorps* & Other Postwar Activities

Last Name	*First Name*	Rank	*Freikorps* Unit	*1914-18 Combat Units*
Biselli	Emil Rudolf	Frw.	FA Haas	
Bismarck, von		Oblt.		FA 26
Bitterle	Johann	Krftfhr.	FA Haas	
Bittrich	Wilhelm	Ltn.	Grenzschutz Ost	-
Blank	Boleslaw	Vzfw.	FA 401 (P)	FA(A) 210
Blanket			FA 424 (P)	
Bleher	Ernst	Gefr.	FA Haas	
Blessing	Ernst	Frw.	FA Haas	
Blessing	Johannes	Frw.	FA Haas	
Blöcker	W. Hein	Ltn.	FA 424 (P, Tech. Officer)	
Blume	Walter	Ltn.	Fliegerhorst Döberitz(Flugleiter), Kurierstaffel der RW Brigade 10	FFA 65, FA(A)280,Js 26, 9(CO)
Boddien, von	Hans-Helmut	Oblt.	FA 424 (P, CO-stv.), 3. PADiv	Js 11, 59 (CO)
Bodenschatz	Karl	Oblt.	Frw. Detachment Probstmayer	FA 3b, Jasta B,JG 1
Boehmer	Hans	Rittm.	Freikorps Dohna (Eskadronführer)	FFA 30, FA 43 (CO), FA 224 (CO),
Boenigk, von	Oskar	Oblt.	FA 418 (CO), Jagdstaffel 202 (CO)	Kasta 32, Js 4, Js 21(CO),JG II(CO)
Boettge	Kurt	Oblt.	FA 433 (O)	FA 37
Bogert	Bruno		FA 409 (waffenmeister-gehilfen)	Js 5
Böhm		Ltn.	FA 426	
Böhmer		Ltn.	FA 424, 3. PADiv	
Bohne	Fritz	Vzfw.	FA 417	
Bohnenkamp	Karl	Vzfw.	Freikorps Maercker	FA(A) 209, Js 22
Bolle		Ltn.	Kofl AOK Sud	
Bolle	Karl	Rittm.	Friedensfliegertruppe	FA 1,KG 4, Kasta 23,Jasta B(CO)
Bonatz	Ernst	Oblt.	FA 420 (P)	FFA 36, FA 25 (O), Schlasta 36 (CO),Schlasta 18 (CO)
Bongardt		Ltn. Frhr.	FA 424, 3.PADiv	
Bongartz	Heinrich	Ltn.		FA 5,KG 5/Ks 27,Schl'sta 9,Js 36(CO)
Böning	Walter	Oblt.	Freikorps von Epp	FFA6b (P), Js 19 (P),Js 76 (CO)
Borghardt	Emil	Flieger	Det. Tüllmann	
Bornhauser	Emil	Krftfhr.	FA Haas	

German Flyers In *Freikorps* & Other Postwar Activities

Last Name	*First Name*	Rank	*Freikorps* Unit	*1914-18 Combat Units*
Bosch	Emil	Frw.	FA Haas	
Bosselmann		Uffz.	KG Sachsenberg (P)	
Bossert		Gefr.	KFSta Würzburg (Mech)	
Bothke	Kurt	Gefr.	FA 409	FFA 31, FFA 61, Js 30
Bovakht	-	Ltn.	FA 422 (O)	
Bracht	Heinrich		FA 424	
Bracht		Ltn.	FA 426 (O - Abt V)	
Brämsch	Karl		FA 401	
Braun		Feldw.	FA 424, 3. PADiv	
Braun	Friedrich Walter	Frw.	FA Haas	
Braun	Karl	Frw	FA Haas	
Bredow, von		Rittm.	Kampfgeschwader von Bredow (CO)	
Brembach	Hellmuth	Lt. z. S.	FA 417	SEE I
Brocke	Alfred?	Lt. d. R.	FA 434	Jasta 6
Brockmann		Vzfw.	FA 431 (P)	
Brockmann	Gustav	Oblt.	FA 418	Kest 8 (CO)
Bruch		Flg.	FA 424	
Bruch	Hermann	Oblt. z. S.		II. Marineflieger Abt., I. Seeflieger Abt.
Bruder		Uffz.	FA 426 / TFst 38, 2. PADiv (FA VIII-Werkmeister)	
Bruns		Dr.	FA 423 (Oberarzt)	
Brunst	Erich	Off.-stv.	FA Haas	
Brüssow		Lt. d. R.	Freiwillige Korps Brüssow	FA 257
Brustle		Vzfw.	KFSta Würzburg (G)	
Brzenk	Hans	Lt. d. R.	FA 415	Js 67, 81, 74, 67, 36, 67
Bubek	Eugen	Frw.	FA Haas	
Buchholz	H.	Flieger	AFst 101	
Büchner	Franz	Ltn.	FA 418, FA Grossenhain, F'horst Grossenhain (P)	Fl. Abt.(A) 270, Js 9, 13(CO)
Buchorn	Albert Michael	Gefr.	FA Haas	
Buck	Emil August	Frw.	FA Haas	FA(A) 224
Bühling	Helmut	Oblt.	FA 409 (O)	
Bulcher		Oblt.	FA 424 (O)	

German Flyers In *Freikorps* & Other Postwar Activities				
Last Name	***First Name***	**Rank**	***Freikorps* Unit**	***1914-18 Combat Units***
Bülow, von	Hilmer Frhr.	Hpt.	FA A1, FK Epp, TFst 21	FFA 11 (CO), FFA 9 ©-stv), FA(A0223 (CO)
Bülowius	Alfred	Oblt.	FA 420 (O), Fl'staffel des AOK Nord, Polizei Fliegerstaffel Ostpreussen	BOGOHL 3 (CO)
Bürgin	Richard	Frw.	FA Haas	
Burianeck	Erwin	Uffz.	FA Haas	
Burkhardt	Wilhelm	Uffz.	FA Haas	II. Seeflieger Abt.
Busch	Johannes		Eiserne FA 8 (P)	
Buschmann		Ltn.	FA 424 (P)	
Busse, von	Joachim		FA 429 (P)	FFA 12, Kasta 22, Js 3, Js 20 (CO)
Camphausen	Franz	Lt. d. R.	FA 432 (O)	
Carle	Karl	Feldw.	FA Haas	
Chorbacher		Off.-stv.	KFSta Würzburg (P)	
Christ	Adolf Ruppert	Frw.	FA Haas	
Christiansen	Friedrich	Kptlt. d. R.	FA Löwenfeld (CO)	IC Staffel, Zeebrügge(CO), Marinefeld JG (CO)
Claussen	Gustav	Vzfw.	FA 424 (P)	
Crailsheim, von	Hanns Frhr.	Rittm	Gruppe Hierl, Detachment Probstmayer (Flivo)	FA 8b, Schusta 22b (CO), Bogohl 25/8 (CO)
Cranz	Eberhard	Ltn.	FA 416 (CO)	Seefront Staffel, Marine Js 4(CO)
Cranz	Karl	Oblt. z. S.	KG Sachsenberg (O.z.b.V.)	
Creydt	-	Hpt.	FA 419 (CO)	FA(A) 234?, FA(A) 216
Dahlmann		Flugmeister	FA 416 (P)	
Dahlmann	Hermann	Oblt.	FEA 3 (O.z.b.V.)	FFA 63, FA(A) 252, Jasta 29, JG III, Jasta 26
Dambacher	Erwin	Frw.	FA Haas	
Dame	Ernst	Asst. Artzt d. R.	FA Haas	
Dams	Wilhelm	Feldw.	Fa 433	
Danhuber	Franz Xaver	Lt. d. R.	AFst 121 (P)	Kg 6/Ks 35b (P) Js 26, 78 (P), Js 79 (CO)
Decker		Gefr.	FA 424	
Degelow §	Carl	Lt. d. R.	Hamburger Zeitfreiwilligen-Korps	Jasta 40(CO)

German Flyers In *Freikorps* & Other Postwar Activities

Last Name	*First Name*	Rank	*Freikorps* Unit	*1914-18 Combat Units*
Deichmann	Paul	Ltn.	Staffel Libau (O), Frw. Fl. Abt. 427 (O), FA 433 (O), AFP Kurland, TFst 2, Fliegerhorst Stolp	FA 8
Deissner		Gefr.	FA 426	
Delling	August	Ltn.	FA Schauer	Jasta 34b
Demnig		Oblt.	FA 423	
Dessloch	Otto	Oblt.	FA Dessloch, TFst 21b (CO)	FA 8b, FA9b,KEK Ensis'm(CO),Js16b,35b(CO)
Deunert	Heinrich	Hpt.	Glogau Reg't. Grenzschutz Ost	FEA 12?
Diebold	Johannes	Uffz.	FA Haas	Js 10?
Dietrich	Paul	Oblt.	FA 425, Lithuanian Air Force	-
Dippon	Emil	Feldw.	FA Haas	
Dittus	Gottlieb Friedrich	Frw.	FA Haas	
Doerstling	Egon	Hpt.	Grenzjäger FA 1 (CO)	Kagohl 4, Kagohl 1, Bogohl 5 (CO)
Donhauser	Hans Christian F.	Vzfw.	AFst 121	FA 10, Js 17
Dörffler	Ernst	Hpt.	FA 425 (CO), Flugpark Kurland (CO), Polizei Fl'staffel Mitteldeutschland, Gotha (CO)	FFA 9, FA(A) 282, FA(A) 249 (CO)
Dreschel		Hpt.	FA Haas	
Drescher	Hans	Vzfw.	FA Haas	
Dressler	Franz Xaver	Frw.	FA Haas	
Drum	Karl	Oblt.	Kofl XVII (Adjutant)	FA 23,48, FA(A) 216,276, Schusta 18
Dukek	Hans	Gefr.	FA Haas	
Düring		Gefr.	FA 424	
Dürr	Hermann	Frw.	FA Haas	
Ebel		Ltn.	FA 426/TFst 38, 6, 5. PADet (CO)	
Eberg	Nikolaus	Flg.	FA 427, AFst 101	
Eberle	Otto	Frw.	FA Haas	
Eberstein	Ernst Otto	Hpt.	FA 431 (CO), Polizei Fliegerstaffel Paderborn	
Ebert		Gefr.	FA 424	

German Flyers In *Freikorps* & Other Postwar Activities

Last Name	*First Name*	Rank	*Freikorps* Unit	*1914-18 Combat Units*
Ehmann	Gottfried	Vzfw.	FA Haas	Schlasta 15 (Gunner)
Eichmayer	Gottfried	Vzfw.	FA Haas (Driver)	
Eifer	Karl	Frw.	FA Haas	
Eise			KG Sachsenberg	
Eitel	Karl	Frw.	FA Haas	
Ellensohn	Albert	Frw.	FA Haas (Driver)	
Emter	Heinrich	Flg.	FA 424	
Engelfried		Ltn.	FA 424, 3 .PADiv	
Engelfried	Karl	Flugobermaat	KG Sachsenberg (P)	Seefrosta, MFJ V
Engelhardt	-	Vzflugm.	KG Sachsenberg	
Englert		Pion.	KFSta Würzburg (Mech)	
Eppinger	Erich		AFst 101 (P)	
Eppler	August	Frw.	FA Haas	
Erdenberger		Ltn	Westrussian Armee	
Erdmann	Friedrich	Hpt.	Eiserne FA 8, FA 427 (CO), Polizei-Fl'staffel Hamburg	FA(A) 210, Kofl 8 (Adj)
Erhardt	August	Frw.	FA Haas	
Ernst		Off-Stv.	FA 424, 3.PAD	
Ertl		Uffz.	FA 423	
Erve			FA 424 (Kraftfahrer)	
Escher	Paul Ernst	Frw.	FA Haas	
Esebeck	Rudolf	Ltn. Frhr.	FA 424	[FA 15, Js 8, KEST 7 (CO),Js 17 (CO)]?
Ewaldt	-	Lt. d. R.	FA 422 (O)	
Exss	Ernst	Ltn.	FA 423 (O, CO - Abt IV)	Kagohl 3,FFA 22, FA(A) 211
Faber	Kurt	Ltn.	FA 429 (P), Schutzstaffel FA Gotha-Erfurt-Halberstadt-Zilly	
Fahrig	Hermann	Rittm.	Fliegerhorst Kowno	FFA 64, FFA 27, FA(A) 244, FA 300
Falkenhayn, von	Erich	Hpt.	Freikorps Faupel	FFA 39
Fast	Friedrich		FA 426	
Fastnacht		Ltn.		
Feist		Fl.	FA 410/ AFst 120, HFP 302 Süd,g, HFP 301Nord,	
Feld		Ltn.	FA 423 (CO - Abt VI)	

German Flyers In *Freikorps* & Other Postwar Activities				
Last Name	***First Name***	**Rank**	***Freikorps* Unit**	***1914-18 Combat Units***
Feller		Ltn.	FA Haas	
Femppel	Karl Friedrich	Frw.	FA Haas	
Fenske	Meinhard?	Hpt.	Kofl XX	
Feucht	Karl	Uffz.	FA Haas	
Feuersenger		Sergt.	AFst 101	
Fick	Walter	Lt. d. R.	FA 427, Staffel Fick, FA 433, Frw. Rgt. Libau, Westrussian Armee	Fl. Abt. (A)250? II. MFLA?
Fiesel	Paul	Frw.	FA Haas	
Filbig	Josef	Lt. d. R.	Eiserne Schar Berthold	FFA 3b, FA 46b, Js 60 (80b?)
Fincke		Gefr.	FA 424	
Fink	Carl	Hpt.	FA 407 (CO)	FFA 7, BAO, Kagohl 1
Fischer		Lt. d. R.	Eiserne Fl. Abt. 8 (P)	
Fischer	Albert	Frw.	FA Haas	
Fischer	Eberhard	Ltn.	FA 401 (O.z.b.V.), AFst 129, Fliegerhorst Brieg (O.z.b.V.)	Kagohl 3, FA(A) 218
Fischer	Fritz		FLA 101 (CO- Photo Abt.)	
Fischer	Josef	Frw.	FA Haas (Driver)	
Fischer	Pankrats	Flg.	FA 426 (P)	
Fischer	Veit	Oblt.	Flugplatz-Kommandanten Sonthofen, Scharfschützentruppe des Freikorps Schwaben	FA 9b, BAO, Kasta 34, Kasta 38 (CO), FA(A) 288, FA Pascha II, FA 304b
Fischer	Wilhelm		FA 426	
Fisser	Johann-Volkmar	Oblt.	FA 432 (O)	FFA 26, FFA 39, FFA 29, FA(A) 222,FA 8, FA 26 (C)-stv), FA 3
Fitsch		Flg.	FA 424	
Fitzau	Erich-Erdmann	Ltn.	FA 423 (P), AFst 116	FA 39, FA(A) 268, FA 40, FA 261
Flach, von		Ltn.	FA 426	
Flecken			Jasta Flecken (CO)	
Fleischauer		Ltn.	FA 426/TFst 38/2. PADiv (Transportation Officer)	
Fleischer	Alfred		Luftpolizei	Js 17
Fliege	Alfred		FA 424	
Flinke	Edmund		FA 426	

German Flyers In *Freikorps* & Other Postwar Activities				
Last Name	***First Name***	**Rank**	***Freikorps* Unit**	***1914-18 Combat Units***
Förster	Helmuth	Maj.	FA 426 (P), FA 433	FA 31 (OzbV, P)
Francke	Rudolf	Vzfw.		Js 8
Franke		Oblt.	Grenzjäger FA 1 (O)	
Frankl	Kurt	Uffz.	FA Haas	FA(A) 262, FA(A) 284, FA 22
Frantzius, von	Gerhard	Lt. d. R.	FA 421, FA 426 O, Flivo), Eis Div.-Intel Off, Stab. Der Kofl.	FA(A) 270 Lb
Freden, von	Hans	Ltn.	FA 414, Stettiner Sicherheitspolizei	FFA 48, Jasta 1, 50(CO)
Freund	Erich	Oblt.	FA 433 (O)	
Freyberg-Eisenberg-Allmendingen, von	Egloff Frhr.	Hpt.	FA 432 (CO)	FFA 11, BAO, FFA 5 (CO), FA 11 (CO), FA(A) 239 (CO)
Fricke	Hermann	Oblt.	FA 432 (CO)	FA 2, FFA 11, FA 3Lb(CO), FA 2Lb(CO)
Frieberg, von	-	Hpt. Frhrr.	FA 424, TFst.37 (Staffelführeer 9)	
Friedewald	Ludwig		FA 423 (P)	FA(A) 261
Friedrich		Feldw.	FA 418	
Frince	Edmund		FA 426	
Fritz	August Karl		FA Haas	
Fritzenschaft	Max	Frw.	FA Haas	
Frohlich	Melchior Georg	Frw.	FA Haas	
Frohlich	Paul	Sergt.	FA Haas (Korporatsschaftsführer 6)	
Frommherz	Hermann	Lt. d. R.	Fliegerstaffel Hassfeld (CO) - Luftpolezei?	Kaghl 4, Kagohl 1, Js 2, 27(CO)
Frott	Arno	Luftsch.	FLA 101 (Photographer)	
Fruhner	Otto	Vzfw.	FA 409	
Fruhner	Otto	Ltn. d. R.	FA 432 (P)	FA 51, Js 26
Funck	Werner	Hpt.	Freischützen-Korps Meyn	FFA 13, FA(A) 209 (CO), FFA 55 (CO), FA(A) 251 (CO)
Funk	Adam	Vzfw.	FA Haas (Werkmeister-stv.)	Js 11, JG II
Funk	Josef	Vzfw.	FA Haas	FA(A) 237, Js 30
Fürst	Joseph	Flieger	FA 424	
Fütterer	Kuno Heriberet	Ltn.	AFst 115 (O)	FFA 45, Bogohl 6

German Flyers In *Freikorps* & Other Postwar Activities

Last Name	*First Name*	Rank	*Freikorps* Unit	*1914-18 Combat Units*
Gablenz, von	Carl-August Frhr.	Ltn.	Freikorps Lüttwitz (Kofl)	FFA 42, KEA 1, Kagohl 5/1 (CO), Bogohl 7
Gaedicke		Oblt.	AFst 101	
Gallen		Ltn.	FA 426 (P - Abt V)	
Gandenberger von Moisy	Hans	Ltn.	FA 434 (O)	FA(A) 255 (O), Turkey?
Gansin	P.	Flieger	FA 418	
Garback			FA 424	
Garbart			FA 424	
Gebbert		Ltn.	FA 418	Js 43
Geerdtz	Franz	Hpt.		
Gehring	Edward		FA 426	
Gehrmann	Bernhard	Luftsch.	FLA 101 (Typist)	
Gehrmann	Georg	Flieger	KG Sachsenberg	
Geisler		Uffz.	Grenzjäger FA 1 (P)	
Genée	Paul	Hpt.		
Genth	Adolf	Ltn.	FA 410 / AFst 120 (O)	
Gerick		Gefr.	AFst 101	
Gertenfeldt	Edm.	Flieger-O-M	FA Löwenfeld	
Geserick	Paul	Uffz.	FA Haas	FFA 52
Gessler	Josef	Gefr.	FA Haas	
Gewert	Fritz	Ltn.	FA 418 (P)	Js 19
Gillmann		Feldw.	FA 427 (Staffel Fick)	
Glöcklen		Ltn.	FA Haas	Js 15
Göbel		Sergt.	FA 423	
Gobel	Wilhelm	Frw.	FA Haas	
Gödeke		Vzfw.	FA 426 (P)	
Goerne, von	Günther	Rittm.	Art. Flugpark 8	
Goerth	Hans	Vzflugm.	KG Sachsenberg	Marine Js 3
Goetz	Xaver		FA Haas (Driver)	
Gölchar		Ltn.	FA 424, 3.PADiv	
Golowin			FA 426 (O)	
Goltz				
Goltz	Konrad	Korvetten-kapitän	Garde-Kavallerie-Schützen-Div.	Frw.-Marine-Fliegerkorps, Marine-Landfl.-Abt., Kdr. der Seeflieger Türkei

German Flyers In *Freikorps* & Other Postwar Activities				
Last Name	***First Name***	**Rank**	***Freikorps* Unit**	***1914-18 Combat Units***
Gordon, von		Kapt. Z. S.	Ftaffel Libau (O)	
Grad	Heinz	Frw.	FA Haas	
Grade	Wilhelm	Major	Gardekorps (Kofl)	
Graf	Carl	Vzfw.	FA Haas	Schlasta 5
Grauert	Ulrich	Hpt.	FA 420 (CO)	FFA 56, Kg5/25, Kasta 28 (CO), Kasta 9 (CO), FA 17 (CO), FA(A) 256 (CO-stv.)
Greim, von	Robert Ritter	Oblt.	Kurierstation Schleissheim	FFA 3b, FA(A) 204, FA 46, Js 34b(CO)
Greiser	Arthur		Grenzschutz Ost	SEE I, II, Küstenfliegerstaffel I & II, Seefrontstaffel, MFJ IV (O)
Grenzisa		Feldw.	FA 418	
Gretz	Heinz		KG Sachsenberg	
Griebsch	Wilhelm	Ltn.	FA 431(P)	FA(A) 250, FA(A)213
Grober	Klemens	Frw.	FA Haas	
Groeckel	Hermann	Hpt. d. R.	FA 425 (CO)	
Gross	Friedrich	Frw.	FA Haas	
Grude	Eberhard	Ltn.	-	FA (A) 290, BOGOHL IV/19
Grün		Off.-stv.	KFSta Würzburg (P)	
Grund	-	Uffz.	FA 427 (P)	
Grünfeld	Alfred		Iron Division	Kasta 109(?)
Gruundmann	Bernhard	Flg.	FA 426	
Gürtler	-	Maj.	FA 431(?)	
Gussmann	Siegfried	Ltn	(P)	Js 11
Gutknecht	Adolf	Hpt.	FA 412 (CO)	Js 43 (CO)
Haagen		Lt. z. S.	FA 416	
Haas	Friedrich		FA Haas (Driver)	
Haase		FlgObmatr	6. R.K.	
Habermann	Alfred	Gefr.	FA Haas	
Habermas		Ltn.	Kofl AOK Sud	
Hachenburg, von	Johannes Ludwig Graf	Hpt.	Iron Brigade / AOK 8, Garde-Kavallerie-Schützen-Div.	Fest.-FA 2, FFA 39, FFA 38, Brieftauben-Abt. M, Kagohl 7/2, Kagohl 11/2 (CO), Kagohl 2/1 (CO), Kagohl 14/3 (CO), FA 204 (CO), FA 248 (CO)
Hackbusch	Hermann	Vzflugm.	FA 417 (P)	Marine Js 1
Hafner	Gustav	Frw.	FA Haas	I. DFA, Putzig
Häfner	Konrad	Hpt.	FA A2/AFst 121, FA Schleissheim	Fl. Abt. 6b, (A)295(CO)
Hähnchen	Martin	Ltn.	FA 407 (P)	Js 13, 53 (CO-stv)

German Flyers In *Freikorps* & Other Postwar Activities				
Last Name	***First Name***	**Rank**	***Freikorps* Unit**	***1914-18 Combat Units***
Hagen		Ltn. z. S.	FA 413	
Hahn		Uffz	FA 417 (P)	
Hallen		Oblt.	FA 426 (P)	
Haller	Georg	Frw.	FA Haas	
Hämmerle	Paul Adolf	Frw.	FA Haas	
Hamscher (Hamster)	Ernst	Ltn.	KG Sachsenberg or FA 433 (P)	Js 37
Hanesse	Friedrich-Carl	Oblt.	FA 400 (O.z.b.V.),Fl'horste Brieg, Paderborn (O.z.b.V.),	FFA 225, Schusta 9 (CO), Schlasta 2 (CO)
Harborth		Feldw.	FA 424/3.PADiv	
Hartmann		Oblt.	FA 431	
Hartmann	A		Erfurt	
Hassel	August	Uffz.	AFst 103 (P)	
Hasselmann	Hermann	Vzfw.	FA 423, AFst 116 (P)	
Hassenmitter	-	Uffz.	FA 431 (P)	
Haupt		Vzfw.	FA 424 (P)	
Haussermann	Paul	Off.-stv.	FA Haas	
Haussler	Friedrich	Frw.	FA Haas	
Haussmann	Friedrich	Uffz.	FA Haas	
Haussmann	Karl Ludwig	Frw.	FA Haas	
Havenit	Pavel	Privat	FA 424/3.PADiv	
Havenith	Leo		FA 424	
Hawelik		Feldw.	Libau FA (P)	
Hecker	Willi		FA 426, Abt. VI	
Heidenreich	Anton	Ltn. d. R.	Technischen Komp., Flugplatz des Gouvernements Libau	FA 289
Heidepriem	-	Lt.	FA 431	
Heimburg, von	Erik	Oblt.	FA 422 (CO)	FFA 40, FA 300
Heinlein		Lt. d. R.	AFP Kurland	
Heising	Fritz	Hpt.	Grenzschutz-Bombenstaffel Ost, Freikorps Severin, Hacketau (Abt.-Führer)	FFA 6, FFA 10, Kasta 23 (CO), Jasta 20 (CO)
Hellbarth	Julius	Wehrm.	FLA 101 (Photographer)	

German Flyers In *Freikorps* & Other Postwar Activities				
Last Name	***First Name***	**Rank**	***Freikorps* Unit**	***1914-18 Combat Units***
Hengl, von	Georg Ritter	Ltn.	FA Schmalschlager, TFst 21a (O)	FA(A) 295 b
Hentschell	Arthur	Gefr.	FA 427	
Herber	Robert?	Maj.	Kofl IV	
Herbst			FA 423 (P)	
Hermann	B.	Ltn.	FA 427 (O)	
Hermann	Eugen Albert	Frw.	FA Haas	
Hertle		Ltn.	FA 423	
Hess	Rudolf	Lt. d. R.	FK Regensburg, FK von Epp, Polizeifliegerstaffel Schleissheim	Js 35b
Heuduck, von		Oblt.	FA 424	FA 18
Heutsch			FA 424	
Heyden-Rinsch		Ltn. Frhr.	FA 424, 3.PADiv	
Heyder	Walter	Hpt.	Fliegerhorst Liegnitz (CO)	FFA2, FFA 57, FFA 44 (CO)
Heydt		Ltn.	FA 434 (O)	
Hillebrandt		Uffz.	FA 424, 3.PADiv	
Hillmann	-	Uffz.	Staffel Libau, FA 433/FP Kurland, FA 424	
Hilpert	Otto	Off. St.	AFst 101 (FA 427)	FFA 28, 48, Kasta 32, Js 12, 2, 19
Hinneberg		Ltn.	FA 426 (Staff), TFst 38, 2. PADiv	
Hinz		Gefr.	FA 424, 3. PADiv	
Hitz	Anton		FA Haas (Driver)	
Hödecke		Ens.	FA 426	
Hoffmann		Ltn.	FA 434 (P)	
Hoffmann	E.		KG Sachsenberg (P)	
Hoffmann	Gustav	Gefr.	FLA 101 (Cook)	
Höhler	Friedrich	Ltn.	FA 426	
Holbach		Ltn.	FA 412 (O)	
Holch	Karl		FA Haas (Driver)	
Holst		Gefr.	FA 424	
Holtzmann	Robert	Hpt.	FA Haas (CO)	Fl. Abt. 1(CO-stv), 16(CO),22(CO)
Hommel	Karl	Frw.	FA Haas	Luftschifftrupp 7
Höper	Heinrich	Ufz.	FLA 101 (Typist)	
Horn		Ltn.	FA 426 (P - Abt V)	

German Flyers In *Freikorps* & Other Postwar Activities

Last Name	*First Name*	Rank	*Freikorps* Unit	*1914-18 Combat Units*
Horn	Hans-Jürgen	Ltn.	FA 401 (CO)	KG 2,Schusta 11, Fl. Abt. (A)221,39
Horster	Waldemar			Seeflabt. In Flandern (P)
Hoth	Hans		FA 426	
Hubrich	Gerhard	Obermaat	FA 417 (P)	Seefrosta 1 (Marine Js 4)
Hückstädt	Albert		FA 426	
Hug	Andre	Hpt.		FFA 26, Brieftauben Abt. "O"
Hummel	Ludwig	Frw.	FA Haas	
Hunninghaus			FA 423 (P)	
Hupp		Pion.	KFSta Würzburg (Mech)	
Huth	Joachim-Friedrich	Lt. d. R.	Grenadier Regt. 11, Freikorps Küttner	Js 14
Hüttner	Xaver	Uffz.	Bayr. Inf. Regt. 45	Js 76
Ilian		Uffz.	FA 423	
Illige		Gefr.	FA 424, 3.PADiv	
Jacob	Kurt	Ltn.		Jasta 36, Jasta 12
Jacobs	Josef Carl	Ltn.	FA 413 (P, Deputy CO)	FFA 11, KEK West, Js 22s,7 (CO)
Jacobsen	Fritz John	Lt. d. R.	Freikorps Lettow-Vorbeck, Hamburger Sipo	FA 1, Js 17, 9, 31, 73
Jagdfeldt	Walter	Ltn.	FA 426 (P)	
Jäger	Erwin	Uffz.	FA Haas	
Jahn	-	Hpt.	FA 431	
Jakob	Paul	Frw.	FA Haas	
Jander		Ltn.	FA 426 (O - Abt V)	
Jankowiak			FA 424 (P)	
Janson		Gefr.	FA 424	
Janzen	Johannes	Lt. d. R.	FA 410/AFst 120 (P)	Kasta 12, Js 23, 6, 4(CO-stv), 6(CO)
Jaser	Albert	Frw.	FA Haas	
Jedwill	-	Vzfw.	Staffel Libau, FA 433 (P)	
Jehle	Benedikt	Ltn.	FA FK Schwaben (CO, P)	FFA 1b, Js 16b
Jena, von		Ltn.	FA 426 (Transportation Officer)	
Jeschinsky	Max	Flg.	FA 404	
Jeschonnek	Hans	Ltn.	FA 401, AFst 129 (P)	JS 40
Jodkow(?)			FA 417 (O)	
Joeden	Helmut	Ltn.	FA 404 (O)	FA(A) 216, FA 303

German Flyers In *Freikorps* & Other Postwar Activities				
Last Name	***First Name***	**Rank**	***Freikorps* Unit**	***1914-18 Combat Units***
Joesting		Ltn.	Westrussian Armee	
Johannesson		Ltn.	FA 426 (O), Abt. VI (CO-stv)	
Joos	Friedrich	Frw.	FA Haas	
Jordan	Georg		FA 426	
Jordan	Otto	Ltn.	FA 432 (O), AFst 103 (O), AFst 116 (O)	FA 41, FA 256, FA 282
Juhnke	Hermann	Vzfw.	Inf-Rgt. 34 (Verpflegungsoffizier)	FA(A) 238, Js 41, Kest 5, Js 52
Junck	Werner	Ltn.	FA 400, TFst 8 (P)	FFA 33, Js 8 (CO)
Jurkschat			FA 424, 3.PADiv	
Kaffanke	-	Hpt.	FA 431	
Kahläne		Hpt. d. R.	Fa 415 (O.z.b.V)	
Kaiser	Paul		FA 426	
Kalähne		Hpt. d. R.	TFst 2 (O.z.b.V.)	
Kalbreier	Fritz		FA 406 (P)	
Kanisch	-	Vzfw.	Staffel Libau, FA 433, Fa 427	
Kasper	Eugen Christian	Vzfw.	FA Haas	
Kaspereit		Oblt.	Westrussian Armee	
Käubler		Ltn.	FA 431	FA 26
Kaumann	Gottfried		KG Sachsenberg	
Kayser		Vzfw.	FA Haas	
Kegel	Max		Luftpolizei Kassel	
Keidel	Eberhardt Heinrich	Frw.	FA Haas	
Keinath	Erwin	Frw.	FA Haas	
Keller	Josef	Ltn.	Selbstschutz Oberschlesien	Js43
Keller	Paul		FA 426	
Kellner		Ltn.	Grenzjäger FA 1 (P)	
Kemmer ?	-	-	Lithuanian Air Force?	-
Kemmler	Eugen Friedrich	Frw.	FA Haas	
Kerber	Eugen	Frw.	FA Haas	
Kern		Vzfw.	KFSta Würzburg (G)	
Kern	Christian	Gefr.	FA Haas	
Kerner	Karl	Lt. d. R.	FA 415, TFst 2	FA(A) 282 (O)

German Flyers In *Freikorps* & Other Postwar Activities				
Last Name	***First Name***	**Rank**	***Freikorps* Unit**	***1914-18 Combat Units***
Kessler	Ernst	Flg.	FA 433 / AFst 101	
Kessler	Ulrich	Lt. z. S.	Garde-Kavallerie-Schützen-Div., Gkdo Lüttwitz	Seefliegerstationen Helgoland, Zeebrügge, & Norderny, Rfa?
Kientach	Otto Wilhelm	Frw.	FA Haas	
Kiesecker	Albert	Frw.	FA Haas	
Kieske		Sgt.	FA 401 (TFst 129)	(G, O)
Kinkelin	Alwin	Oblt.	KFSta Ingolstadt	Kasta 36
Kinzel		Uffz.	FA 424, 3.PADiv	
Kirsch		Hpt.	FA Haas	FA(A) 218
Kirschstein	Walter	Oblt.	Eiserne FA 8 (O), FA 427 (O, CO-Stv), Afst 101 (O)	FFA 47, FFA 37
Kisling	Wilhelm		FA 424	
Kitzinger	Seb. (astian?)	Vzfw.	FA Krug, TFst 21a (G)	
Klaiber	Otto	Feldw.	FA 418 (P)	FA 232, Js 12
Klaile	Paul	Frw.	FA Haas	
Kleb		Ltn.	FA 411 (O)	
Klee	Adolf	Gefr.	AFst 101	
Klein		Flg,	FA 426	
Klein		Ltn.	FA 425	
Klein		Ltn.	FA 426	
Kleinath	Erwin	Frw.	FA Haas	
Kleinath	Wilhelm	Gefr.	FA Haas	
Kleinschmidt	Erich	Uffz.	FA 101	
Klempert		Gefr.	FA 426	
Klink	Ernst	Frw.	FA Haas	
Klöwer		Ltn.	FA 426 / TFst 38, 8. PADiv (O)	
Knabel	Viktor		Fa 429 (Mechanic)	
Knöringer		Oblt.	FA Haas	
Koch		Hpt.	FA 410 (CO)	
Köckler	Friedrich Wilhelm	Hpt.	FA 411(O.z.b.V), FA 408 (CO), TFst30 (O.z.b.V)	
Köhl	Hermann	Ltn?	Polizei-Fliegerstaffel Stuttgart (O)	Kasta 19, Bogohl 4(CO), Bogohl 7 (CO)

German Flyers In *Freikorps* & Other Postwar Activities

Last Name	*First Name*	Rank	*Freikorps* Unit	*1914-18 Combat Units*
Koehler	Fritz	Ltn. d. R.	FA Haas (Ordonnanz-Offz. bei Stab), FA Dessloch, TFst 21b (P)	FA 48, Js 25, Js 38, Kest 9
Köhler	Fritz	Ltn.d. R.	TFst 21b (P)	
König	Otto	Frw	FA Haas	
Koepsch	Egon	Oblt. d. R.	Oberschleisischen Freiwilligenkorps	FA 256, Jg 11
Kohlbach	Wilhelm	Ltn.	Frw. Landesjägerkorps Maercker	FFA 4, FA(A) 255, Js 50, Js 10
Kohne			FA 423 (P)	
Kohnle	Matthaus	Frw.	FA Haas	
Koller	Karl	Vzfw.	Polizei-Fliegerstaffel 1, Schleissheim (P)	Js 34, FA 273, FA 47, Js 76
Kollmar	Hermann	Frw.	FA Haas	FFA 59
Konpoka	Kosta		FA 426, Abt. VI	
Köpf	Karl	Frw.	FA Haas	
Koppel		Gefr.	KFSta Würzburg (Mech)	
Koppenhäfer	Heinrich Georg	Uffz.	FA Haas (Korporalsschaftsführer)	
Korfay		Feldw.	FA 418	
Korner		Vzfw.	FA 412 (P)	
Kortüm	Joachim	Ltn.	FA 412, AFst 117 (P)	FA 216, Js 10, FA(A) 242
Kost	Erwin Karl	Frw.	FA Haas	
Kotz	V.		FA 426	
Kraft	Herbert		Grenzschutz Ost (O)	
Kraft	Johann	Sergt.	FA Haas	
Krahmer	Eckhard	Oblt.	Fliegerstaffel der Polizei Berlin (P)	FA 236
Kramer	August Wilhelm	Frw.	FA Haas	
Krämer	Richard	Sergt.	FA Haas	
Krampitz	Paul	Gren.	KG Sachsenberg	
Kratz		Oblt.	FA 426 (P, CO - Abt VIII)	
Krausser, von	Friedrich Ritter	Oblt.	FA A1 (CO)	Schusta / Schlasta 23b(CO), FA 45b (CO), Schlasta 2 (CO)
Kraut	Richard	Lt. D. R.	FA 434 (P)	Js 4, Js 63
Kredel	Ernst	Ltn.	FA 417 (O, CO)	FA(A)202, FA(A)237
Kretschmar			FA 424 (Fitter)	

German Flyers In *Freikorps* & Other Postwar Activities				
Last Name	***First Name***	**Rank**	***Freikorps* Unit**	***1914-18 Combat Units***
Kriegbaum	Oskar	Oblt.	FA der Mecklenburgischen-Sicherheits-Polizei (Flugleiter)	FFA 57 (P)
Kriegelsteiner	Reinhard	Hpt.	FA Kriegelsteiner (CO), TFst 21b	Kasta 35b (O), FA(A) 293b (CO)
Krieger		Ltn.	FA 406 (CO)	
Krocker	Victor	Hpt.	FA 423 (CO)	FFA 28, FA(A) 243, FA(A) 263, FA(A) 233, FA(A) 210, FA(A) 211
Krogmann	Ernst?		FA 423 (P)	FFA 63?
Kroll	Heinrich	Lt. d. R.	FA 426/TFst 38 (O.b.z.V), 2. PADiv/8. PADet (O.z.b.V.), Polizei FSt Hamburg	FFA 17 (CO), Js 9, 24s(CO)
Krotz	F.	Oblt.	FA 426 (Abt VIII)	
Krotzky		Flg.	TFst 37, 1. PADet	
Krug	Michael	Hpt.	Kurierfliegerstation Schleissheim (CO), FA Fürth (CO), TFst 21a (CO)	FFA 2b, Kasta 35, Bogohl 8, FA 45b
Krull	Ernst	Ltn.	FA Sturmabteilung Rossbach (P)	
Krummel	Alfons		FA 427	
Kübler	Karl Georg	Gefr.	FA Haas	Schlasta 16 (?)
Kuhn	Hermann	Frw.	FA Haas	
Kühne			Grenzjäger FA 1 (Harness maker)	
Kuke		Ltn.	FA 418	
Kulikowski	Erich	Off -stv	FA 424 (P)	
Kulikowski	Erich	Oblt.	FA 425, Lithuanian Air Force	
Kumme	Hans	Hpt.	Kofl I	
Kunze		Oblt.	Westrussische Armee	
Künzel	Georg		AFst ?	
Kupfer	Karl	Uffz.	FA Haas	
Küpker		Vzfw.	AFst 129 (O)	
Kuppinger	Oskar	Oblt.	FA 421(O)	FFA 64, KAGOHL VII/41,Schusta 2(CO),KG I/3(CO), BOGOHL IV/19(CO), FA(A) 216
Kuring	Carl	Flugmeister		
Küsgen		Ltn.	FA 423 (CO - Abt V)	

German Flyers In *Freikorps* & Other Postwar Activities				
Last Name	***First Name***	**Rank**	***Freikorps* Unit**	***1914-18 Combat Units***
Kuttlau			Fa 407	
Lahmer	Paul	Flg.	FA 427	
Laib	Alfred	Frw.	FA Haas	
Laischke			FA 433	
Lang		Oblt.	Schlasta Anschütz	
Lang	Karl	Frw.	FA Haas	
Lashowski	Franz	Flieger	FA 414, Afst 102	
Lässle	Johannes	Vzfw.	FA Haas	
Lautenbach			FA 424 (G)	
Lebdiez	Fritz	Vzfw.	FA 424 (P), Poizei FSt Seerappen	FFA 56, FA(A) 212
Lebedev		Ltn.	FA 424/3. PADiv	
Lechner	Hermann	Ltn.	FA ?, Grenzschut Ost (O)	FA(A) 297b
Leesemann	Friedrich	Lt. d. R.	FA 406 (CO - Staffel A), AFst 127	FA(A) 230, FA 11, Jagdgruppe 3, FA(A) 281
Lehmann		Lt. d. R.	FA 427	
Lehmann	F.	Off. St.	FA 411 (C-P)	
Leichner	Hermann		FA 425 (O)	
Lemmle	Andreas	Uffz.	FA Haas	II. See FA
Lenz	Alfred	Lt. d. R.	FA 431(O.z.b.V.), Selbstschutz Kreis Kreuzburg (CO)	FFl. Abt. 61,23,Js 4,14,22(CO)
Lenz	Willy	Flg.	TFst 37, 2. PADet.	
Lepa	Alfred	Luftsch.	FLA 101 (Cook)	
Lerche		Wachmeister AFST 101	AFst 101	
Lesermoser			FA 426 (P)	
Lessmann		Flg.	FA 423	
Leusch	Wilhelm	Ltn.	FA 418 (P)	FFA Metz, FA 19, Js 13, 19 (stv CO)
Lichtenberg		Ltn.	FA 426 (P - Abt VIII)	
Lidle	Otto Hermann	Frw.	FA Haas	
Liebermann, von		Hpt.	Kofl XVII	
Liebig	Friedrich-Wilhelm	Ltn.	FA 431 (P)	Js 22
Lilienfeld, von	Alex	Ltn.	FA 433	
Link	Oswald	Frw.	FA Haas	

German Flyers In *Freikorps* & Other Postwar Activities

Last Name	*First Name*	Rank	*Freikorps* Unit	*1914-18 Combat Units*
Lipp	Christian	Frw.	FA Haas	
Lippmann		Ltn.	FA 412 (CO - Schlachtstaffel)	
Lippmann		Ltn.	FA Sturmabteilung Rossbach (CO)	
Lippmann	Max Walter		Frw. Jägerbattalion Schneider, Freikorps von der Goltz	FFA 18, FFA 62 (CO)
Litzbarski		Flg.	FA 429	
Loerzer	Bruno	Hpt.	FA 427 (P)	FA25,60,(A)203,KEK Jametz, FA71(Jasta 17),Js 26(CO),JG III(CO)
Loewen			FA 433	
Logan		Rittm.	FA Haas	
Lohmann	Gunther	Ltn.	FA 409 (O), FA. 427 (O), FA 410 (O), Fliegerhorst Königsberg	FFA 26, FA 15, FA 233
Lorenz	Heinrich	Oblt.	Stab des Fliegergruppen-Kommandeurs, Libau	Kagohl 1, Js 19, Js 1, Js 33, Js 10, Js 7
Ludwig		Ltn.	KG Sachsenberg	
Ludwig		Ltn. d. R.	FA 417 (O)	
Ludwig	Wilhelm Karl	Frw.	FA Haas	
Luge	Walter	Hpt.	Heimat Schützpolizei	
Lunt		Vzfw.	FA 411 (C-P)	
Lutz	Ernst Gottlob	Frw.	FA Haas	
Lux			(P)	
Maass		Hpt.	FA 431 (Faulbrücke detachment)	
Maass	Joseph	Gefr.	FA Haas	
Mack	Friedrich	Frw	FA Haas	
Madel		Flg.	VI. R.K.	
Madlener	Max	Ltn.	FA FK Schwaben (O)	
Mahn	Herbert	Ltn.	FA 418 (P)	Js 72
Mahnke	Alfred	Hpt.	Frw.-Regt. Reinhardt, F'horst Königsberg	FA 14, FA(A) 231
Mai	Josef	Ltn.		Staffel 29/KAGOHL 5, Js 5
Maier	Adolf	Frw.	FA Haas	
Maier	Ernst	Gefr.	FA Haas	

German Flyers In *Freikorps* & Other Postwar Activities				
Last Name	***First Name***	**Rank**	***Freikorps* Unit**	***1914-18 Combat Units***
Maier	Friedrich	Vzfw.	FA Haas	
Maier	Lorenz	Frw.	FA Haas	
Maier	Nikolaus	Lt. d. R.	FA Krug, TFst 21a, FA Miller	Schusta 27
Mailer		Ltn	FA, II. Schützenbataillon, Memmingen (O)	
Majewski	Bruno	Ltn.	FA 417 (CO)	Marine Feld Jasta I
Malsfey	Johann	Frw.	FA Haas	
Mammele	Heinrich Otto	Frw.	FA Haas	
Manteuffel, von	Eitel-Friedrich Roediger	Ltn.	AFst 103	FFA 26, FFA 29, FFA 27, Js 14,Kest 9, Kest 1b, Js 2
Markert	Wilhelm	Lt. d. R.	FA Sorg (CO)	FA(A) 294b
Marschalk		Lt. d. R.	Russian Westarmee	
Marschall von Bieberstein	Wilhelm	Rittm. Frhr.	FA 429, FA 424, Russian Westarmee	Kampfgeschwader I, BOGOHL I/2(CO)
Martin		Ltn.	FA 431(2 O.z.b.V.)	
Martin		Oblt.	FA 429	
Martin	Erwin	Oblt.	FA 424 (CO)	FFA 19
Martin	Hans	Ltn.	FA 424	
Marx	Johannes	Gefr.	FA Haas	
Mascus	Hellmut	Oblt.	FA 402 (O)	FA 71, FA 305
Maucher	Georg		FA Haas (Driver)	
Mauelshagen	-	Ltn.	FA 431	
Mauer	Eugen		FA Haas (Werkmeister-stv.)	FFA 20, FA(A) 284, FA 12
Mauler	Otto	Lt. d. R.	FA FK Schwaben (O)	FF(A) 291b
Mauritius, von		Ltn.	FA 426	
Mauscherning		Lt. d. R.	FA 427 - Staffel Libau (O)	
Mayer		Off.-stv.	FA Haas (II Werkmeister)	
Mayer		Vzflugm.	KG Sachsenberg	Marine Js 3
Mayer	Karl Georg Wilhelm	Frw.	FA Haas	FFA 73
Mecklenburg		Vzfw.	FA 426	
Meder		Gefr.	FA 424	

German Flyers In *Freikorps* & Other Postwar Activities

Last Name	*First Name*	Rank	*Freikorps* Unit	*1914-18 Combat Units*
Meder		Flg.	FA 426	
Medera			FA 426 (P - Abt VII)	
Meidel	Baron		FA 433	
Meinberg		Vzfw.	AFst 103	
Meinecke	Emil	Off. St.	TFst 11	Turk. FA 1, FA 6
Meinhardt	Karl	Frw.	FA Haas	
Meissner	Gustav		FK Oberland	
Meister	Rudolf	Ltn.	FA 420 (O)	FA(A) 244, FA(A) 285
Meller		Ltn.		
Mendele	Paul	Frw.	FA Haas	
Menhofer	Martin	Frw.	FA Haas	Js 5
Mentele	Albert hans	Frw.	FA Haas	
Mern	Firdolin(?)	Uffz.	FA Haas	Js 71
Mertens		Ltn.	FA 424	
Mertscher			FA Grenzschutz Ost	
Merz	Horst	Ltn. z. S.	Grenzschutz Ost	FFA 36, 2. Staffel, Marine JG II
Mesch	Christian	Vzfw.	FA 426 / TFst 38 (P-Abt VIII)	Js 26, Jg III
Meser		Ltn.	FA 426	
Messinger		Ltn.	118 Brigade Nachrichten Abt, Ferner Abt. 301(O?)	
Messow		Oblt.	FA 404 (CO)	
Metzger	Hans	Lt. d. R.	FA Haas	
Metzger		Vzfw.	Grenzjäger FA 1 (P)	
Meuer	Walter	Lt. d. R.	FA 427 (O)	
Meyer		Ltn.	FA 431	
Meyer	Albert	Gefr.	FA Haas	
Meyer	Karl	Oberflugm.		
Miet		Ltn.	FA 426 (P - Abt V)	
Mikulski		Hpt.	FA 412 (CO)	
Milch	Erhard	Hpt.	FA 412 (CO), FA 429 (O),Fliegersonderstaffel des Kofl XVII. Armeekorps (CO), FA 412 / AFst 117 (CO), Polizei-Fst Königsberg (CO)	FA(A) 204, FA(A) 205, FA 5, Jagdgruppe 6
Milkovsky			FA 424	
Miram		Kornet	FA 433	

German Flyers In *Freikorps* & Other Postwar Activities

Last Name	*First Name*	Rank	*Freikorps* Unit	*1914-18 Combat Units*
Missfelder	Johannes	Hpt.	FA 408 (CO)	FFA 35, Schlasta 12, Schlagru D (CO)
Möbius			Grenzjäger FA 1 (Radio operator)	
Moehn	-	-	FA 421(CO)	
Möhrle	Karl	Frw.	FA Haas	
Möller		Uffz.	FA 424, 3.PADiv	
Monnington	Kurt	Ltn. d. R.	FA Haas (P)	FA 62, Js 15, Js 18
Moosmayer		Hpt.	FA 407 (CO)	
Morzik	Friedrich-Wilhelm	Offz.-stv.	FA 418 / TFstl 27 (P), TFst 6 (P)	FFA 17, FA 300, Js 26, Kest 5, Kest 8
Muchinow			FA 423 (P)	
Muggenthaler	Hermann	Oblt.	FA Dessloch, TFst 21b (P)	FA(A) 286, Schusta 30, FA 46b
Mühlich		Sergt.	AFst 101	
Mühlig-Hofmann	Albert	Hpt.	FA 414 / AFst 102 (CO)	FFA 17, FFA 38, FA(A) 21
Müller		Off -stv	FA 426	
Müller	Albert	Frw.	FA Haas	FA(A) 228
Müller	Christian	Frw.	FA Haas	
Müller	Gottlob	Ltn.	FA A2, AFst 121 (P)	
Müller	Robert	Gefr.	KG Sachsenberg (?)	
Müller			FA 423 (P)	
Müller		Ltn.	FA 424 (P)	
Müller-Arles	Aribert	Hpt.	Grenzjäger FA 1 (O), Flugabt.F'horst Grossenhain (CO)	FFA 23
Müller-Kahle	Albert	Oblt.	FA 409 (P, CO), Fliegerhorst Stolp	
Muller		Ltn.	FA 412 (O)	
Muller			FA 423 (P)	
Münch, Frhr. von		Oblt.	TFst 37, 2. PADet	
Münch, von		Oblt.	FA 424	
Murizk			AFst 121 (P)	
Müthel			FA 433	
Naase		Uffz.	FA 423	
Nafe	Max	Frw.	FA Haas	
Nagel	Karl	Frw.	FA Haas	

German Flyers In *Freikorps* & Other Postwar Activities				
Last Name	***First Name***	**Rank**	***Freikorps* Unit**	***1914-18 Combat Units***
Nägele	Max Emil	Frw.	FA Haas	
Näther	Max	Lt. d. R.	FA 415 (CO)	Js 62 (CO)
Neetzel		Ltn.	FA 424, 3.PADiv	
Neger	Alfred		FA 426	
Neidhardt	Wilhelm	Frw.	FA Haas	
Neitsch		Vzfw.	Grenzjäger FA 1	
Neubauer		Ltn.	FA 426 (Abt VI)	
Neyman			FA 424	
Niebling	Wilhelm	Frw.	FA Haas	FA(A) 216
Niederer	Karl	Gefr.	FA Haas	
Niedmayer	Friedrich Johann	Frw.	FA Haas	
Niemann	Robert	Vzfw.		Fl. Abt. (A)250
Niethammer	Werner	Ltn.	FA 418 (P)	Js 13
Noak	Karl	Vzfw.	FA 426 (P)	
Noltenius	Friedrich	Ltn.		Fl. Abt. (A)234, Js 27, 6, 11
Nosik	Adolf		FA 426	
Nülle	Moritz	Flugzeug-matrosen	FA 415, TFst 2, AFst 115 (P)	2. Marine Feld Jasta
Obel		Ltn.	FA 426 (Abt V)	
Obermeier	-	Lt. d. R.	HFP 302	
Ochs	Albrecht	Vzfw.		Bogohl 4/20, 7/22
Odebrett	Karl	Ltn.		FA 47, FA(A) 215, Js 16, 42(CO)
Oelsnitz, von der		Hpt.	FA 402 (CO)	
Oertzen, von	Jasper?	Hpt.	Kofl X	
Oheimb, von		Ltn.	FA Grenzschutz Ost & Ostpr	FA(A) 239
Ompteda, von	Ernst-Albrecht Frhr.	Ltn.	FA 417 (O)	Fl. Abt. 3, 33
Opel		Vzfw.	Grenzjäger FA 1 (P)	
Ossig	Richard	Offz.-stv.	FA 403/TFst 6 (P)	FFA 26, FFA 63, FFA 29
Osterkamp	Theo	Lt. d. R	FA 413 (CO)	Marine LFJ I,II (CO)
Osterroht, von		Ltn.	FA 426/TFst 38 (O.z.b.V.), 2. PADiv	
Ostrop	Paul		FA 426 (Staff Quarters)	
Otto		Gefr.	FA 424, 3.PADiv	

German Flyers In *Freikorps* & Other Postwar Activities

Last Name	*First Name*	Rank	*Freikorps* Unit	*1914-18 Combat Units*
Ozinek			FA 423 (P)	
Paech		Vfw.	FA 426/TFst 38, 2. PADiv (Werkmeister)	
Pallmer		Flg.	FA 424	
Palmer	Hermann	Hpt.	FA 415 (CO)	
Pape	Gerhard	Ltn.	FA 423 (O)	FA(A) 226
Passelot			AFst 121 (P)	
Patze	Friedrich	Oblt. z. S.	KG Sachsenberg	
Pätzhold			FA 423 (Werkmeister)	
Pausser	Otto	Ltn.	FA 416	
Peckmann	Werner	Lt. d. R.	TFst 15	FFA 65, KEK A, JS 14, 15, 31, KEST 3, JS 9
Pelzer		Off-Stv.	FA 431 (Werkmeister)	
Peterch	-	Major	FA 431(?)	
Peterkowsky, von		Ltn.	Grenzjäger FA 1 (O)	
Petersen		Vzfw.	FA 415	
Pfänder	Albert	Frw.	FA Haas	
Pfeifer	Karl		FA 426	
Pfeiffer	Alfred	Lt. d. R	FA 434	
Pfeiffer	Herbert	Ltn.	FA 421 (P)	FFA 226
Pfeil	Friedrich Jakob	Frw.	FA Haas (Radio operator)	
Pfister	Otto	Frw.	FA Haas	
Pflüger	Hermann	Frw.	FA Haas	
Pietsch	Arthur	Luftsch.	FLA 101 (Photographer)	
Plehn, von		Oblt.	FA 424, 3.PADiv	
Polla	Eugen Johann	Frw.	FA Haas	
Polte			KG Sachsenberg	
Preis		Vzfw.	FA 413 (P)	
Preusch			AFst 121 (P)	
Preuss	Werner	Lt. d. R.	FA Freikorps Schleswig-Holstein	Js 66 (CO)
Prim	Karl	Uffz.	FA FK Schwaben (P)	Schlasta 27b, Js 32b
Prinz		Flg.	FA 417 (P)	
Prüssner		Ltn.	FA 426/TFst 38, 2. PADiv/VI. PADet (O.b.z.V)	

German Flyers In *Freikorps* & Other Postwar Activities				
Last Name	***First Name***	**Rank**	***Freikorps* Unit**	***1914-18 Combat Units***
Qoos	August	Uffz.	FA 431(P), Fliegerhorst Brieg (P)	FFA 30, FA 47, FA 14, FA 13
Quade	Erich	Hpt.	FA 434 (CO)	
Raab	Georg	Uffz.	FA Haas	
Raben	August	Lt. d. R.	Schutztruppe Regt. v. Lettow-Vorbeck, Westrussian Armee	Js 36, 39 (CO),15(CO), 18(CO)
Radlik	Wenzel (Waczlaw)	Uffz.	FA 403 / TFst 6	
Rahm	Gotthilf Max	Frw.	FA Haas	
Rahn	Otto	-	Lithuanian Air Force	
Raikowski	Stanislaus	Gefr.	FA Haas	
Raithel	Johann	Ltn.	Kurierfliegerstation Bamberg (P,Tech. Off.)	FA(A) 295, FA(A)291, Js 34
Rall	Adolf Christian		FA Haas (Mstr. Zahlmeister)	
Rapp	Hermann		FA Haas (Driver)	
Rath	Gottlob Johann	Uffz.	FA Haas	
Rath	Richard Paul	Uffz.	FA haas	
Rattey		Ltn.	FA 412 (P)	
Rätzer		Gefr.	FA 423	
Rauch	Heinrich	Ltn.	FA 410 (O), FA 433 / TFst / FP Kurland1 (O), FP Graf Keller, TFst 38, Abt. VI (O)	Fl. Abt. (A)265, FA(A) 234, FA(A) 260, FA 37
Rauch	Ullrich	Uffz.	TFst 37, 2. PADet	
Rauscher	Albert Jakob	Gefr.	FA Haas	
Reck		Ass. Arzt	FA 426 / TFst 38 (Medical Officer), 2. PADiv	
Rehl		Oblt.		
Rehm	Heinrich Adolf	Frw.	FA Haas	
Reiber	Eugen	Frw.	FA Haas	
Reich			FA 424 (Fitter)	
Reichardt	Karl	Frw.	FA Haas	
Reichert	-	Ltn.	FA 431	
Reicherzer	Michael	Frw.	FA Haas	

German Flyers In *Freikorps* & Other Postwar Activities				
Last Name	***First Name***	**Rank**	***Freikorps* Unit**	***1914-18 Combat Units***
Reichle	Richard	Frw.	FA Haas	
Reihling	Hermann Robert	Frw.	FA Haas	
Rein(l?)		Hpt.	TFst 6 (CO)	
Reinhard		Ltn. Frhr.	TFst 37, 3. PADiv	
Reinhard	Josef	Uffz.	FA Krug, Ffst 21a (P), Fliegerhorst Schleissheim	
Reinhardt	Karl Christian	Frw.	FA Haas	FA 9
Reinhold		Vize-Wachmeister	TFst 37, 3. PADiv	
Reinicke		Hpt.	FA 401 (CO)	
Reisch		Hpt.	FA 404 (CO)	
Reiske			FA 423 (Geometer)	
Rempe	Leonhard	Lt. d. R.	FA 423, AFsl 116 (P)	FA(A)261
Rempfer	Konrad August	frw.	FA Haas	
Renz	Max Otto	Frw.	FA Haas	
Resch	Ergart		FA 424, 3. PADiv	
Reteuch	Reinhold		FA 424	
Reutter	Robert Richard	Frw.	FA Haas (Radio operator)	
Richard	Johann		FA 426, Abt. VI	
Richter	-	Ltn.	FA 418	
Richter		Vzfw.	AFst 101	
Richter		Gefr.	FA 433	
Rieckhoff	Herbert	Ltn.	FA 414 (O), FA. 400 (O)	FA(A) 219
Rieldorf		Ltn. (d.R.?)	Afst 103 (O)	
Rienau	Rudolf	Ltn.	FA 418 (P)	Js 1, 19
Ripberger	Eugen	Sergt.	FA Haas	Js 17
Riteleki		Ltn.	FA 427 (O)	
Rodenberg	Paul		FA 427 (P)	
Rödiger		Lt. d. R.	AFst 101 (Tech. Off.)	
Roehre			FA 430 (OzbV)	
Rohn	Richard	Frw.	FA Haas	
Röhn, von		Ltn.	FA 424	
Rohrmiser		Gefr.	AFst 101	
Römer	Robert	Frw.	FA Haas	

German Flyers In *Freikorps* & Other Postwar Activities

Last Name	*First Name*	Rank	*Freikorps* Unit	*1914-18 Combat Units*
Roosen	Otto	Vzfw.	FA 404 (P), TFst 20 (P)	FA 2
Rosenstein	Willy	Lt. d. R.		FFA 19, Js 9, 27, KEST 1a, 1b, Js 40
Roth	Heinrich	Frw.	FA Haas	
Rothe	Paul	Uffz.	Grenzjäger FA 1, Luftpolizei Dresden (P)	FFA 62, Js 14
Rother	Harry	Oblt.	Lithuanian Air Force	
Rothfelder	Karl Eugen	Frw.	FA Haas	
Rüdenberg	Friedrich	Lt. D. R.	Garde-Kavallerie-Schützenkorps	FA(A) 209, FFA 26, FA(A) 259, Js 10, Js 75
Rüdiger		Ltn.	AFst 101	
Rüdikeit		Gefr.	AFst 101	
Ruff	Alfred	Hpt.	Kofl XIII	
Rühle	Otto	Vzfw.	Grenzjäger FA 1	
Rundel	Anton	Frw.	FA Haas	
Rupprecht			FA 417 (P)	
Rusche	Helmut	Lt.	Art. Fl. Staffel 129	
Ryon	Karl	Ltn.	FA 424	
Sachsenberg	Gotthard	Oblt. z. S.	KG Sachsenberg (CO)	M.Fl.Abt.2,M.Jasta1,M'feldJG(CO)
Salewski	Karl	Lt. d. R.	FA 427 (Flivo - Aviation liaison officer), AFst 101	FA 26
Salewsky	Karl	Ltn.	FA 424	
Sallerau, von		Ltn.	FA 424 (P)	
Säuberlich	Helmuth		FA 427 (P)	
Sawatzki	Horst	Flugmaat	KG Sachsenberg	Kusta 3, Marine Js 1
Schach	Pius	Uffz.	FA Haas	
Schaefer	Erich	Lt. z. S.	FA 416 (P)	
Schaefer	Friedrich	Ltn.	Probstmayer, Bogendörfer or 17. I. D. ?	FA 38, Jasta 10 (Tech. Off.)
Schaefer			FA 424	
Schäfer	Richard	Ltn.	Baltische Landeswehr	
Schaff		Ltn.	FA 426 (P)	
Schaich	Robert		FA Haas (Driver)	
Schäppel		Uffz.	FA Haas (Korporalsschaftsführer IV)	
Scharon	Karl	Vzflugm.	KG Sachsenberg	Marine Js 2

German Flyers In *Freikorps* & Other Postwar Activities				
Last Name	***First Name***	**Rank**	***Freikorps* Unit**	***1914-18 Combat Units***
Schars	Alfred	Flg.	FA 424, 2. PADet	
Schatz	Martin	Oblt.	FA (CO) - Grenzschutz Ost	
Schauer	Ludwig	Hpt.	Flugplatz Kommandant Fl'horst Bamberg (Frw. FA Schauer)	FFA 6, FA(A) 297
Scheel	Friedrich	Lt. d. R.	AFst 103 (P)	
Scheel, von		Ltn.	Grenzjäger FA 1 (P)	
Scheibe	Kurt	Ltn.	FA 407 (CO)	KG1/Ks 3, BG 1.Bs 3
Schele, von		Oblt. Frhr.	FA 426/TFst 38 (Staff Quarters), 2. PADiv	
Schell, von			FA 426	
Schelling	Hermann Christian	Sergt.	FA Haas	
Schiemann	Fritz	Vzfw.	FA 401	
Schinzing	Alfred	Hpt.	Fliegerhorst Freiburg (CO)	FFA 4, FFA 47(FA 6)-CO
Schirmer	Otto	Frw.	FA Haas	FA(A) 206, Js 75
Schleich, von	Eduard Ritter	Hpt.	Fliegerstaffel der Polizeiwehr Bayern	FA 2, SS 28, Js 21(CO),32b(CO)
Schleiff	Franz	Oblt.	Kampfgeschwader von Bredow	FA 54,300, Js 41, Js 56(CO)
Schlekweis	Friedrich	Frw.	FA Haas	
Schlenker	Jakob		FA Haas (Driver)	
Schleswig	Adolf		FA 425	
Schlicker		Uffz.	FA 424, 3. PADiv	
Schlieben, von	Richard	Hpt.	FA Schlieben (CO), FA 409 (CO)	Js 21(CO)
Schmalschläger	Benedikt	Oblt.	FA Schmalschläger (CO), TFst 21a	Fl. Abt. 2b, (A)294(CO), Schl 26b(CO)
Schmecker		Gefr.	FA 424	
Schmid	Gotthold Johannes	Frw.	FA Haas	FA(A) 224
Schmid	Jakob	Sergt.	FA Haas	
Schmid	Karl	Sergt.	FA Haas	
Schmidt		Vzfw.	FA 426 (P)	
Schmidt	August		FA Schauer	Bayr. Inf. Regt. 18
Schmidt	Eugen	Flg.	FA 427	
Schmitt		Ltn.	FA 426 (P, CO - Abt. VIII)	

German Flyers In *Freikorps* & Other Postwar Activities				
Last Name	***First Name***	**Rank**	***Freikorps* Unit**	***1914-18 Combat Units***
Schmückle	Karl	Off.-stv.	FA Haas	Kagohl I/18, Schusta 1, FA(A) 254, Js 21 JG II/Js 15
Schnäble		Vzfw.	KFSta Würzburg (G)	
Schnöppel	Otto	Uffz.	FA Haas	
Schnotz	Otto		FA Haas	
Schoenebeck,von	Karl August	Ltn.	FA 424 (P)	Fl. Abt. (A)203, Js 11, 59, 33(CO)
Schoy	Joseph	Frw.	FA Haas	
Schramm			Grenzjäger FA 1 (Gunner)	
Schrank	Erhard	Frw.	FA Haas	
Schranz	Wilhelm	Frw.	FA Haas	
Schreiber		Flg.	FA 423	
Schreyer	Bruno	Sergt.	FA 426	
Schröder	Felix	Flg.	TFst 37, 2. PADiv	
Schröder, von	Joachim	Hpt.	FA 422/ TFst 14, Polizei Fliegerstaffel Hamburg (CO)	Bombengeschwader 7 (CO)
Schroeder	Walter	Vzfw.	FA 404/TFst 20 (P)	FA 17
Schroll		Sergt.	FA 423 (Küchen)	
Schroth	Günther	Ltn.	FA 415 (P, Tech. Off.)	FFA 42, Js 19, FA 303, Js 2/ Heeresgruppe F
Schtomprowsky			FA 426 (P - Abt VII)	
Schubart	-	Ltn.	FA 417	
Schubert			FA 426 (P - Abt VI)	
Schubert	Theodor	Hpt.	FA 429 (CO)	Fest.-FA 4, FA 36, Kagohl 38/7, Schusta 14
Schüssler		Ltn.	KG Sachsenberg	
Schuldes	Erich		FA 426, Abt. VI	
Schultz	Fritz	Oblt.	FA 425, Lithuanian Air Force	-
Schulz	Ferdinand	Lt. d. R	Grenzschutz Ost	Schlasta 10 (P)
Schulz	Julius	Hpt.	FA 409 (CO), Iron Division	FFA 5 (O), Fliegerschue Krefeld, Fliegerhorst Insterberg 12/18-1/19
Schulz	Paul Josef	Feldw.	FA Haas	
Schumm	Werner	Vzfw.	FA 417 (P)	
Schupan		Ltn.	FA 412 (O)	
Schurr	Josef		FA Haas (Driver)	
Schuster		Ltn.	FA 427	

German Flyers In *Freikorps* & Other Postwar Activities				
Last Name	***First Name***	**Rank**	***Freikorps* Unit**	***1914-18 Combat Units***
Schuster		Sergt.	FA 423	
Schuz	Klaus Willi Julius	Sergt.	FA Haas (Korporats-schaftsführer)	
Schwab	Karl		FA Haas (Driver)	
Schwabedissen	Walter	Oblt.	FA 432	FA 28, FA 20, FA 3 (Lb) (O)
Schwamm			FA 423 (P)	
Schwarz		Uffz.	FA 413 (P)	
Schwarzkopf		Ltn..	FA 426/TFst 38, 2. PADiv (Tech Off),	
Schwartzkopff	Günther	Ltn.	FA 406 (Tech. Off.)	FFA 4, FA 13
Schwenk	Wilhelm	Uffz.	FA Haas (Korporalsschaftsführer)	Schlasta 1
Schwenzner		Gefr.	FA 424	
Seebald	Josef	Ltn.	FA 411 (P)	FA(A) 247, Js 21s
Seidel	Hans Joachim	Lt. d. R.	FA 410 (O)	
Seiffert				FA(A) 227
Seifried	Albert	Frw.	FA Haas (Driver)	
Seitz	Friedrich Paul	Uffz.	FA Haas (Korporatsschaftsführer II)	FA 9
Selzer		Ltn.	FA 426 (Tech. Offz. - Abt VI	
Seyfert	-	Vzfw.	Iron Division	FA 21
Seywald	Heinrich	Ltn.	FA A2 (P)	Kasta 33, FA 274, FA 48b, FA 1b, Js 23
Sieberg			FA 426	
Silberzahn	Adam Christian	Frw.	FA Haas	
Sillich, von		Hpt.	FA 406 (CO)	
Skribanowitz		Feldmeister	FA 433	
Sölter	Gustav	Uffz.	FA 434	FA(A) 250
Sommer	Otto		KG Sachsenberg	
Sorg	Max	Hpt	FA Kitzingen (CO), Korps von Möhl (OzbV)	FFA 1b
Spang	Willibald	Hpt.	FA Haas (P, Grufl. Gruppe West)	FFA 20, FFA 5, Kagohl 3, FA(A) 225, FA(A) 230
Speidel	Hermann	Frw.	FA Haas	
Sperrle	Hugo	Hpt.	FA des Freikorps Lüttwitz (Kofl), Rw Gruppenkdo	FFA 4, FFA 42, FFA 60, FA(A) 249

German Flyers In *Freikorps* & Other Postwar Activities

Last Name	*First Name*	Rank	*Freikorps* Unit	*1914-18 Combat Units*
Spiegel	Leonhard	Frw.	FA Haas	
Spille	Karl	Flieger	FA 415	
Spobert	Otto	Off. St.	FA 415	Fl. Abt.7, KEST 8, Js 27, 62
Stähle	Wilhelm	Frw.	FA Haas	
Stahr		Hpt.	Kofl AOK Sud	
Stams	Walter			
Stapel	Erich	Luftsch.	FLA 101 (Radioman)	
Stark	Rudolf		Polizei-Fliegerstaffell1 Schleissheim	FA(A) 296, Js 34b, 77b, 35Bb
Starke	Friedrich	Hpt.	Landesjägerkorps Maercker	FFA 21, FA(A) 275, Bgohl 4,
Steffen		Hpt.	8. A. K.	
Stehr		Ltn.	FA 426 (P - Abt V)	
Stein	Erich?	Hpt.	Kofl III	
Steiner	Franz	Vzfw.	FA Krug	
Steiner	Friedrich	Uffz.	FA Haas	
Steinkopf-Hartig	Ernst	Hpt.	Kofl XIV	
Steinmann	Heinrich		FA 400/FA A z. V. (P)	
Steinmann			FA 426/TFst 38, 2. PADiv (Werkmeister)	
Steinnagel		Gefr.	AFst 101	
Stengele	Joseph	Frw.	FA Haas	
Stephan		Ltn.	FA 426/TFst 38, 2. PADiv/8. PADet (CO)	
Stephan		Uffz.	FA 423	
Stephanski			AFst 121 (P)	
Steppat	Otto	Off -stv	FA 408 (O)	
Stepphuhn	Hugo	Ltn.	FA 431	
Sterz		Ltn.	FA 426 (P - Abt V)	
Stihl		Gefr.	FA 424	
Stör	Wilhelm	Vzfw.	FA 426 (P)	
Stolf		Ltn.	FA 426/TFst 38 (Tech Offz), 2. PADiv (Werkmeister)	
Stracke	Hermann	Vzfw.	FA 426	
Streccius	Alfred	Major	Kofl AOK Nord	FFA 13 (CO), Kofl 9, KG 7 (CO)
Strecker	Karl	Uffz.	FA Haas (Driver)	

German Flyers In *Freikorps* & Other Postwar Activities				
Last Name	***First Name***	**Rank**	***Freikorps* Unit**	***1914-18 Combat Units***
Stremple	Kurt	Vzfw	FA 416?	Fl. Abt. 1 (probably)
Strobel	Karl	Frw.	FA Haas (Driver)	
Strobel	Lorenz Gustav	Frw.		
Svezyg?	-	Major	Lithuanian Air Force, FA 425?	
Taubes		Kapitan	FA 413	
Tautz		Hpt.	FA 408 (CO)	
Teubern, von	Walther	Hpt.	Kofl XII	
Theurer	Karl	Frw.	FA Haas	
Thieme	Wolf	Hpt.	Kofl II	
Thiemicken	Max	Flieger	FA 426	
Thoene	Willy	Lt. z. S.	Grenzschutz Ost	Marine JG Flandern
Thom	Karl	Ltn.	FA 411	FA(A) 216, FA 48, FA(A) 234, Js 21s
Thuy	Emil	Lt. d. R.		FA 53, Js 21s, 28(CO)
Tillmann		Flg.	FA 427	
Tirkschleid		Vzfw.	FA 411 (C-P)	FA(A) 235, FA(A) 211, FA(A) 206
Tödheide	Fritz	Vzfw.	FA 417 (P)	
Tolle	Hermann		FA Schlieben (P)	Schlachtstaffel 19
Töpert	Günther	Gefr.	AFst 101	
Treuhaupt		Ltn.	FA 418 (O)	
Triendl	Theodor	Oblt.	AFst 121 (P)	FA 6, Kasta 31, Schlasta 26, FA(A) 287
Tsikmer	Emil	Ltn.	FA 424	
Tull	Franz		FA 426	
Turban	Karl		Kdo der Flieger-Bomben Versuchs Abt, 3 Kurlandisches Inf Regt. der Eisernen Division (P)	
Uebelmesser	Eugen	Frw.	FA Haas	
Uhler	Alfons Karl	Frw.	FA Haas	
Veith	Maximilian	Ltn.	Abteilung Lütz, Iron Division	
Veltjens	Josef	Lt. d. R.	Freikorps Lüttwitz, Freikorps Gerstenberg	Fl.Abt.23, Js14,15(CO),JG II(CO-stv)
Verheyen		Sergt.	AFst 101	
Vielstich	Fritz			FA 305
Viereck			FA 423 (P)	
Vilgis		Flg.	FA 424	

German Flyers In *Freikorps* & Other Postwar Activities

Last Name	*First Name*	Rank	*Freikorps* Unit	*1914-18 Combat Units*
Vinz		Offz.-stv.	FA 426 (Paymaster),TFst 38, 2. PADiv	
Vogel	Anton Benedict	Gefr.	FA Haas	
Vogt		Feldw.	FA 423	
Vogt		Major	FA 431(?)	
Voigt		Gefr.	FA 424	
Volck	Herbert	Ltn.	Lüneberg-Volk Freikorps	FFA 31?
Volkmann	Hellmuth	Hpt.	Garde-Kavallerie-Schützenkorps	FFA 25, FFA Metz, FFA 71, Js 6, Js 10, FA(A) 211
Wacker	Karl	Frw.	FA Haas	
Wadehn	Walther	Ltn.	KG Sachsenberg (O)	FA(A) 272
Wagner		Hpt.	FA 429 (CO)	
Wagner	Karl	Frw.	FA Haas	
Wahl	Eugen Michael	Frw.	FA Haas	
Wahl	Fritz Louis	Gefr.	FA Haas	II. SFA, FA Borkum, I. SFA
Wahl	Irwin	Gefr.	FA Haas	
Waldbauer	Wilhelm	Uffz.	FA Haas	
Walter	Gustav	Hpt.	Kofl V	FFA 26 (CO), Stofl. 2, Kofl. 2
Walter	Karl		FA Haas (Driver)	Kagohl Staffel 2
Walter	Max	FlgObmatr.	6. R.K.	
Walter	Wilhelm	Frw.	FA Haas	
Waltgerr		Off-Stv.	FA 427	
Walz	Franz Josef	Hpt.	Fliegerhorst Schleissheim, Polizei-Fliegerstaffel Kitzingen	BrAbt C,Kasta 2 (CO),Kg 1 Js 19(CO),2(CO),34b(CO),FA 304b(CO)
Walz	Lorenz	Frw.	FA Haas	
Walz	Paul	Frw.	FA Haas	
Wappenhans	Waldemar	Ltn.	Silesia	FA 305 , Schlasta 27 (CO)
Warther	Willy	Luftsch.	FLA 101 (Radioman)	
Weber	Emil		FA 426	
Weber	Vitalis	Gefr.	FA Haas	
Wedge	-	Oberstlt.	FA 431(?)	
Weiner	Georg	Ltn.	AFst Grossenhain (P)	Kagohl 38/7, Js 30, Kest 3, Js 3
Weingarth		Hpt.	FA Haas	
Weinhold	Otto		FA 429 (Mechanic)	

German Flyers In *Freikorps* & Other Postwar Activities

Last Name	*First Name*	Rank	*Freikorps* Unit	*1914-18 Combat Units*
Weinschenck	Hellmuth	Oblt.	FA 426 (CO), Westrussian Armee	Js 18, 16?
Weinschenk	Alfred	Oblt.	FA 425, Lithuanian Air Force	
Weisheit	Georg		8. PADet - Graf Keller (O)	
Weiss		Lt. d. R.	FA 401 / AFst 129 (O)	
Weiss		Vzfw.	FA 424 (P)	
Weisshaar	Otto	Lt. d. R.	FA Haas (P)	FA(A) 279, Js 65
Weitkus	Paul	Lt. d. R.	FA 432	
Welke		Vzwachm.	TFst. 37, 2. PADet.	
Welkoborsky	Norbert		FA 423 (P)	FA(A) 211?
Weller	Hermann	Vzfw.	FA Haas (1 Werkm.-stv.)	FA(A) 269
Wenig	Erwin	Oblt.	TFst 21b (P)	FFA 7b, FFA 8b, KEK Ensisheim, Js 2, 76 (P), Js 80 (CO)
Wenzl	Richard	Lt. d. R.	FK von Epp	Js 11, Js 6
Wenzler	Otto	Frw.	FA Haas	
Werner	Erich	Hpt.	FA 101 (CO)	
Westphal	-	Hpt.	FA 431	
Wichmann		Gefr.	FA 424	
Wiedemann		Feldw.	FA 418	
Wiedemann	Karl	Flg.	TFst 37, 2. PADet	
Wiedennauer	Christian	Frw.	FA Haas	FA 112
Wieland	Günther	Ltn.	FA 428 (O)	FA(A) 273
Wildgans		Uffz.	KFSta Würzburg (Mech)	
Wilhelm	Erwin		FA 424	
Wilhelmi		Oblt.	Grenzjäger FA 1 (O)	
Wilke	Felix	Gefr.	FA 425 (?)	
Wimmer	Max	Vzfw.	FA Krug, TFst 21a	
Winckler	Paul	Hpt.	FA 411 (CO)	FFA 43, Schlasta 10 (CO), FA(A) 217(Stv. CO)
Winkler		Gefr.	FA 424	
Winter	-	Lt. (z. S.?)	FA 417 (O)	
Winterfeld, von	Alexander	Ltn.	FA 429 (P), TFst 37, 3. PADiv (P)	Js 20
Wippermann	Karl	Ltn.d. Marine	FA 403	
Wirth		Gefr.	FA 424	

German Flyers In *Freikorps* & Other Postwar Activities				
Last Name	***First Name***	**Rank**	***Freikorps* Unit**	***1914-18 Combat Units***
Wittemayer		Lt. d. R.	FA 427 (O)	
Wörmer	Wilhelm	Frw.	FA Haas	
Wolf		Gefr.	FA 433 (P)	
Wrensch			FA 423 (P)	
Wurner	Albert	Frw.	FA Haas	
Zaak		Ltn.	FA 426 (P - Abt VI)	
Zaiosa	Ernst	Gefr.	FA Haas	
Zalevsky		Ltn.	Fa 427	
Zastrow, von	Theobald	Ltn.	FA 420	FFA 31
Zech	Otto	Hpt.	Landesjägerkorps Maercker	FFA 9, Kagohl 4
Zell	Georg		FK von Epp	Bogohl 4/19 (P)
Zeller	Karl	Uffz.	FA Haas	
Zenzes	Alexandre	Vzflugm.	KG Sachsenberg	Marine Js 2
Zernik		Ltn.	FA 409 (CO)	
Ziefle	Johann	Frw.	FA Haas	
Ziegler	Jakob		FA Haas (Driver)	
Zielinsky	Lorwnz	Frw.	FA Haas (Driver)	
Zikmer	Emil		FA 424	
Zschochlitz		Uffz.	FA 423	
Zühlke	Bruno	Hpt.	Kofl VI	
Zunke	Arnold	Vzfw.	FA 415 (P)	

Appendix 2

Freikorps/Provisional *Reichswehr* Flying Unit Profiles

This appendix contains Profiles for all the flying detachments known to have operated during the *Freikorps* & Provisional *Reichswehr* era. These Profiles are capsule summaries of information about each unit's personnel, bases, aircraft types, key dates and reporting relationships, in addition to details like uniform insignia, document stamps and *Feldpost* numbers. Regrettably, very little information about some units has come to light; consequently, their Profiles are relatively sketchy -- additional information is always welcome.

All dates are presented in military style, i.e.: day, month, year.

A date preceded by a right arrow (>) should be interpreted as "before or by this date".

A date followed by a right arrow (>) means "on and from this date forward".

A date not preceded or followed by an arrow means "on this date", but possibly earlier and/or later, as well.

A double right arrow (>>) following the date means "at some unknown time following this date".

Unit Names. The name(s) listed at the top of the page are, as far as known, their "official" names; i.e., the names given to them by to their superior authorities. If multiple names appear here, they are placed in chronological order; most of the Prussian and Bavarian *Flieger Abteilungen* were redesignated *Reichswehr Truppen-* or *Artillerie-Fliegerstaffeln* in May or June 1919.

Alternate Names are unofficial ones, usually given to them by unit veterans or historians. These often appear in memoirs or biographies. Appendix 3 contains an alphabetized cross-reference of these names.

Uniform Insignia. A few units like *Geschwader* Sachsenberg and *Flieger Abteilung* 424 had their own unique insignia. More often, though, they shared the sleeve or collar insignia of their parent unit. In those cases, the name of the parent unit is placed in parentheses below the illustration of the insignia.

Unit Stamp. Facsimile of the stamp used on the unit's documents.

***Feldpost* Numbers**, where known, are listed. As a rule, aviation units were covered by the same *Feldpost* number as their parent unit. Except for *Grenzschutz Ost* detachments, units stationed within Germany's borders usually did not have *Feldpost* numbers, as their permanent bases were served by the *Reichspost*.

Date established. The date orders authorizing the unit were issued.

Location established. The base, usually a *Flieger Ersatz Abteilung* or *Fliegerschule*, where the unit was assembled and equipped. Some units remained at these bases for combat duty; others were sent elsewhere.

Founding Commander. The officer placed in charge at the unit's formation. Some of these commanders appear to be placeholders until a flying officer could assume command.

Reported to. All aviation units reported to a tactical ground unit, and some, more than one during their existence. In those cases, the date of handover from one command to the other is noted, if known.

Bases. Air bases known to have been occupied by the unit, in chronological order. If known, dates of arrival at those bases are listed in parentheses after the base name.

Disbanded. Date the unit was to be wound down, along with the *Idflieg* order or memo authorizing it. Delays in ratification of the Versailles Treaty allowed this date to be postponed, and many units continued to function for many months after their planned official demobilization date.

Commanding Officers. Officers known to have led the unit, in chronological order. If known, dates these officers assumed command are added in parentheses after their name. *Stv.*, an abbreviation for *stellvertreter* (acting, or literally, standing in place) indicates the officer held temporary command in the unit leader's absence.

Personnel. Personnel names have been obtained from many sources. Although they served with the unit at some time in its existence, starting and ending service dates are not included because, in many cases, they're simply not known. Some names of personnel who joined the West Russian Army are probably phonetic misspellings; they were obtained from Russian documents after they had already been transliterated into the Russian language.

Where known, personnel functions are presented in abbreviated form after their names:

CO - Flying Detachment Commanding Officer (*Flieger Abteilungsführer* or *Fliegerstaffelführer*)
OzbV - *Offizier zur besonderer Verwendung* (Adjutant)
P - Pilot (*Flugzeugführer*)
O - Observer (*Beobachter*)
G - Aerial gunner *(Bordschützen)*
Tech. Off. - *Technische Offizier* (Technical Officer)
Flivo -- *Fliegerverbindungs Offizier* (Aviation Liaison Officer)
Abteilungsführer - Detachment or Department Commander. Not to be confused with *Flieger Abteilungsführer* or *Fliegerstaffelführer* (Flying Detachment Commanding Officer). This refers to the leader of a component department of the *Abteilung*, like ordinance, finance or transportation. It may also apply to a sub-section of the flying troops.
Artzt - Medical Officer
Werkmeister -- Equipment Officer
Zahlmeister -- Paymaster

Other functions, such as mechanic or driver, are specified if they are known.

Aircraft. As with any military unit, aircraft inventories changed in number and type over time. All aircraft types known to have been used by the unit during its existence are listed, along with their quantities and military serials, if known. Not all these aircraft may have been on strength at the same time.

A/C Unit Marking. Unlike the Great War, when colorful, unique German aircraft markings abounded, most of the aircraft serving during the *Freikorps*/ Provisional *Reichswehr* period bore standard factory colors without embellishments. A few units did, however, apply a distinctive unit marking to their machines, and some pilots also added their own personal markings. In addition, aircraft gleaned from flying schools and disbanded units often brought with them markings from their previous service.

Above: Another photo of the Fokker D.VII (OAW) of *FA* 418/ *Truppen-Fliegerstaffel* 27 shown on page 81. (Greg VanWyngarden)

Unit Name: *Grenzschutz Flieger Abteilung 400/Truppen-Fliegerstaffel 8*

Alternate Names: *Flieger Abteilung* A
Feldpost No.:
Successor Unit: *Truppen-Fliegerstaffel* 8

Uniform Insignia **Unit Stamp**

Date Established: 2 December 1918,
per 26991.18 A7L. IV. C., *Idflieg* 407/18 pers. A3
Location Established: *Flieger Ersatz Abteilung* 11, Brieg
Founding Commander: *Ltn.* Werner Junck, Hpt. Paul Backhaus

Reported to: *Genkdo.* VI. *Armee Korps,* Breslau, *Rw.* 8. Brigade, Oppeln
Bases: Oppeln
Disbanded: *Fliegerhorst* Brieg, per *Idflieg Nr.* 1157/8.19 *Abt.* I of 15.8.19

Commanding Officers: *Ltn.* Werner Junck
Personnel known to have served with unit:

Hpt. Paul Backhaus - CO
Oblt. Friedrich-Carl Hanesee - *O.z.b.V.*
Ltn. Werner Junck - CO
Ltn. Herbert Rieckhoff - O
Heinrich Steinmann - P

Aircraft:

Aircraft Unit Marking:

Unit Name: *Grenzschutz Flieger Abteilung 401/Artillerie-Fliegerstaffel 129*

Alternate Names: *Flieger Abteilung* B, *Flieger Abteilung bei der* 117. *I-D.*, Gleiwitz *Grenzschutz Fliegerstaffel*

Uniform Insignia **Unit Stamp**

Feldpost No.:
Successor Unit: *Artillerie-Fliegerstaffel* 129

Date Established: 2 December 1918, per 26991.18 A7L. IV. C., *Idflieg* 407/18 pers. A3
Location Established: *Flieger Ersatz Abteilung* 11, Brieg
Founding Commander: *Ltn.* Hans-Jürgen Horn, Hpt. Reinicke (>23.5.19)

Reported to: *Gkdo.* VI. *Armee Korps*, Breslau, *Rw.* 29 Brigade, Bernstadt
Bases: Gleiwitz
Disbanded: *Fliegerhorst* Brieg, per *Idflieg Nr.* 1157/8.19 *Abt.* I of 15.8.19

Commanding Officers: *Ltn.* Hans-Jürgen Horn
Personnel known to have served with unit: 18 Officers, 140 NCOs & enlisted men

Vzfw. Boleslaw Blank - P
Karl Brämsch
Ltn. Eberhard Fischer - *O.z.b.V.*
Ltn. Hans-Jürgen Horn - CO
Ltn. Hans Jeschonnek - P
Sgt. Kieske - G, O
Vzfw. Küpker - P
Hpt. Reinicke - CO
Ltn. Helmuth Rusche - P
Vzfw. Fritz Schiemann
Lt. d. R. Weiss - O

Aircraft:

Aircraft Unit Marking:

Unit Name: ***Grenzschutz Flieger Abteilung 402/Artillerie-Fliegerstaffel 106***

Alternate Names: *Flieger Abteilung* C
Feldpost No.:
Successor Unit: *Artillerie-Fliegerstaffel* 106

Uniform Insignia **Unit Stamp**

Date Established: 3 January 1919,
per 27559.18 A7L. IV. C., *Idflieg* 34/1.19 Ia
Location Established: *Fliegerschule* Hundsfeld, Breslau
Founding Commander:

Reported to: *Gkdo.* VI. *Armee Korps*, Breslau, *Rw.* 6. Brigade, Breslau
Bases: Brieg
Disbanded: *Fliegerhorst* Brieg, per *Idflieg Nr.* 1157/8.19 *Abt.* I of 15.8.19

Commanding Officers: *Hpt.* von der Oelsnitz
Personnel known to have served with unit:
Oblt. Helmut Mascus - O
Hpt. von der Oelsnitz - CO

Aircraft:

Aircraft Unit Marking:

Unit Name: *Grenzschutz Flieger Abteilung 403/Truppen-Fliegerstaffel 6*

Alternate Names: *Heimatschutz Schlesien Flieger Abteilung* D*

Feldpost No.:

Successor Unit: *Truppen-Fliegerstaffel* 6

Date Established: 12 December 1918, per 27559.18 A7L. IV. C., *Idflieg* 34/1.19 Ia

Location Established: *Fliegerschule* Hundsfeld, Breslau

Founding Commander:

Reported to: *Gkdo.* VI. *Armee Korps,* Breslau, *Rw.* 6. Brigade, Breslau

Bases: Breslau-Klein Gandau, Breslau-Hundsfeld

Disbanded: 27.9.19 or later**

Commanding Officers: *Hpt.* Reisch (19.8 - 27.9.19>)

Personnel known to have served with unit:

Offz-stv. Richard Ossig - P
Uffz. Wenzel (Waczlaw) Radlik
Hpt. Reisch - CO
Ltn. d. Marine Karl Wippermann

Uniform Insignia **Unit Stamps**

Aircraft:

Aircraft Unit Marking:

Notes: **FA* D was already in existence by 6 December 1918.

**Originally scheduled for demobilization at *Fliegerhorst* Brieg, per *Idflieg Nr.* 1157/8.19 *Abt.* I of 15.8.19.

Unit Name: *Grenzschutz Flieger Abteilung 404/Truppen-Fliegerstaffel 20*

Alternate Names: **Uniform Insignia** **Unit Stamp**
Feldpost No.:
Successor Unit: *Truppen-Fliegerstaffel* 20

Date Established: 2 December 1918,
per *K. M. Nr.* Ia 26991.18 A 7 L.IV.C
Location Established: *FliegerErsatz Abteilung 2,* Schneidemühl
Founding Commander: *Oblt.* Messow

Reported to: *Gkdo.* II. *Armee Korps,* Stettin, *Rw.* 20. Brigade, Allenstein
Bases: Schneidemühl, Graudenz
Disbanded: 1920*

Commanding Officers: *Oblt.* Messow
Personnel known to have served with unit: 28 officers & enlisted men, including:
Flieger Max Jeschinsky
Ltn. Hellmut Joeden - O
Oblt. Messow - CO
Vzfw. Otto Roosen - P
Vzfw. Walter Schroeder -P

Aircraft:

Aircraft Unit Marking:

* **Note:** Originally scheduled to be demobilized at *Fliegerhorst* Insterburg, per *Idflieg Nr.* 1203/8.19 *Abt.* I of 15.8.19. It was still active 19.9.19.

Unit Name: *Grenzschutz Flieger Abteilung 405*

Alternate Names: *Aviation-Abteilung*
Feldpost No.:
Successor Unit:

Uniform Insignia **Unit Stamp**

Date Established: 3 January 1919,
per *Idflieg Nr.* 34/1.19 Ia
Location Established: *Fliegerschule* Gandau, Gandau
Founding Commander:

Reported to: *Gkdo.* VI. *Armee Korps*, Breslau
Bases: Gandau, Brieg
Disbanded:

Commanding Officers:
Personnel known to have served with unit:

Aircraft:

Aircraft Unit Marking:

Unit Name: *Grenzschutz Flieger Abteilung 406/Artillerie-Fliegerstaffel 127*

Alternate Names: **Uniform Insignia** **Unit Stamp**
Feldpost No.:
Successor Unit: *Truppen-Fliegerstaffel* 127

Date Established: 15 November 1918,
per *Idflieg Nr.* 407/18 A2 *pers.*
Location Established: *Bomben-Lehr-Abteilung,* Frankfurt a. O.
Founding Commander: *Ltn.* Krieger

Reported to: *Gkdo.* V. *Armee Korps,* Stettin, *Rw.* 27. Brigade, Frankfurt a. O.
Bases: Glogau, Meseritz
Disbanded: *Fliegerhorst* Neuruppin, per *Idflieg Nr.* 1203/8.19 *Abt.* I of 15.8.19

Commanding Officers: *Ltn.* Krieger, *Hpt.* von Sillich, *Lt. d. R.* Friedrich Leesemann (15.1.19)
Personnel known to have served with unit:

Fritz Kalbrier - P
Ltn. Krieger - CO
Lt. d. R. Friedrich Leesemann - *Staffelführer*
Ltn. Günther Schwarzkopff - Tech. Off.
Hpt. von Sillich - *CO*

Aircraft:

4 DFW C.V
Fokker D.VII
Halberstadt CL.II
Halberstadt CL.IV (Rol) 9494/18
Junkers CL.I
LVG C.VI
Roland D.VIb 1535/18
SSW D.IV

Aircraft Unit Marking:

Unit Name: *Grenzschutz Flieger Abteilung 407/Truppen-Fliegerstaffel 5*

Alternate Names: *Freiwillige Flieger Abteilung* **Lissa** **Uniform Insignia** **Unit Stamp**
Feldpost No.:
Successor Unit: *Truppen-Fliegerstaffel* 5

Date Established: 15 November 1918,
per *Idflieg Nr.* 407/18 A3 *pers.*
Location Established: *Bomben-Lehr-Abteilung,* Frankfurt a. O.
Founding Commander: *Ltn.* Kurt Scheibe

Reported to: *Gkdo.* V. *Armee Korps,* Glogau, Rw. 5. Brigade, Lissa (?)
Bases: Lissa
Disbanded: *Fliegerhorst* Döberitz, per *Idflieg Nr.* 1203/8.19 *Abt.* I of 15.8.19

Commanding Officers: *Ltn.* Kurt Scheibe, *Hpt.* Carl Fink (16.1.19), *Hpt.* Moosmayer
Personnel known to have served with unit:
- *Hpt.* Carl Fink- CO
- Ltn. Martin Hähnchen - P
- Kuttlau
- *Hpt.* Moosmayer - CO
- *Ltn.* Kurt Scheibe - CO

Aircraft:

Aircraft Unit Marking:

Unit Name: *Grenzschutz Flieger Abteilung 408/Truppen-Fliegerstaffel 30*

Alternate Names: **Uniform Insignia** **Unit Stamp**
Feldpost No.: 3072
Successor Unit: *Truppen-Fliegerstaffel* 30

Date Established: 15 November 1918,
per *Idflieg Nr.* 407/18 *pers.* A3
Location Established: *Flieger-Beobachter-Schule,* Königsberg
Founding Commander: *Hpt.* Tautz

Reported to: *I. Armee Korps, G.K.Z.b.V. 52* - Brigade Schaulen, *Freikorps* Diebitsch (26 Jul. - Nov. 19)
Bases: Tauroggen, Kielmy (Jan. 19), Schaulen/Ginkuniai (Mar. 19), Kielmy (Oct. 19), Königsberg (4 November 19)
Disbanded: 4 November 19 or later* *Fliegerhorst* Insterburg, per *Idflieg Nr.* 1203/8.19 *Abt.* I of 28 August 19

Commanding Officers: *Hpt.* Tautz, *Hpt.* Johannes Missfelder (1 Jan. - 18 July 19), *Hpt.* Friedrich Köckler (19 July - November 19)

Personnel known to have served with unit:
Hpt. Friedrich-Wilhelm Köckler - CO
Hpt. Johannes Missfelder - CO
Off-stv. Otto Steppat - O
Hpt. Tautz - CO

Aircraft: 4 aircraft on strength beginning March 1919.

Aircraft Unit Marking:

***Note:** Originally scheduled for demobilization at *Fliegerhorst* Insterburg, per *Idflieg Nr.* 1203/8.19 *Abt.* I of 28 August 19

Unit Name: *Grenzschutz Flieger Abteilung 409/Schlacht-Fliegerstaffel 201*

Alternate Names:
Feldpost No.: 3072
Successor Unit: *Schlacht-Fliegerstaffel* 201

Uniform Insignia **Unit Stamp**

Date Established: 15 November 1918,
per *Idflieg Nr.* 407/18 *pers.* A3
Location Established: *Flieger-Beobachter-Schule,* Königsberg
Founding Commander: *Ltn.* Zernik

Reported to: *I. Armee Korps, G.K.Z.b.V. 52; Gruppe Deetjen/Korps Oven* (25 April 19)
Bases: Tauroggen, Kielmy (Jan. 19), Schaulen/Ginkuniai (Mar. 19), Schleissheim (25 April 19)
Graudenz (8 May 19)
Disbanded: *Fliegerhorst* Jüterbog, per *Idflieg Nr.* 1203/8.19 *Abt.* I of 28 August 19

Commanding Officers: *Ltn.* Zernik, *Hpt.* Julius Schulz (Jan. 19), *Oblt.* Richard von Schlieben (5 May 19), *Oblt.* Albert Müller-Kahle (9 September - 12 November 19)

Personnel known to have served with unit:
Ltn. Otto Abernetty - O
Bruno Bogert - *Waffenmeister Hilfer*
Gefr. Kurt Bothke - P
Oblt. Helmut Bühling - O
Vzfw. Otto Fruhner
Ltn. Günther Lohmann - O
Oblt. Albert Müller-Kahle - P, CO
Oblt. Richard von Schlieben - CO
Hpt. Julius Schulz - CO
Ltn. Zernik - CO

Aircraft: 3 aircraft on strength beginning March 1919.
12 May 1919: DFW C.V 6025/17, 17196/17, Fok. D.VII 6823/18

Aircraft Unit Marking:

Unit Name: *Grenzschutz Flieger Abteilung 410/Artillerie-Fliegerstaffel 120*

Alternate Names: *Feld-Flieger-Abteilung* Koch **Uniform Insignia** **Unit Stamp**
Feldpost No.: 3054
Successor Unit: *Artillerie-Fliegerstaffel* 120

Date Established: 15 November 1918,
per *Idflieg Nr.* 407/18 *pers.* A3
Location Established: *Flieger Ersatz Abteilung* 8, Graudenz
Founding Commander: *Hpt.* Koch

Reported to: II. *Armee Korps,* Stettin, *Rw.* 20. Brigade, Allenstein
Bases: Graudenz, Lyck, Allenstein-Deuthen
Disbanded: *Fliegerhorst* Insterberg, May 1920*
Commanding Officers: *Hpt.* Koch, Walter Aschoff

Personnel known to have served with unit:
Walter Aschoff - O, CO
Oblt. Karl Barlen - P
Flieger Feist
Ltn. Adolf Genth - O
Lt. d. R. Johannes Janzen - P
Hpt. Koch - CO
Ltn. Günther Lohmann - O
Ltn. Heinrich Rauch - O
Ltn. d. R. Hans Joachim Seidel - O

Aircraft: Alb. C.III, Alb. C.VII, 3 Fok. D.VII**, including Fok. D.VII (OAW) 8426/18 (?) & 8453/18, Fok. D.VIII, 6 Halb. C.V**, Han. CL.II or CL.III, 3 LVG C.VI, Pfalz D.IIIa 6014/17, 2 Ru. C.I, Ru. C.IV.
**Per Polish intelligence report dated 6.6.1919.

Aircraft Unit Marking:

***Note:** Originally scheduled for demobilization per *Idflieg Nr.* 1203/8.19 *Abt.* I of 28 August 19.

Unit Name: *Grenzschutz Flieger Abteilung 411*

Alternate Names: *Freiwillige FA Winkler*
Feldpost No.: 992
Successor Unit:

Date Established: 18 December 1918,
per *Idflieg Nr.* 121/12.18 Ia
Location Established: *Flieger Ersatz Abteilung* 8, Graudenz
Founding Commander: *Hpt.* Paul Winkler

Uniform Insignia (Collar tabs or upper left sleeve)

(Grenzschutztruppen XVII. A.K.)

Unit Stamp

Reported to: XVII. *Armee Korps,* Danzig
Bases: Graudenz, Freistadt-Langenau (6 June 19)
Disbanded:

Commanding Officers: *Hpt.* Paul Winkler

Personnel known to have served with unit:

Ltn. Kleb - O
Hpt. Friedrich-Wilhelm Köckler - O.z.b.V.
Off-stv. F. Lehmann - C-P
Vzfw. Lunt - C-P
Ltn. Josef Seebald - P
Ltn. Karl Thom - P
Vzfw. Tirkschleid - C-P
Hpt. Paul Winckler - CO

Aircraft: 3 Fok. D.VII, 3 Halb. C.V, 2 Alb. C.I, Alb. C.XII 1911/17, 1 Ru. C.IV

Aircraft Unit Marking:

Unit Name: *Grenzschutz Flieger Abteilung 412/Artillerie-Fliegerstaffel 117*

Alternate Names:
Feldpost No.:
Successor Unit: *Artillerie-Fliegerstaffel* 117

Uniform Insignia (Collar tabs or upper left sleeve) **Unit Stamp**

(Grenzschutztruppen XVII. A.K.)

Date Established: 18 December 1918, per *Idflieg Nr.* 121/12.18 Ia
Location Established: *Flieger Ersatz Abteilung* 8, Graudenz
Founding Commander: *Hpt.* Mikulski

Reported to: XVII. *Armee Korps,* Danzig, *Rw.* 17. Brigade, Danzig
Bases: Graudenz, Thorn, Gut Dietrichsdorf (before 12.3.19 through at least 6.6.19); *Schlachtstaffel*: Graudenz/Sturmback Station (as of 6.6.19)
Disbanded: *Fliegerhorst* Schwerin, per *Idflieg Nr.* 1203/8.19 *Abt.* I of 28.8.19

Commanding Officers: *Hpt.* Mikulski, *Hpt.* Erhard Milch (30.4 - 5.9.19), *Hpt.* Adolf Gutknecht

Personnel known to have served with unit:

Vzfw. Auersbach - P
Hpt. Adolf Gutknecht - CO
Ltn. Holbach - O
Vzfe. Körner - P
Ltn. Joachim Kortüm - P
Ltn. Lippmann - CO, *Schlachtstaffel*
Hpt. Erhard Milch - CO
Hpt. Mikulski - CO
Ltn. Müller - O
Ltn. Rattey - P
Ltn. Schupan - O

Aircraft: 7 Fok. D.VII (May 1919), 3 Halb. CL.IV, incl. CL.IV (Rol) 9401/18 & 9522/18

Aircraft Unit Marking:

Unit Name: *Kampfgeschwader* Sachsenberg (*Flieger Abteilungen* 413, 416 & 417)

Alternate Names: *Gemischtes* [Mixed] *Geschwader Sachsenberg*
Feldpost No.: 2134
Successor Units: (FA 413) -*Truppenfliegerstaffel* 34
(FA 416 & FA 417) - Artillerie-Fliegerstaffel 108

Uniform Insignia (Upper left sleeve)

Unit Stamp

Date Established: January 1919
Location Established: *Flieger Beobachter Schule*, Jüterbog
Founding Commander: *Oblt. z. S.* Gotthard Sachsenberg

Reported to: VI. Reserve *Korps*/1. *Garde Reserve* Division
Bases: Jüterbog, Wainoden (28.2.19), Alt-Auz (18.3.19), Peterfeld (4.19), Weissenhof (23.5.19), Peterfeld (18.6.19), Stettin (30.6.19), Graudenz (7.19), Jüterbog (14.8.19)
Disbanded: Truppenstaffel 34 and Artillerie-Fliegerstaffel were ordered to be demobilized in Jüterbog, per *Armee-Tagsbefehl* 145 of 14 August 1919. Demobilization was officially complete 15 October 1919, but by then, many of the unit's personnel had returned to the Baltic to rejoin *Oblt. zur See* Sachsenberg in a reconstituted *Kampfgeschwader* Sachsenberg, which joined the West Russian Army.

Commanding Officer: *Oblt. z. S.* Gotthard Sachsenberg

Personnel known to have served with unit: 50 officers & 650 NCOs and enlisted men. Those known to have served with FAs 413, 416 & 417 are listed under those units; Geschwader personnel and those whose affiliation is uncertain are listed below.

Lt. z. S. Paul Achilles - P
Vzflugm. Hans Antonious - P
Karl Barnow
Ltn. z. S. Philip Becht - P
Uffz. Bosselmann - P
Oblt. z. S. Karl Cranz - *O.z.b.V.*
Flugmeister Dahlmann
Vzflugm. Engelhardt - P
Eise
Flieger Georg Gehrmann
Vzflugm. Hans Goerth
Uffz. Hahn - P
E. Hoffmann - P
Flugmeister Carl Kuring =- P
Gren. Paul Krampitz
Ltn. Ludwig
Vzflugm. Mayer
Gefr. Robert Müller
Oblt. z. S. Friedrich Patze - P
Flugmaat Horst Sawatzki - P
Ltn. Schüssler
Vzflugm. Karl Scharon - P
Otto Sommer (?)
Ltn. Walther Wadehn - O
Vzflugm. Alexandre Zenses - P
Oblt. z. S. Gotthard Sachsenberg - CO, P

Aircraft: Initially, 70. Additional aircraft continued to arrive in succeeding months. See individual entries for *FA*s 413, 416 & 417.

A/C Unit Marking: See individual entries for *FA*s 413, 416 & 417.

Unit Name: *Kampfgeschwader* Sachsenberg/4. *Plastun* Aviation Division

Alternate Names:
Feldpost No.:
Successor Unit: 4. *Plastun* Aviation Division (7.19)

Date Established: 2 July 1919 or earlier
Location Established: Peterfeld
Founding Commander: *Oblt. z. S.* Gotthard Sachsenberg

Uniform Insignia
(Upper left sleeve)

Unit Stamp

(Staffel Majewski)

Reported to: VI. Reserve *Korps*; *Deutsche* Legion,
West Russian Army (7.19)
Bases: Peterfeld
Disbanded: December 1919

Commanding Officers: *Oblt. z. S.* Gotthard Sachsenberg

Personnel known to have served with unit:
Karl Gripekoven
Ltn. Wilhelm Kauth
Flieger Ernst Kessler
Loewen
Ltn. Bruno Majewski - *Staffelführer, Staffel* Majewski
Müthel
Oberflugmaat Prenzlau
Oblt. z. S. Gotthard Sachsenberg - CO
Flugmaat Schäfer
Feldw. Erich Schaefer - P
Feldmeister Skribanowitz

Aircraft:
Fokker D.VII
Fokker D.VIII
Junkers D.I
SSW D.IV - at least 2

A/C Unit Marking:

Note: After the repatriation of the 1. *Garde* Reserve Division and with it, *Truppen-Fliegerstaffel* 34 and *Artillerie-Fliegerstaffel* 108, to Germany in late July or early August 1919, *Oblt. z. S.* Sachsenberg remained at Peterfeld with a few key personnel. After the demobilization of the original Sachsenberg *Staffeln* in August, many of their personnel returned to Kurland to rejoin their old commander in a reconstituted *Kampfgeschwader* Sachsenberg. Technically an entirely new unit, it joined Colonel Bermondt-Awaloff's West Russian Volunteer Army in July, where it was assigned the name 4. *Plastun* Aviation Division. Like the original, it was comprised of three components :

9. *Plastun* Aviation Detachment, the *Schlachtstaffel,* also known as *Staffel* Majewski after its commander, *Ltn.* Bruno Majewski, former *Staffelführer* of FA 417, and

10. & 11. *Plastun* Aviation Detachments, the fighter and reconnaissance units, mainly populated with personnel from the recently disbanded *Truppen-Fliegerstaffel* 108.

Unit Name: *Flieger Abteilung* 413/*Truppen-Fliegerstaffel* 34

Alternate Names: *Jagdgeschwader des Kampfgeschwader* Sachsenberg
Feldpost No.: 2134
Successor Unit: *Truppen-Fliegerstaffel* 34

Uniform Insignia (Upper left sleeve)

Unit Stamp

Date Established: January 1919
Location Established: Jüterbog
Founding Commander: *Ltn. d. R.* Theo Osterkamp
Ltn. Josef Jacobs (deputy commander)

Reported to: 1. *Garde* Reserve Division; *Reichswehr* Brigade 34 (6.19)
Bases: Jüterbog, Wainoden (28.2.19), Alt-Auz (18.3.19), Peterfeld (4.19), Weissenhof (23.5.19), Peterfeld (18.6.19), Stettin (30.6.19), Graudenz (7.19), Jüterbog (14.8.19)
Disbanded: Jüterbog, per *Armee-Tagsbefehl* 145 of 14 August 1919. Demobilization was officially complete 15 October 1919, but by then, many of the unit's personnel had returned to the Baltic to rejoin *Oblt. zur See* Sachsenberg in a reconstituted *Kampfgeschwader* Sachsenberg.

Commanding Officers: *Ltn. d. R.* Theo Osterkamp

Personnel known to have served with unit:

Lt. z. S. Hagen
Ltn. Josef Jacobs - P, Deputy CO
Ltn. d. R. Theo Osterkamp - CO, P
Vzfw. Preis - P
Uffz. Schwarz
Kapitan Taubes

Aircraft:

Fokker D.VII
Junkers D.I

A/C Unit Marking:

Unit Name: *Grenzschutz Flieger Abteilung 414/Artillerie-Fliegerstaffel 102*

Alternate Names: **Uniform Insignia** **Unit Stamp**
Feldpost No.: 2134
Successor Unit: *Artillerie-Fliegerstaffel* 102

Date Established: 4 January 1919,
per *Idflieg Nr.* 319/12.18 Ia
Location Established: *Flieger Ersatz Abteilung* 13, Bromberg
Founding Commander: *Hpt.* Albert Mühlig-Hoffman

Reported to: II. *Armee Korps*, Stettin, *Rw.* 2. Brigade, Stettin
Bases: Bromberg, Hohensalza, Nackel
Disbanded: *Fliegerhorst* Neuruppin, per *Idflieg Nr.* 1203/8.19 *Abt.* I of 28 August 19

Commanding Officers: *Hpt.* Albert Mühlig-Hoffman

Personnel known to have served with unit:
Ltn. Hans von Freden
Flieger Franz Lashowski
Hpt. Albert Mühlig-Hoffman - CO
Ltn. Herbert Rieckhoff - O

Aircraft:

Aircraft Unit Marking:

Unit Name: *Grenzschutz Flieger Abteilung 415/Truppen-Fliegerstaffel 2*

Alternate Names: *Fliegerstaffel zur Befreiung der Ostmark, Freiwillige Flieger Abteilung Näther*
Feldpost No.:
Successor Unit: *Truppen-Fliegerstaffel* 2

Uniform Insignia

Unit Stamp

Date Established: 8 January 1919, per *Idflieg Nr.* 77/1.19 Ia
Location Established: *Flieger Ersatz Abteilung* 2, Schneidemühl
Founding Commander: *Ltn.* Max Näther

Reported to: II. *Armee Korps,* Stettin, *Rw.* 2. Brigade, Stettin
Bases: Schneidemühl, Kolmar, Wissek, Fitzeri, Margonin, Stolp-West (19.9.19), Schneidemühl (31.12.19), Döberitz
Disbanded: *Fliegerhorst* Döberitz, 5 May 1920 or later
Commanding Officers: *Ltn.* Max Näther, *Hpt.* Hermann Palmer, *Hpt.* Lothar Baas

Personnel known to have served with unit: 7 officers, 15 NCOs, 30 enlisted men

Hpt. Lothar Baas - CO
Lt. d. R. Hans Brzenk - P
Paul Deichmann - O
Hpt. d. R. Kalähne - O.z.b.V.
Lt. d. R. Karl Kerner - O
Ltn. Max Näther - P, CO
Flugmaat Moritz Nülle - P
Hpt. Hermann Palmer - CO
Vzfw. Petersen
Ltn. d. R. Günther Schroth - P, Tech. Off.
Flieger Karl Spille
Off-stv. Otto Spobert
Vzfw. Arnold Zunke - P

Aircraft:

Aircraft Unit Marking:

***Note:** Demobilization ordered per *Idflieg Nr.* 1203/8.19 *Abt.* I of 28 August 19.

Unit Name: *Flieger Abteilung* 416/*Artillerie-Fliegerstaffel* 108

Alternate Names: *Aufklarungs-Staffel des Kampfgeschwader* Sachsenberg
Feldpost No.: 2134
Successor Unit: Merged with *FA* 417 on 25 June 1919 to form *Artillerie-Fliegerstaffel* 108

Date Established: January 1919
Location Established: Jüterbog
Founding Commander: *Ltn.* Eberhard Cranz

Uniform Insignia
(Upper left sleeve)

Unit Stamp

Reported to: 1. *Garde* Reserve Division; *Reichswehr* Brigade 34
Bases: Jüterbog, Wainoden (28.2.19), Alt-Auz (18.3.19), Peterfeld (4.19), Weissenhof (23.5.19), Peterfeld (18.6.19), Stettin (30.6.19), Graudenz (7.19), Jüterbog (14.8.19)
Disbanded: Jüterbog, per *Armee-Tagsbefehl* 145 of 14 August 1919. Demobilization was officially complete 15 October 1919, but by then, many of the unit's personnel had returned to the Baltic to rejoin *Oblt. zur See* Sachsenberg in a reconstituted *Kampfgeschwader* Sachsenberg.

Commanding Officers: *Ltn.* Eberhard Cranz, *Oblt.* Adolf Averdam (25.6.1919)

Personnel known to have served with unit:

Oblt. Adolf Averdam - CO
Ltn. Eberhard Cranz - CO
Flugmeister Dahlmann - P
Lt. z. S. Haagen
Ltn. Otto Pausser
Erich Schaefer - P

Aircraft:

DFW C.V
LVG C.V, including 9037/17
LVG C.VI
Rumpler C - types

A/C Unit Marking:

Unit Name: *Flieger Abteilung* 417/*Artillerie-Fliegerstaffel* 108

Alternate Names: *Schlachtstaffel des Kampfgeschwader* Sachsenberg
Feldpost No.: 2134
Successor Unit: Merged with *FA* 416 on 25 June 1919 to form *Artillerie-Fliegerstaffel* 108

Uniform Insignia (Upper left sleeve)

Unit Stamp

Date Established: January 1919
Location Established: Jüterbog
Founding Commander: *Ltn.* Bruno Majewski

Reported to: 1. *Garde* Reserve Division; *Reichswehr* Brigade 34
Bases: Jüterbog, Wainoden (28.2.19), Alt-Auz (18.3.19), Peterfeld (4.19), Weissenhof (23.5.19), Peterfeld (18.6.19), Stettin (30.6.19), Graudenz (7.19), Jüterbog (14.8.19)
Disbanded: Jüterbog, per *Armee-Tagsbefehl* 145 of 14 August 1919. Demobilization was officially complete 15 October 1919, but by then, many of the unit's personnel had returned to the Baltic to rejoin *Oblt. zur See* Sachsenberg in a reconstituted *Kampfgeschwader* Sachsenberg. *Ltn.* Ernst Kredel, former *Führer* of *FA* 417, succeeded Averdam as commander of *Artillerie-Fliegerstaffel* 108 and was recorded as holding that command until 15 January 1920, demobilization date notwithstanding.

Commanding Officers: *Ltn.* Bruno Majewski, *Ltn.* Ernst Kredel (5.6.19), *Oblt.* Adolf Averdam (25.6.19)

Personnel known to have served with unit:

Oblt. Herbert-Armin Angelroth - O
Oblt. Adolf Averdam - O, CO
Vzfw. Fritz Bohne
Lt. d. R. Hellmuth Brembach
Vzflugm. Hermann Hackbusch - P
Uffz. Hahn
Obermaat Gerhard Hubrich - P
Jodkow (?) - O
Ltn. Ernst Kredel - CO
Ltn. Ludwig - O
Ltn. Bruno Majewski - CO
Ltn. Ernst-Albrecht *Frhr.* von Ompteda - O
Flieger Prinz - P
Ruppricht - P
Ltn. Schubart
Vzfw. Werner Schumm - P
Vzfw. Fritz Todheide -P
Lt. z. S. Harry Winter - O

Aircraft:

Hannover CL.III
Halberstadt C.V, including 4237/18, 6902/18, 6905/18, 6911/18
Junkers CL.I, including 12600/18, 12921/18, 12925/18

A/C Unit Marking:

Some Ju. CL.I bore a white band centered on the gunner's cockpit. At least two also carried a tactical letter on the band, including "M" and"V" (CL.I 12921/18).

Unit Name: *Grenzschutz Flieger Abteilung 418/Truppen-Fliegerstaffel 27*

Alternate Names:
Feldpost No.:
Successor Unit: *Truppen-Fliegerstaffel* 27

Date Established: After 9 January 1919
Location Established: *Bomben-Lehr-Abteilung*, Frankfurt a. O.
Founding Commander: *Oblt.* Oskar *Frhr.* von Boenigk

Reported to: V. *Armee Korps*, Glogau, *Rw.* 27. Brigade, Frankfurt a. O.
Bases: Glogau
Disbanded: *Fliegerhorst* Neuruppin, per *Idflieg Nr.* 1203/8.19 *Abt.* I of 28 August 19

Uniform Insignia
(Epaulet)

(Red on feldgrau)

Unit Stamp

Commanding Officers: *Oblt.* Oskar *Frhr.* von Boenigk

Personnel known to have served with unit:
- *Ltn.* Hermann Becker - P
- *Off-stv.* Friedrich Bernhardt - P
- *Oblt.* Bilke
- *Oblt.* Oskar *Frhr.* von Boenigk - CO
- *Oblt.* Gustav Brockmann
- *Ltn.* Franz Büchner - P
- *Feldw.* Friedrich
- *Flieger* P. Gansin
- *Ltn.* Gebbert
- *Ltn.* Fritz Gewert - P
- *Feldw.* Grenzisa
- *Feldw.* Otto Klaiber - P
- *Feldw.* Korfay
- *Ltn.* Kuke
- *Ltn.* Wilhelm Leusch - P
- *Ltn.* Herbert Mahn - P
- *Off-stv.* Friedrich-Wilhelm Morzik - P
- *Ltn.* Werner Niethammer - P
- *Ltn.* Richter
- *Ltn.* Rudolf Rienau - P
- *Flieger* T. Gansin
- *Ltn.* Treuhaupt - O
- *Feldw.* Wiedemann

Aircraft:

Aircraft Unit Marking:

Unit Name: *Flieger Abteilung 419/Artillerie-Fliegerstaffel 104*

Alternate Names: *FA der 3. Landesschützen Brigade,* *Staffel* Creydt, *Jagdstaffel der Landesschützenkorps,* *Freiwillige Flieger Abteilung Cadets der Landesschützenkorps*

Uniform Insignia **Unit Stamp**

Feldpost No.:
Successor Unit: *Artillerie-Fliegerstaffel* 104

Date Established: 8 January 1919, per K. M. Nr. 400.19 A 7 L. IV.C.
Location Established: Marburg
Founding Commander: *Hpt.* Creydt

Reported to: 3. *Landesschützen* Brigade, Potsdam, *Rw.* 4. Brigade, Magdeburg
Bases: Döberitz, Bremen (2.19), Cuxhaven, Emden, Norderny
Disbanded: *Fliegerhorst* Döberitz, per *Idflieg Nr.* 1203/8.19 *Abt.* I of 28 August 19

Commanding Officers: *Hpt.* Creydt

Personnel known to have served with unit:
Oblt. Karl Barlen - *O.z.b.V.*
Hpt. Creydt - CO

Aircraft:

Aircraft Unit Marking:

Notes: *FA* 419 was directly descended from the disbanded *FA(A)* 216, commanded by *Hauptmann* Creydt and demobilized at Marburg, Hessen.

Unit Name: *Flieger Abteilung 420/Artillerie-Fliegerstaffel 115*

Alternate Names: *FA der 7. Garde-Kavallerie-Schutzen-*Division, *Gruppe Heuduck, Flieger Abteilung* Grauert

Uniform Insignia **Unit Stamp**

Feldpost No.:
Successor Unit: *Artillerie-Fliegerstaffel* 115

Date Established: 8 January 1919, per K. M. Nr. 400.19 A 7 L. IV.C.
Location Established: Potsdam
Founding Commander: *Hpt.* Ulrich Grauert

Reported to: *7. Garde-Kavallerie-Schutzen*-Division, *Gruppe* Heuduck, Potsdam, *Rw.* 15. Brigade, Berlin
Bases: Potsdam, Döberitz (3.19), Bremen (?), Lübeck-Blankensee, Schneidemühl
Disbanded: Schneidemühl, 31 March 1920*

Commanding Officers: *Hpt.* Ulrich Grauert

Personnel known to have served with unit:
Oblt. Karl Angerstein - P
Oblt. Ernst Bonatz - P
Oblt. Alfred Bülowius -P
Ltn. Kuno Heribert Fütterer - O
Hpt. Ulrich Grauert - CO
Lt. z. S. Ulrich Kessler
Ltn. Rudolf Meister - O
Lt. d. R. Werner Peckmann - P
Ltn. Theobald von Zastrow - O

Aircraft: 23, including Halb. C.V

Aircraft Unit Marking:

***Note:** *Artillerie-Fliegerstaffel* 120 was originally planned to be demobilized at *Fliegerhorst* Döberitz, per *Idflieg Nr.* 1203/8.19 *Abt.* I of 28 August 19.

Name: *Flieger Abteilung 421*

ıate Names: *FA* Möhn *des Deutsche Schutzen-* Division

Uniform Insignia **Unit Stamp**

ɔst No.:
ssor Unit:

:stablished: 8 January 1919,
per K. M. Nr. 400.19 A 7 L. IV.C.
on Established: Potsdam
ling Commander: Möhn

ted to: *Deutsche Schutzen* Division, Korps Oven, *Deutsche Schutzen* Division
Potsdam, Döberitz, Berlin-Johannisthal (3.19), Schleissheim (3.5.19), Döberitz (9.5.19>>), Bremen(?)
nded:

ıanding Officers: Möhn

ınel known to have served with unit:
Lt. d. R. Gerhard von Frantzius - P
Möhn - CO
Ltn. Herbert Pfeiffer - P
Oblt. Oskar Kuppinger (O)

ft: 9 aircraft in March 1919, including:
Halb. C.V 8771/18. 8777/18 & 8780/18, Ru. C 3977/18 & 6420/18 & Alb. B.II 1309

ft Unit Marking:

Unit Name: *Flieger Abteilung 422/Truppen-Fliegerstaffel 15*

Alternate Names: *Jagdstaffel Löwe, Flieger Abteilung Löwe*
Feldpost No.:
Successor Unit: T*ruppen-Fliegerstaffel* 15

Date Established: 8 January 1919,
per K. M. Nr. 400.19 A 7 L. IV.C.
Location Established: Potsdam
Founding Commander: *Hpt.* Joachim von Schröder

Uniform Insignia
(Collar tabs & surrounding cap cockade)

(Freiwillige Landesschützenkorps)

Unit Stamp

Reported to: Division Lettow; *Korps* Oven; Division Lettow; *Rw.* 15. Brigade
Bases: Potsdam, Döberitz, Ludwigslust, Ingolstadt, Oberwiesenfeld (4.5.19), Döberitz (9.5.19>>), Blankensee bei Lübeck (6.19), Blankenese (25.7.19), Ludwigslust (>3.7.19)
Disbanded:

Commanding Officers: *Hpt.* Joachim von Schröder, *Oblt.* Erik von Heimburg
Personnel known to have served with unit:

Ltn. Bovakht - O
Lt. d. R. Ewaldt - O
Oblt. Johann-Volkmar Fisser - P
Oblt. Erik von Heimburg - CO
Lt. d. R. Fritz John Jacobsen
Lt. d. R. Werner Peckmann - P
Ltn. Werner Preuss - P
Hpt. Joachim von Schröder - CO

Aircraft:

Hannover CL.V
Fokker D.VII

Aircraft Unit Marking:

Unit Name: *Flieger Abteilung 423/Artillerie Fliegerstaffel 116*

Alternate Names: *FA der Freiwillige Landesjägerkorps* Maercker, *Staffel* Weimar, *FA* Krocker des *Landesjäger-Korps*
Feldpost No.:
Successor Unit: *Artillerie-Fliegerstaffel 116* (2 May 19)

Date Established: 8 January 1919, per K. M. Nr. 400.19 A 7 L. IV.C.
Location Established: *Geschwaderschule* Paderborn
Founding Commander: *Hpt.* Victor Krocker

Uniform Insignia
(Collar tabs)

(*Freiwillige Landesjägerkorps*)

Unit Stamp

Reported to: *Freiwillige Landesjägerkorps, Rw.* 16. Brigade, Weimar
Bases: Paderborn, Tetlow (14.1.19), Weimar (22.2.18), Paderborn
Disbanded: *Fliegerhorst* Paderborn, 27 September 1919

Commanding Officers: *Hpt.* Victor Krocker

Personnel known to have served with unit: 126 Officers, NCOs and enlisted men, including:

Hpt. von Apell - CO, *Abt.* Ia
Off.-Stv. Paul Aue - P
Uffz. Bading
Oblt. Hans Behrendt - O, *Flivo*
Ltn. Bindner - CO, *Abt.* VII
Vzfw. Karl Bihnenkamp
Dr. Bruns - *Oberarzt*
Oblt. Demnig
Uffz. Ertl
Ltn. Ernst Exss - CO, *Abt.* IV, O
Ltn. Erich-Erdmann Fitzau - P
Ltn. Feld - - CO, *Abt.* VI
Ludwig Friedewald - P
Sgt. Göbel - *Kammer*
Uffz. Gutermann - *Kassenwesen*
Vzfw. Hermann Hasselmann - P
Herbst - P
Ltn. Hertle
Huninghaus - P
Uffz. Ilian
Ltn. Otto Jordan - O
Kohne - P
Ltn. Wilhelm Kohlbach
Hpt. Victor Krocker - CO
Ernst(?) Krogmann - P
Ltn. Küsgen - CO, *Abt.* V
Flieger Lessmann
Flieger Müller
Muller - P
Munichow - P
Uffz. Naase
Ozinek - P
Oblt. Gerhard Pape - O
Pätzhold - *Werkmeister*
Gefr. Rätzer
Geom. Reiske
Lt. d. R. Leonhard Rempe - CO, *Abt.* III, P
Flieger Schreiber
Sergt. Schroll - Kitchen
Sergt. Schuster
Schwamm - P
Hpt. Friedrich Starke (?)
Uffz. Stephan
Viereck - P
Feldw. Vogt
Norbert Welkoborsky - P
Wrensch - P
Hpt. Otto Zech
Uffz. Zschochlitz

Aircraft: 6 Fokker D.VII, incl. D.VII F 5136/18 & 8448/18
13 LVG C.VI, incl 7692/18 & 7766/18 as of 15 April 1919.

A/C Unit Marking: White numeral on fuselage sides of LVGs, below gunner's cockpit.
White streamers on lower wings of Fok. D.VIIs

Unit Name: *Freiwillige Flieger Abteilung* 424/*Truppen-Fliegerstaffel* 37

Alternate Names:
Feldpost No.: 3066
Successor Units: *Reichswehr Truppen-Fliegerstaffel* 37, later 3. Plastun Aviation Division, West Russian Army
Date Established: 18 December 1918
Location Established: *Flieger Beobachter Schule*, Königsberg
Founding Commander: *Oblt.* Erwin Martin

Uniform Insignia
(Upper left sleeve)

Unit Stamp

Reported to: *10. Armee, GKdo. der Zusammengesetzt Reservekorps; VI. Reservekorps; Deutsche* Legion, West Russian Army, *Kofl. VI. Reservekorps*
Bases: Mariampol (18.12. 18), Alt-Auz / Gut Weitenfeld (10.5.19), Ginkuniai (* 1. *Plastun* Av. Detach., end 8.19)*
Disbanded: Stolp & Seerappen / Königsberg, January 1920 and Döberitz, March 1920

Commanding Officers: *Oblt.* Erwin Martin, *Oblt.* Hans-Helmut von Boddien (stv. 22.8.19 - ?)

Personnel known to have served with unit:

Flieger Alfred
Ltn. Gerhard Bassenge - P
Flieger Hans Beer
Uffz. Behrens
Gefr. Bene
Ltn. Benke
Flieger Rudolf Bickel
Blanket - P
Ltn. W. Hein Blöcker - P & Tech. Officer
Oblt. Hans-Helmut von Boddien - P, CO - stv.
Ltn. Böhmer
Ltn. Frhr. von Bongart - O
Heinrich Bracht
Feldw. Braun
Flieger Bruch
Oblt. Bulcher - O
Ltn. Buschmann -P
Ltn. Gustav Claussen - P
Gefr. Decker
Gefr. Düring
Gefr. Ebert
Flieger Heinrich Emter
Ltn. Engelfried
Off.-stv. Ernst
Kraftfahrer Erve
Ltn. Frhr. Rudolf Esebeck
Captain Filatov
Gefr. Fincke
Flieger Fitsch
Alfred Fliege
Hpt. Frhrr. von Frieberg - *Staffelführer*
Flieger Joseph Fürst
Garback
Garbart
Ltn. Gölchar
Feldw. Harborth
Vzfw. Haupt - P
Privat Pavel Havenit
Leo Havenith
Oblt. von Heuduck
Heutsch
Ltn. Frhr. von der Heyden-Rinsch
Uffz. Hillebrandt
Gefr. Hinz
Gefr. Holst
Gefr. Illige
Gefr. Janson
Ltn. Jenkowiak - P
Jurkschat
Ltn. Krabbe
Wilhelm Kisling
Kretschmar - Fitter
Off-stv. Erich Kulikowski - P
Lautenbach - MG
Vzfw. Fritz Lebiedz - P
Lieutenant Lebedev
Flieger Willy Lenz
Off.-stv. Lewa
Uffz. Kinzel
Flieger Kütner
Rittm. Frhr. Wilhelm Marschall von Bieberstein
Oblt. Erwin Martin - CO
Ltn. Hans Martin
Gefr. Meder
Ltn. Mertens
Milkovsky
Uffz. Möller
Ltn. Müller - P
Oblt. Frhr. von Münch
Ltn. Neetzel - P
Neyman
Gefr. Otto
Flieger Pallmer
Oblt. von Plehn
Reich - Fitter
Ltn. Frhr. von Reinhardt
Vize-Wachmeister Reinhold
Ergart Resch
Reinhold Reteuch
Ltn. von Röhn
Ltn. Karl Ryon
Ltn. Karl Salewsky
Schaefer
Flieger Alfred Schars
Uffz. Schlicker
Gefr. Schmecker
Ltn. Carl-August von Schoenebeck - P
Flg. Kurt Schönemann
Flieger Felix Schröder
Ltn. von Sallerau - P
Gefr. Schwenzner
Gefr. Stihl
Ltn. Emil Tsikmer
Vize-Wachmeister Welke
Gefr. Wichmann
Flieger Karl Wiedemann
Gefr. Winkler
Ltn. Alexander von Winterfeld
Gefr. Wirth
Flieger Vilgis
Gefr. Voigt
Vzfw. Weiss - P
Erwin Wilhelm
Emil Zikmer

Aircraft: 12 aircraft at the end of February 1919. At various times, equipment included:
10 Fokker D.VII
1 Friedrichshafen G.IIIa
Halb. CL.IV, incl. CL.IV(Rol) 8167/18
LVG C.VI
SSW D.IV, probably 6177/18
A/C Unit Marking: None known

Unit Name: *Freiwillige Flieger Abteilung* 425/*Truppen-Fliegerstaffel* 1

Alternate Names: *Freiwillige Fl.-Abt. der X. Armee, Flugpark Kurland* **Uniform Insignia** **Unit Stamp**
Feldpost No.:
Successor Units: *Truppen-Fliegerstaffel* 1 (6.19), *Flugpark Kurland,* after merger with *FA* 433 (28.7.19)

Date Established: 18 December 1918
Location Established: *Flugplatz* Alexsotas, Kowno
Founding Commander: *Hpt. d. R.* Hermann Gröckel

Reported to: Armee HQ, *10. Armee (18.12.18), GKdo. der Zusammengesetzt Reservekorps (1.19); Freiwilligen Reservekorps (1.5.19)*
Bases: Kowno/Aleksotas, Alt-Auz (7.7.19)
Disbanded: Alt-Auz, 14 August 1919, per AOK Nord Tagsbefehl 145

Commanding Officers: *Hpt. d. R.* Hermann Gröckel (12.18 - 3.5.19), *Hpt.* Ernst Dörffler (3.5.19 - 17.8.19)

Personnel known to have served with unit:
- *Oblt.* Paul Dietrich - O
- *Hpt.* Ernst Dörffler - CO
- *Hpt. d. R.* Hermann Gröckel - CO
- Klein
- *Oblt.* E. Kuligowski - P
- Hermann Leichner - O
- Erich Schulz - O
- *Oblt.* Fritz Schultz - P
- Adolf Schwesig - P
- *Oblt.* Albrecht Weinschenk - P
- *Gefr.* Felix Wilke

Aircraft: 3 aircraft at the end of February 1919. At various times, equipment included:
- Alb. B.II, incl. 326, 1349, 2128, & 2274
- Alb. CXII 1144/17
- Halb. Cl.IV, incl. 9432
- LVG C.VI, incl. 7626/18 & 8942/18
- Ru. C.I, incl. 2699 & 8414

A/C Unit Marking: None known

Unit Name: *Freiwillige Flieger Abteilung* 426/*Truppen-Fliegerstaffel* 38

Alternate Names: 2. *Freiwillige Flieger Division, Freiwillige FA* Weinschenck
Feldpost No.: 737
Successor Units: *Truppen-Fliegerstaffel* 38, 2. *Plastun* Aviation Division
Date Established: December 1918
Location Established: *Flieger Ersatz Abteilung* 3, Gotha
Founding Commander: *Oblt.* Hellmuth Weinschenck

Unit Insignia (Upper left sleeve)

Unit Stamp

Reported to: *Landwehrkorps, GKdo. der Zusammengesetzt Reservekorps*, Reinforced Regiment *von* Klüfer (3-9.5.19), Wirgolitsch *Korps*,West Russian Army (13.8.19)
Bases: Suwalki (2.19), Radziwilishki (12.8.19), Mesoten (Abt. VIII) (4. 10.19), Grossenhain (>6.2.20)
Demobilized: Stolp, January 1920*

Commanding Officers: *Oblt.* Hellmuth Weinschenck, *Ltn.* Stephan (>6.2.20)
Personnel known to have served with unit: Approximately 250 men, including:

Hermann Aberglaub
Ltn. d. R. Adomeit - *Stabs Abt.* V
Ahling - P, *Abt.*VII
Vzfw. Karl Albrecht
Ltn. Altenburg
Ltn. Hans Arimond - CO, *Abt.* IV
Fw.. Bachert - P, *Abt.*VII
Ltn. Bahl - P, *Abt.*VII
Baumgartner - O, *Abt.* V
Beitat - P
Off.-Stv. Karl Berwaldt
Ltn. Böhm
Ltn. Bracht - O, *Abt.* V
Uffz. Bruder, *Werkmeister, Abt.* VIII
Gefr. Deissner
Ltn. Ebel - P, CO, *FA* VI, *FA* V
Friedrich Fast
Flieger Pankrats Fischer - P
Wilhelm Fischer
Ltn. von Flach
Ltn. Fleischauer - Transport *Offizier, Stabs Abt.* I
Edmund Flinke
Maj. Helmuth Förster
Lt. d. R. Gerhard von Frantzius - O, *Flivo*
Edmund Frince
Ltn. Gallen - P, *Abt.*V
Edward Gehring
Vzfw. Gödeke - P, *Abt.* VI
Golowin - O
Flieger Bernhard Grundmann
Oblt. Hallen - P
Willi Hecker - *Abt.* VI
Ltn. Hinneberg- *Stabs Abt.* III
Ens. Hödecke
Ltn. Friedrich Höhler
Ltn. Horn - P, *Abt.*V
Hans Hoth
Albert Hückstädt
Ltn. Walter Jagdfeldt -P
Ltn. F. von Jena - Transport *Offizier, - Stabs Abt.* II
Ltn. Jander - O, *Abt.* V
Ltn. Johannesson - O, *Abt.* V, CO (*stv.*), *Abt.* VI
Georg Jordan
Paul Kaiser
Paul Keller
Flieger Klein
Ltn. Klein
Gefr. Klempert
Ltn. Klöwer - O
Oblt. Kratz - P, CO, *Abt.*VIII
Kosta Konpoka - *Abt.* VI
V. Kotz
Lt. d. R. Erich Kroll - *O.z.b.V., Abt.*VIII
Ltn. F. Krotz, *Abt.* VIII
Lesermoser - P
Ltn. Lichtenberg - P, *Abt.* VIII
Ltn. von Mauritius
Vzfw. Mecklenburg
Flieger Meder
Medera - P, *Abt.* VII
Ltn. Meser
Vzfw. Christian Mesch - P, *Abt.*VIII
Ltn. Miet - P, *Abt.*V
Off.-Stv. Müller
Alfred Neger
Vzfw. Karl Noak - P
Adolf Nosik
Ltn. Obel - *Abt.* V
Ltn. von Osterroht - *O.z.b.V.*
Paul Ostrop
Vzfw. Paech - *Werkmeister*
Karl Pfeiffer
Ltn. Prüssner - *O.z.b.V., Abt.* VI
Ltn. Heinrich Rauch - O
Uffz. Ullrich Rauch
Dr. Reck - *Ass. Arzt*
Johann Richard, *Abt.* VI
Ltn. Schaff - P
Oblt. Frhr. von Schele - Staff Quarters
von Schell
Vzfw. Schmidt - P
Ltn. Schmitt - P, CO, *Abt.* VIII
Schtomprowsky - P, *Abt.* VII
Schubert - P, *Abt.* VI
Serg. Bruno Schreyer - P
Eric Schuldes, *Abt. VI*
Ltn. Schwarzkopf - *Tech. Offz*
Ltn. Selzer - *Tech. Offz., Abt.* VI
Sieberg
Werkm. Steinmann, *Werkmeister*
Ltn. Stephan - CO, *Abt.* VII, *Abt.* VIII
Ltn. Stehr - P, *Abt.* V
Ltn. Sterz - P, *Abt.* V
Vzfw. Wilhelm Stör - P
Ltn. Stolf - *Tech. Offz.*
Vzfw. Hermann Stracke
Flieger Max Thiemicken
Franz Tull
Off. Stv. Vinz - *Zahlmeister, Stabs Abt.* IVa
Emil Weber
Oblt. Hellmuth Weinschenck - CO
Ltn. Zaak - P, *Abt.* VI

Aircraft:
6 C-types and 4 D-types at the end of February. Specific types included Alb. C.XII, Alb. C.XV, DFW C.V, Fok. D.VII, Halb. Cl.II, Halb. Cl.IV (Rol) , incl. 8176/18

A/C Unit Marking: White noses on Halb. Cl.IV & DFW C.V

***Note:** Originally scheduled for demobilization at *Fliegerhorst* Jüterbog per *Idflieg Nr.* 1203/8.19 *Abt.* I of 28.8.19.

Unit Name: *Freiwillige Flieger Abteilung 427*

Alternate Names: *Eiserne Flieger Abteilung* (8)
Feldpost No.: 383
Successor Unit: *Artillerie-Fliegerstaffel 101**

Date Established: 9 December 1918
Location Established: Weissenhof, Kurland
Founding Commander: *Hpt.* Friedrich Erdmann

Reported to: *Eiserne* Division; West Russian Army (9.19)
Staffel **Fick:** *Baltische Landeswehr*
Bases: Weissenhof, Insterburg (12.1.19), Wainoden (9.2.19), Alt-Auz (18.3.19), Mitau/Paulsgnade (6.19), Ramenhof (18.6.19),Wiebersholm (6.19), Lutzausholm (6.19), Mitau/Paulsgnade (3.7.19), Radziwilischki (21.11.19).
Staffel **Fick:** Katzdangen, Tuckum (23.4.19), Weissenhof (22.5.19)
Disbanded: Stolp, January 1920

Uniform Insignia

(Red on feldgrau)

Unit Stamp

Commanding Officers: *Hpt.* Friedrich Erdmann (12.18 - 12.5.19, *Oblt.* Walter Kirschstein (stv.-12.5 - 20.7.19), Kirschstein (20.7.19 - 2.1.20)
Staffel **Fick:** *Lt. d. R.* Walter Fick

Personnel known to have served with unit:

Johannes Ausmus Busch - P
Ltn. Paul Deichmann - O
Flieger Nikolaus Eberg - P
Hpt. Friedrich Erdmann - CO
Lt. d. R. Walter Fick - CO, *Staffel* Fick
Lt. d. R. Fischer - P
Feldwebel Gillmann - *Staffel* Fick
Alfred Grünfeld
Uffz. Grund - P
Hpt. Johannes Ludwig *Graf* von Hachenburg?
Gefr. Arthur Hentschel
Ltn. B. Hermann - O
Off. Stv. Otto Hilpert - P
Oblt. Walter Kirchstein - CO
Vzfw. Kanisch
Alfons Krummel
Flg. Paul Lahmer
Lt. d. R. Lehmann
Hpt. Bruno Loerzer - P
Ltn. Gunther Lohmann - O
Lt. d. R. Mauschering - O
Ltn. Walter Meuer - O
Ltn. Meyer
Ltn. Riteleki - O
Paul Rodenberg - P
Lt. d. R. Karl Salewski - *Flivo*
Helmuth Säuberlich - P
Hpt. Julius Schulz
Flg. Eugen Schmidt
Hpt. Julius Schulz
Ltn. Schuster
Vzfw. Seyfert
Flg. Tillmann
Karl Turban
Off-stv. Waltgerr
Georg Weisheit - O
Lt. d. R. B. Wittemayer - O
Lt. Zalevsky

Aircraft:

Junkers
Halberstadt Cl.IV

A/C Unit Marking:

***Note:** *FA* 427 was to become *Artillerie-Fliegerstaffel* 101 in the Provisional *Reichswehr*. However, *it* continued to exist after *Artillerie-Fliegerstaffel* 101 was established 10 August 1919. The two units operated in close association until September, when *FA* 427 merged into *Artillerie-Fliegerstaffel* 101.

Unit Name: *Artillerie-Fliegerstaffel 101**

Uniform Insignia (Upper left sleeve)

(Yellow on feldgrau)

Unit Stamp

Alternate Names:
Feldpost No.: 383
Successor Unit:

Date Established: 10 August 1919
Location Established: Mitau/Paulsgnade
Founding Commander: *Hpt.* Friedrich Erdmann

Reported to: *Eiserne* Division, West Russian Army (9.19)
Bases: Mitau/Paulsgnade (22.5.19 or later), Königsberg & Stolp (1.20)
Disbanded: Stolp, February 1920

Commanding Officers: *Hpt.* Friedrich Erdmann, *Oblt.* Walter Kirschstein (9.9.19)

Personnel known to have served with unit:

Flugzeugmatrose Max Blumberg
Flieger H. Bucholz
Flieger Nikolaus Eberg - P
Erich Eppinger - P
Hpt. Friedrich Erdmann - CO
Sgt. Feuersenger
Oblt. Gaedicke
Gefr. Gerick
Flieger Ernst Kessler
Oblt. Walter Kirchstein - CO
Gefr. Adolf Klee
Wachmeister Lerche
Sgt. Mühlich
Vzfw. Richter
Gefr. Rohrmiser
Ltn. Rüdiger - Tech. Off.
Gefr. Rüdikeit
Gefr. Steinnagel
Ltn. Günther Töpert
Sgt. Verheyen

Aircraft:

A/C Unit Marking:

***Note:** *Artillerie-Fliegerstaffel* 101 was to be the designation of the former *FA* 427 when it became part of the Provisional *Reichswehr*. However, *FA* 427 continued to exist, while a new unit named *Artillerie-Fliegerstaffel* 101 was established 10 August 1919. The two units operated in close association until September, when *FA* 427 was absorbed into *Artillerie-Fliegerstaffel* 101.

Unit Name: ***Grenzschutz Flieger Abteilung* 428/*Truppen-Fliegerstaffel* 50(?)**

Alternate Names: **Uniform Insignia** **Unit Stamp**

Feldpost No.:
Successor Units: *Truppen-Fliegerstaffel* 50(?)

Date Established: By 24 Feb. 1919
Location Established:
Founding Commander:

Reported to: Generalkommando. II Armee Korps, Stettin

Bases: Köslin
Disbanded: After October 1919

Commanding Officers:

Personnel known to have served with unit:
Ltn. Günther Wieland - O

Aircraft:

A/C Unit Marking:

Unit Name: *Grenzschutz Flieger Abteilung* 429/*Truppen-Fliegerstaffel* 31

Alternate Names:
Feldpost No.:
Successor Unit: *Truppen-Fliegerstaffel* 31
Date Established: 12 January 1919 or earlier
Location Established: Langführ
Founding Commander: *Hpt.* Wagner

Uniform Insignia

Unit Stamp

Reported to: Volunteer Reserve *Korps* (5.19), 10. A.K.
Bases: Danzig/Langfuhr (27.2.19), Wilkowischki (15.3.19), Deutsch-Eylau (Schloss Schönberg) (10 (18?).5.19),
Disbanded:

Commanding Officers: *Hpt.* Wagner (12.1.19), *Hpt.* Theodor Schubert (9.3.19* - 22.10.19)

Personnel known to have served with unit:

Joachim von Busse
Ltn. Kurt Faber - P
Viktor Knabel - Mechanic
Flieger Litzbarski
Oblt. Martin
Hpt. Erhard Milch - O
Hpt. Theodor Schubert - CO
Hpt. Wagner - CO
Otto Wienhold - Mechanic
Ltn. Alexander von Winterfeldt - P

Aircraft:

Halb. C.V

A/C Unit Marking:

Note: * 12.4.19 per some sources.

Unit Name: *Grenzschutz Flieger Abteilung* 430/*Truppen-Fliegerstaffel* 52(?)

Alternate Names:
Feldpost No.:
Successor Unit: Truppen-Fliegerstaffel 52 (?)

Uniform Insignia

Unit Stamp

Date Established:
Location Established: Breslau-Hundsfeld
Founding Commander: *Hpt.* Dietrich(?) Arndt

Reported to: Gkdo VI. A.K.
Bases: Breslau-Hundsfeld
Disbanded:

Commanding Officers: *Hpt.* Dietrich(?) Arndt

Personnel known to have served with unit:
- *Hpt.* Dietrich(?) Arndt - CO
- *Flieger* Jos. Barthel
- Roehre - *O.z.b.V.*

Aircraft:
- Fokker D.VII
- Fokker D.VIII
- Junkers Cl.I

A/C Unit Marking: None known.

Unit Name: *Grenzschutz Flieger Abteilung* 431/Truppen-Fliegerstaffel 29

Alternate Names:
Feldpost No.: 809
Successor Unit: *Truppen-Fliegerstaffel* 29 (22.5.19)

Date Established: 16 January 1919
Location Established: Breslau-Gandau
Founding Commander: *Hpt.* Ernst Otto Eberstein

Uniform Insignia
(Collar tabs)

(Freiwilliges Jäger-Bataillon Schull)

Unit Stamp

Reported to: Genkdo. VI. A . K., *Freiwillige Korps Schlesien* (9th Inf. Div - Freiwilliges Jäger-Bataillon Schull); *Rw.* 29. Brigade, Bernstadt
Bases: Breslau-Gandau (16.1.19), Faulbrücke bei Reichenbach (15.3.19), Breslau-Hundsfeld (part: 5.5.19), Rudelsdorf (23-25.6.19), Paderborn (16.8.19)
Disbanded: 1 October 1919, Paderborn

Commanding Officers: *Hpt.* Ernst Otto Eberstein (21.1.19 - 1.10.19)

Personnel known to have served with unit:

Vzfw. Brockmann - P
Hpt. Ernst Otto Eberstein - CO
Ltn. d. R. Wilhelm Griebsch - P
Maj. Gürtler?
Oblt. Hartmann
Uffz. Hassenmitter -P
Ltn. Heidepriem
Hpt. Jahn
Ltn. Käubler
Hpt. Kaffanke
Lt. d. R. Alfred Lenz - *O.z.b.V.*
Ltn. Friedrich-Wilhelm Liebig - P
Hpt. Maass (Faulbrücke detachment)
Ltn. Martin - *O.z.b.V.*
Ltn. Mauelshagen
Ltn. Meyer
Off. Stv. Pelzer - *Werkmeister*
Maj. Peterch?
Uffz. August Qoos - P
*Ltn.*Reichert
Ltn. Hugo Steppuhn
Maj. Vogt?
Oberstltn. Wedge?
Hpt. Westphal
Hpt. Bruno Zühlke - *Grufl, 9. I. D.*

Aircraft:

DFW C.V, Fokker D.VII (OAW), Halberstadt CL.IV, Junkers CL.I (Jfa) 12608/18, Junkers J.I 750/17, LVG C.V (at least 3), LVG C.VI (at least 2), SSW D.IV 3087/18 & 6178/18

A/C Unit Marking:

Unit Name: *Flieger Abteilung* 432/*Artillerie-Fliegerstaffel* 103

Alternate Names: *Fliegerabteilung* Behrendt, Döberitz *Jagdstaffel*, *Freiwillige FA des Freikorps* Hülsen, *Reichswehr Flieger Abteilung* Potsdam

Uniform Insignia **Unit Stamp**

Feldpost No.:
Successor Unit: *Reichswehr Artillerie-Fliegerstaffel* 103

Date Established: > 6 January 1919
Location Established: Berlin
Founding Commander: *Oblt.* Hermann Fricke

Reported to: *Freikorps* Hülsen; *Reichswehr* Brigade 3
Bases: Potsdam, Döberitz (>15.8.19)
Disbanded: Döberitz, 28.8.19, per *Idflieg Nr.* 1283/8.19 *Abt. I.II.Ang.*

Commanding Officers: *Oblt.* Hermann Fricke, *Oblt.* Hans Behrendt (11.1.19), *Hpt.* Egloff *Frhr.* von Freyberg-Eisenberg-Allmendingen (18.9.19)

Personnel known to have served with unit:

Oblt. Kurt Baur - O
Oblt. Werner Bautz -O
Oblt. Johannes Behrendt - CO
Lt. d. R. Franz Camphausen - P
Oblt. Johann-Volkmar Fisser - O
Hpt. Egloff *Frhr.*von Freyberg-Eisenberg-Allmendingen - CO
Oblt. Hermann Fricke - CO
Vzfw. (Lt. d. R. ?) Otto Fruhner
Uffz. August Hassel - P
Ltn. Otto Jordan - O
Ltn. Eitel-Friedrich Roediger von Manteuffel - P
Ltn. (d. R.?) Rieldorf - O
Vzfw. Meinberg - P
Lt. d. R. Friedrich Scheel - P
Oblt. Walter Schwabedissen - O
Hermann Tolle - P
Lt. d. R. Paul Weitkus

Aircraft:
Fok. D.VII (at least 8), Fok. D.VIII (at least 2, incl. #5), Halb. CL.IV (Rol), LVG C.VI 7205/18

A/C Unit Marking: At least one Halb. CL.IV fuselage overpainted in black/red color scheme, ace of spades (see profile).

Unit Name: *Grenzschutz Flieger Abteilung* 433

Alternate Names: *FA des Ostpreussisches - Freiwilligenkorps* Memel *(prior to 22 April 1919), 1. Baltische Fliegerabteilung, Feldflieger-Abteilung 433*

Uniform Insignia **Unit Stamp**

Feldpost No.: 3068
Successor Unit: *AFP Kurland, Truppen-Fliegerstaffel* 1 (after merger with *FA* 425)

Date Established: 1 February 1919
Location Established: Memel
Founding Commander: *Hpt.* Erich Alt

Reported to: *Ostpreussisches Freiwilligenkorps*, Baltic *Landeswehr* (22.4.19 - 4.7.19)
Bases: Memel, Tuckum (23.4.19), Weissenhof (23.5.19), Wenden (18.6.19), Weissenhof (22.6.19), Tuckum (4.7.19), Alt-Auz (27.7.19)
Disbanded: Alt-Auz, 14 August 1919

Commanding Officers: *Hpt.* Erich Alt

Personnel known to have served with unit:

Hpt. Erich Alt - CO
Vzfw. von Bartdorff - P
Otto Max Berengrün
Oblt. Kurt Boettge - O
Feldwebel Wilhelm Dams
Ltn. Paul Deichmann - O
Flg. Drether
Lt. d. R. Walter Fick
Maj. Helmuth Förster
Oblt. Erich Freund - O
Ltn. Ernst Hamster - P
Uffz. Hillmann
Vzfw. Jedwill - P
Vzfw. Kanisch - P
Flieger Ernst Kessler - P
Laischke
Alex von Lilienfeld
Loewen
Baron Meidel
Kornett Miram
Müthel
Ltn. Heinrich Rauch - O
Gefr. Richter
Feldmeister Skribanowitz
Gefr. Wolf - P
Wrensch - P

Aircraft:

Alb. C.III, Alb. D.V, DFW C.V 847, Halb. CL.IV, Hannover CL.IIIa 7015/18, LVG C.VI 12027/1

A/C Unit Marking: On some aircraft, a plain white square was painted over the wing, fuselage and rudder crosses after the unit was transferred to The *Baltische Landeswehr.*

Note: For the attack on Riga 22 May 1919, some aircraft and crews from the idled *Geschwader* Sachsenberg were temporarily attached to *FA* 433.

Unit Name: *Grenzschutz Flieger Abteilung* 434/Truppen-Fliegerstaffel 9

Alternate Names: **Uniform Insignia** **Unit Stamp**
Feldpost No.: 3067
Successor Unit: *Truppen-Fliegerstaffel 9*

Date Established: > 23 April 1919
Location Established: Thorn
Founding Commander: *Hpt.* Erich Quade

Reported to: *Genkdo.* XVII *A. K.*, Danzig, *Rw.* 9. Brigade, Schwerin
Bases: Thorn (<4.6 19 - 5.11.19>)
Disbanded: 31 March 1920

Commanding Officers: *Hpt.* Erich Quade

Personnel known to have served with unit:

Lt. d. R. Alfred(?) Brocke
Ltn. Hans Gandenberger von Moisy - O
Ltn. Heydt - O
Ltn. Hoffmann - P
Ltn. d. R. Richard Kraut - P
Lt. d. R. Alfred Pfeiffer
Hpt. Erich Quade - CO
Uffz. Gustav Sölter - O

Aircraft:

Fok. D.VII

A/C Unit Marking: *Ltn. d. R.* Richard Kraut's Fok. D.VII - Dark rectangle on fuselage sides aft of cockpit, with white block letter "K"

Unit Name: *Freiwillige Flieger Abteilung* 435/*Artillerie-Fliegerstaffel* 109(?)

Alternate Names: *Freiwillige Festungsflieger-Abt. Graudenz* **Uniform Insignia** **Unit Stamp**
Feldpost No.:
Successor Unit: *Artillerie-Fliegerstaffel 109(?)*

Date Established: >>23 April 1919
Location Established: Graudenz
Founding Commander:

Reported to:
Bases:
Commanding Officers:

Personnel known to have served with unit:

Aircraft:

A/C Unit Marking:

Unit Name: ***Freiwillige Fliegerstaffel* Libau** **Uniform Insignia** **Unit Stamp**

Alternate Names: Libau *Fliegerstaffel*
Feldpost No.:
Successor Unit: *Staffel* Fick, *FA 427*

Date Established: December 1918
Location Established: Libau
Founding Commander: *Hpt. z. S.* S. Gordon?

Reported to:
Bases: Libau
Disbanded: 9 February 1919, Libau. Absorbed by *Staffel* Fick of *FA* 427.

Commanding Officers: *Hpt. z. S.* S. Gordon?
Personnel known to have served with unit:
Hpt. z. S. S. Gordon - CO?, O
Feldw. Hawelick - P
Uffz. Hillmann - O
Vzfw. Jedwill - P
Flieger Paul Lahmer

Aircraft: 6 Albatros C.III, 4 Seaplanes

A/C Unit Marking:

Unit Name: ***Flieger Abteilung* Schlieben** **Uniform Insignia** **Unit Stamp**

Alternate Names: *Flieger Frei-Korps Horst* Hannover
Feldpost No.:
Successor Unit:

Date Established: 12.1.19
Location Established: *Flieger Ersatz Abteilung (FEA) 5,* Hannover
Founding Commander: *Hpt.* Richard Schlieben

Reported to: *Freikorps* Hülsen
Bases: *FEA* 5, Hannover
Disbanded: By 5.5.19. At least part of its personnel were absorbed into *FA* 409.

Commanding Officers: *Hpt.* Richard Schlieben

Personnel known to have served with unit:

Gefr. Kurt Bothke - P
Oblt. Friedrich Cranz (?)
Lt. d. R. Otto Fruhner
Hpt. Richard Schlieben - CO

Aircraft:

A/C Unit Marking:

Unit Name: ***Fliegerabteilung der Sturmabteilung Rossbach***

Unit Insignia
(Upper left sleeve)

(Sturmabteilung Rossbach)

Unit Stamp

Alternate Names: *Schlachtstaffel der Sturmabteilung Rossbach*
Feldpost No.:
Successor Unit: None
Date Established:
Location Established: Culmsee
Founding Commander: *Ltn.* Ernst Krull

Reported to: *Sturmabteilung* Rossbach
Bases: Culmsee, Sturmbach station by Graudenz (6.6.19), Alt-Auz
Disbanded: Aircraft grounded early October 1919 at Alt-Auz for lack of fuel.

Commanding Officers: *Ltn.* Ernst Krull, *Ltn.* Lippmann (6.6.19)

Personnel known to have served with unit:
- *Ltn.* Ernst Krull - CO
- *Ltn.* Lippmann - CO

Aircraft:
- 2 LVG C.VI
- 2 LVG P.I -- 13/19 & 23/19

A/C Unit Marking:

Unit Name: *Freiwillige Flieger Abteilung* Löwenfeld (3. Marine Brigade)

Alternate Names: **Uniform Insignia** **Unit Stamp**
Feldpost No.:
Successor Unit:

Date Established: 14 April 1919?
Location Established: Kiel
Founding Commander:

Reported to: 3. (Löwenfeld) Marine Brigade
Bases: Kiel, Dorsten (3. 20)
Disbanded: Paderborn, 7 May 1920

Commanding Officers: *Kapitänleutnant d. R.* Friedrich Christiansen (14.4.19 - 21.1.20)

Personnel known to have served with unit:
Kapitänleutnant d. R. Friedrich Christiansen - CO
Flieger-O.-M. Gertenfeldt
Kapitänleutnant d. R. Konrad Goltz

Aircraft:
At least two Halb.CL.IV, incl. CL.IV(Rol.) 9455/18

A/C Unit Marking:
One CL.IV marked with 3-pointed star in triangle and stripes in black & white
CL.IV 9455/18 marked with numeral 2 below gunner's cockpit

Unit Name: ***Flieger Abteilung A1 des Bayerische Schützenkorps/ Truppen-Fliegerstaffel* 21**

Alternate Names: *FA* Krausser,
FA A1 der *Bayerische Schützenkorps*
Feldpost No.:
Successor Unit: *Truppen-Fliegerstaffel* 21

Date Established: 13 April 1919
Location Established: Gotha
Founding Commander: *Oblt.* Friedrich *Ritter* von Krausser

Uniform Insignia (Upper left sleeve)

(*Freikorps* Epp)

Unit Stamp

Reported to: *Bayerische Schützenkorps* (*Freikorps* Epp)
Bases: Gotha, Dornstadt (16.4.19), Lechfeld (3.5.19), Schleissheim (>14.5.19), Paderborn (24.3.20), Schleissheim
Disbanded: Schleissheim, 30 September 1919, per *Bayr. Gruppenkommando Nr.* 4 Ia/0/O.Qu./ *Flieg. Nr.* 5215. Portions of the unit were integrated into *Reichswehr Fliegerhorst* Schleissheim.

Commanding Officers: *Oblt.* Friedrich *Ritter* von Krausser, *Hpt.* Hilmer von Bülow

Personnel known to have served with unit: 21 officers, 28 NCOs, 101 men
Hpt. Hilmer von Bülow - CO
Oblt. Friedrich *Ritter* von Krausser - CO

Aircraft:

A/C Unit Marking:

Unit Name: *Flieger Abteilung A2 des Bayerische Schützenkorps/ Artillerie-Fliegerstaffel* 121

Alternate Names: *FA* Häfner, *FA* A2 der *Bayerische Schützenkorps*
Feldpost No.:
Successor Unit: *Artillerie-Fliegerstaffel* 121

Uniform Insignia

(*Freikorps* Epp)

Unit Stamp

Date Established: 21 April 1919
Location Established: Gotha
Founding Commander: *Hpt.* Konrad Häfner

Reported to: *Bayerische Schützenkorps* (*Freikorps* Epp)
Bases: Gotha, Dornstadt (16.4.19), Lechfeld (3.5.19), Schleissheim (23.5.19). Paderborn (24.3.20), Schleissheim
Disbanded: Schleissheim, 30 September 1919, per *Bayr. Gruppenkommando Nr.* 4 Ia/0/O.Qu./ *Flieg. Nr.* 5215. Portions of the unit were integrated into *Reichswehr Fliegerhorst* Schleissheim.
Commanding Officers: *Hpt.* Konrad Häfner

Personnel known to have served with unit: 10 officers, 19 NCOs, 57 men

Obergefreiter Beck - P
Franz Xaver Danhuber - P
Hpt. Konrad Häfner - CO
Ltn. Gottlob Müller - P
Murizk - P
Passelot - P
Preusch - P
Ltn. Heinrich Seywald - P
Stephanski - P

Aircraft: One *staffel* of 9 Hannover CL.IIIa, including C.7010/18/C.7012/18, C.7018/18, C.7020/18, C.7047/18 & C.7048/18.

A/C Unit Marking: White disc on a black diamond in the former locations of the fuselage crosses.

Unit Name: *Truppen-Fliegerstaffel* 21a

Alternate Names: **Uniform Insignia** **Unit Stamp**
Feldpost No.:
Successor Unit:

Date Established: 1 July 1919, from the merger of *Freiwillige FliegerAbteilungen* Krug & Schmalschläger
Location Established: Fürth
Founding Commander:

Reported to: Reichswehr 21. Brigade
Bases: Fürth
Disbanded: Fürth, 30 September 1919. Portions of the unit were integrated into *Reichswehr Fliegerhorst* Schleissheim.

Commanding Officers:

Personnel known to have served with unit:
Vzfw. Hans Bauer
Ltn. Georg *Ritter* von Hengl - O
Vzfw. Seb(astian?) Kitzinger
Lt. d. R. Nikolaus Maier
Oblt. Benedikt Schmalschlager
Vzfw. Franz Steiner
Vzfw. Max Wimmer

Aircraft:

A/C Unit Marking:

Unit Name: *Truppen-Fliegerstaffel* 21b

Alternate Names: **Uniform Insignia** **Unit Stamp**
Feldpost No.:
Successor Unit:

Date Established: 1 June 1919, from the merger of *Freiwillige FliegerAbteilungen* Dessloch & Kriegelsteiner
Location Established: Gersthofen
Founding Commander: *Hpt.* Friedrich Berthold

Reported to: Reichswehr 21. Brigade
Bases: Gersthofen, Kitzingen
Disbanded: Kitzingen, 30 September 1919. Portions of the unit were integrated into *Reichswehr Fliegerhorst* Schleissheim.

Commanding Officers: *Hpt.* Friedrich Berthold

Personnel known to have served with unit:
- *Hpt.* Friedrich Berthold - CO
- *Lt. d. R.* Fritz Koehler - P
- *Oblt.* Hermann Muggenthaler - P
- *Oblt.* Erwin Wenig - P

Aircraft:

A/C Unit Marking:

Unit Name: ***Schlachtstaffel*** **Anschütz/*****Bayerische Schlachtstaffel***

Alternate Names: *Schlacht-Fliegerstaffel* **Uniform Insignia** **Unit Stamp**
Feldpost No.:
Successor Unit: *Bayerische Schlachtstaffel (25.6.19)*

Date Established:
Location Established:
Founding Commander: *Ltn.* Oskar Anschütz

Reported to:

Bases: Dornstadt (16.4.19), Gersthofen (14.5.19). Schleissheim
Disbanded: Absorbed into *Fliegerhorst* Schleissheim, 30 September 1919, per *Bayr. Gruppenkommando Nr.* 4 Ia/0/O.Qu./*Flieg. Nr.* 5215 (?).

Commanding Officers: *Ltn.* Oskar Anschütz, *Oblt.* Lang

Personnel known to have served with unit:
Ltn. Oskar Anschütz - CO
Oblt. Lang - CO

Aircraft:

A/C Unit Marking:

Unit Name: *Freiwillige Flieger Abteilung* Dessloch

Alternate Names: **Uniform Insignia** **Unit Stamp**
Feldpost No.:
Successor Unit: Merged with *Freiwillige FliegerAbteilung Krieglsteiner* to form *Truppen-Fliegerstaffel* 21b (1.6.19)

Date Established: 23 April 1919
Location Established: Gersthofen
Founding Commander: *Oblt.* Otto Dessloch

Reported to:
Bases: Gersthofen (14.5.19), Kitzingen
Disbanded: See *Truppen-Fliegerstaffel* 21b.

Commanding Officers: *Oblt.* Otto Dessloch

Personnel known to have served with unit:
Oblt. Otto Dessloch - CO
Oblt. Hermann Muggenthaler - P
Lt. d. R. Fritz Koehler - P

Aircraft:

A/C Unit Marking:

Unit Name: *Freiwillige Flieger Abteilung* Kriegelsteiner

Alternate Names: *FA* Kriegelsteiner
Feldpost No.:
Successor Unit: Merged with *Freiwillige FliegerAbteilung* Dessloch to form *Truppen-Fliegerstaffel* 21b (1.6.19)

Uniform Insignia

Unit Stamp

Date Established: 20 April 1919
Location Established: Lechfeld
Founding Commander: *Hpt.* Reinhard Kriegelsteiner

Reported to:
Bases: Lechfeld, Schleissheim, Gersthofen, Kitzingen
Disbanded: See *Truppen-Fliegerstaffel* 21b.

Commanding Officers: *Hpt.* Reinhard Kriegelsteiner

Personnel known to have served with unit:
Hpt. Reinhard Kriegelsteiner - CO
Oblt. Hermann Muggenthaler - P

Aircraft:

A/C Unit Marking:

Unit Name: *Freiwillige Flieger Abteilung* Krug

Alternate Names: *Flieger Abteilung* Fürth, *Kurier-Flieger-Station* Fürth | **Uniform Insignia** | **Unit Stamp**

Feldpost No.:
Successor Unit: Merged with *Freiwillige FliegerAbteilung* Schmalschläger to form *Truppen-Fliegerstaffel* 21a (1.7.19)

Date Established: 21 April 1919, from former *Kurier-Flieger-Station* Schleissheim
Location Established: Fürth
Founding Commander: *Hpt.* Michael Krug

Reported to: *Kurier-Flieger-Staffel,* Bamberg from 21.4.19 to 1.7.19.
Bases: Fürth
Disbanded: 1.7.19 -- See *Truppen-Fliegerstaffel* 21a.

Commanding Officers: *Hpt.* Michael Krug

Personnel known to have served with unit: 3 officers, 15 NCOs & 26 enlisted men, including:

Vzfw. Hans Bauer
Vzfw. Seb(astian?) Kitzinger - G
Hpt. Michael Krug - CO
Lt. d. R. Nikolaus Meier
Uffz. Josef Reinhard - P
Vzfw. Franz Steiner
Vzfw. Max Wimmer

Three pilots were on strength 27 April 1919.

Aircraft: 3 aircraft after 21.4.19, later 6.

A/C Unit Marking:

Unit Name: ***Freiwillige Flieger Abteilung*** **Mühl**

Alternate Names: *Flieger Abteilung* 2, Fürth **Uniform Insignia** **Unit Stamp**
Feldpost No.:
Successor Unit:

Date Established: 21 April 1919
Location Established: Grafenwöhr
Founding Commander:

Reported to:

Bases: Grafenwöhr, Fürth
Disbanded: 21 June 1919

Commanding Officers:

Personnel known to have served with unit: 7 Officers, 15 NCOs, 30 Enlisted men

Aircraft:

A/C Unit Marking:

Unit Name: *Freiwillige Flieger Abteilung* Schmalschläger

Alternate Names: **Uniform Insignia** **Unit Stamp**
Feldpost No.:
Successor Unit: Merged with *Freiwillige FliegerAbteilung* Krug to form *Truppen-Fliegerstaffel* 21a (1.7.19)

Date Established:
Location Established:
Founding Commander: *Oblt.* Benedikt Schmalschläger

Reported to:
Bases: Fürth (28.4.19), Schleissheim (14.5.19)
Disbanded: See *Truppen-Fliegerstaffel* 21a.

Commanding Officers: *Oblt.* Benedikt Schmalschläger

Personnel known to have served with unit:
Ltn. Georg Ritter von Hengl - O
Oblt. Benedikt Schmalschläger - CO

Aircraft:

A/C Unit Marking:

Unit Name: *Freiwillige Flieger Abteilung* Sorg

Alternate Names: **Uniform Insignia** **Unit Stamp**
Feldpost No.:
Successor Unit:

Date Established: 7 March 1919
Location Established: Kitzingen
Founding Commander: *Hpt.* Max Sorg

Reported to:
Bases: Kitzingen, Schleissheim
Disbanded: Schleissheim, 30 September 1919

Commanding Officers: *Hpt.* Max Sorg

Personnel known to have served with unit:

Lt. d. R. Wilhelm Markert
Hpt. Max Sorg - CO

Aircraft:

A/C Unit Marking:

Unit Name: *Kurier-Flieger-Station* Bamberg

Alternate Names: *Flieger Abteilung* Schauer **Uniform Insignia** **Unit Stamp**

Successor Unit: *Freiwillige Flieger Abteilung* Schauer (12.7.19)

Date Established: 21 April 1919
Location Established: Bamberg
Founding Commander: *Hpt.* Ludwig Schauer

Reported to: *Kurier-Flieger-Staffel,* Bamberg

Bases: Bamberg
Disbanded: Became *FA* Schauer, 12.7.19. Disbanded 31 October 1919.

Commanding Officers: *Hpt.* Ludwig Schauer

Personnel known to have served with unit:
Ltn. August Delling
Hpt. Ludwig Schauer - CO
August Schmidt

Aircraft: 6 Aircraft

A/C Unit Marking:

Unit Name: *Kurier-Flieger-Station* Ingolstadt

Alternate Names: *Kurierfliegerstaffel* Kinkeln
Flieger Abteilung Kinkeln
Feldpost No.:
Successor Unit: None

Uniform Insignia **Unit Stamp**

Date Established: April 1919
Location Established: Ingolstadt
Founding Commander: *Oblt.* Alwin Kinkeln

Reported to: *Kofl., Korps* Oven

Bases: Ingolstadt
Disbanded: May 1919

Commanding Officers: *Oblt.* Alwin Kinkeln

Personnel known to have served with unit:
Oblt. Alwin Kinkeln - CO

Aircraft:
2 C-type Aircraft

A/C Unit Marking:

Unit Name: *Kurier-Flieger-Staffel* München

Alternate Names:
Feldpost No.:
Successor Unit: None

Uniform Insignia

Unit Stamp

Date Established: 21 April 1919
Location Established: Munich
Founding Commander:

Reported to: *Kurier-Flieger-Staffel* Bamberg

Bases: Munich
Disbanded: 28 July 1919

Commanding Officers:

Personnel known to have served with unit:

Aircraft:
3 Aircraft

A/C Unit Marking:

Unit Name: *Kurier-Flieger-Station* Schleissheim

Alternate Names: *Kurierfliegerstaffel* Schleissheim, *Flieger Abteilung* Krug, *Truppenfliegerstaffel* Krug

Uniform Insignia **Unit Stamp**

Feldpost No.:
Successor Unit: *Kurier-Flieger-Station* Fürth

Date Established: 12 December 1918
Location Established: Schleissheim
Founding Commander: *Hpt.* Michael Krug

Reported to:

Bases: Schleissheim (vacated about 18 April 1919), Fürth
Disbanded: Merged with *Flieger Abteilung* Schmalschläger 1 July 1919 to form *Truppen-Flieger-Staffel* 21a.

Commanding Officers: *Hpt.* Michael Krug

Personnel known to have served with unit:
Hpt. Michael Krug - CO
Lt. d. R. Nikolaus Maier

Aircraft:

A/C Unit Marking:

Unit Name: *Kurier-Flieger-Station* Würzburg

Alternate Names:

Uniform Insignia **Unit Stamp**

Date Established: 21 April 1919
Location Established: Würzburg
Founding Commander:

Reported to: *Kurier-Flieger-Staffel,* Bamberg

Bases: Würzburg
Disbanded: 12 July 1919

Commanding Officers:

Personnel known to have served with unit:

Gefr. Bossert - Mechanic
Vzfw. Brüstle - G
Offstv. Chorbacher - P
Pionier Englert - Mechanic
Offstv. Grün - P
Pionier Hupp - Mechanic
Vzfw. Kern - G
Gefr. Koppel - Mechanic
Vzfw. Schnäble - G
Uffz. Wildgens - Mechanic

Aircraft:

3 Aircraft

A/C Unit Marking:

Unit Name: *Flugstaffel des Freikorps* Schwaben

Alternate Names: *Flieger Abteilung* Schwaben
Feldpost No.:
Successor Unit:

Date Established: 28 April 1919
Location Established: Memmingen
Founding Commander: *Ltn.* Benedikt Jehle

Reported to: *Freikorps* Schwaben, Gruppe West
Bases: Memmingen (28.4.19), Lechfeld
Disbanded: 12 July 1919

Commanding Officers: *Ltn.* Benedikt Jehle

Personnel known to have served with unit:
- *Ltn.* Benedikt Jehle - CO, P
- *Ltn.* Max Madlener - O
- *Lt. d. R. Otto* Mauler - O
- *Uffz.* Karl Prim - P

Uniform Insignia
(*Upper left sleeve*)

(*Freikorps Schwaben*)

Unit Stamp

Aircraft: 2 Albatros C.Ia (Bayr.)

A/C Unit Marking: At least one Alb. C.Ia carried a white edelweiss on a green background on the normal position of the fuselage cross. The fin and rudder were crudely overpainted in white.

Unit Name: *Jagdstaffel* Flecken

Alternate Names: *Flieger Abteilung Flecken* **Uniform Insignia** **Unit Stamp**
Feldpost No.:
Successor Unit:

Date Established: About 27 April 1919
Location Established:
Founding Commander:

Reported to: *Gruppe Nord, Korps* Oven

Bases: Ingolstadt (1.5.19), Schleissheim (14.5.19)
Disbanded: Schleissheim, about 14 May 1919 per order of *Kofl., Gruppen-Kommando* 4

Commanding Officers:

Personnel known to have served with unit:

Aircraft:
3 D - types

A/C Unit Marking:

Unit Name: *Freiwilligen Fliegerstaffel* 18 **Unit Insignia** **Unit Stamp**

Alternate Names: Freiwilligen Flieger-Geschwader Staffel 327
Date Established: Before 28 February 1919
Location Established:
Founding Commander:

Reported to: 46. (Sächs.) *Landwehr* Div.-*Frw.*, Brigade *Süd-Litauen* (*Reichswehr* Brigade 28)
Bases:
Disbanded: 1 April 1919 or later

Commanding Officers:

Personnel known to have served with unit:

Aircraft:

A/C Unit Marking:

Unit Name: *Grenzjäger Flieger Abteilung 1/Truppen-Fliegerstaffel* Bautzen

Alternate Names:
Feldpost No.
Successor Unit: *Flugabteilung der Reichswehr*
Fliegerhorst Grossenhain (1.10.19)

Date Established: 25 (?) February 1919
Location Established: Grossenhain, Saxony
Founding Commander: *Hpt.* Egon Doerstling

Uniform Insignia (Collar Tabs) **Unit Stamp**

(Freikorps Michaelis)

Reported to: *Freiw. Grenzjäger-Abt.* 1 (*Freikorps* Michaelis), *Reichswehr* Brigade 12
Bases: *Fliegerhorst* Grossenhain, Bautzen (3.19), *Fliegerhorst* Grossenhain (1.10.19)
Disbanded: Grossenhain, 7.5.20

Commanding Officers: *Hpt.* Egon Doerstling, *Hpt.* Aribert Müller-Arles (1.10.19 - 7.5.20)

Personnel known to have served with unit:

- *Hpt.* Egon Doerstling (CO)
- *Oblt.* Franke (O)
- *Uffz.* Geisler (P)
- *Ltn.* Kellner (P)
- Kühne (Harness Maker)
- *Vzfw.* Metzger (P)
- Möbius (Radio Operator)
- *Hpt.* Aribert Müller-Arles (O, CO)
- *Vzfw.* Neitsch
- *Vzfw.* Opel (P)
- *Ltn.* von Peterkowsky (O)
- *Uffz.* Rothe
- *Vzfw.* Paul Rothe (P)
- *Vzfw.* Otto Rühle (P)
- *Ltn.* von Scheel (P)
- Schramm (Gunner)
- *Oblt.* Wilhelmi (O)

Aircraft:

A/C Unit Marking:

Unit Name: *Grenzjäger Flieger Abteilung 2*

Alternate Names: | **Uniform Insignia** | **Unit Stamp**
Feldpost No.
Successor Unit: *Flugabteilung der Reichswehr*
Fliegerhorst Grossenhain (1.10.19)

Date Established: 25 February 1919
Location Established: Grossenhain, Saxony
Founding Commander:

Reported to: *Reichswehr* Brigade 12
Bases: *Fliegerhorst* Grossenhain
Disbanded: Grossenhain, 7.5.20

Commanding Officers:

Personnel known to have served with unit:
Ltn. Werner Anton (O)

Aircraft

A/C Unit Marking:

Unit Name: *Fliegerkommando* Dresden-Kaditz

Alternate Names:
Feldpost No.
Successor Unit: *Flugabteilung der Reichswehr*
Fliegerhorst Grossenhain (1.10.19)

Uniform Insignia **Unit Stamp**

Date Established: 25 February 1919 (?)
Location Established: Dresden-Kaditz, Saxony
Founding Commander:

Reported to: *Reichswehr* Brigade 12
Bases: *Fliegerhorst* Kaditz, *Fliegerhorst* Grossenhain
Disbanded: Grossenhain, 7.5.20

Commanding Officers:

Personnel known to have served with unit:

Aircraft:

A/C Unit Marking:

Unit Name: ***Freiwillige Fliegerabteilung* Haas**

Unit Insignia (Collar tabs)

(Württ. Frw. Abt. Haas)

Unit Stamp

Alternate Names: *Freiwillige Württembergische Fliegerabteilung* Haas, *Grenzschutz Fl. Abt. bei den Truppen des* General Haas
Feldpost No.:
Successor Unit:

Date Established: 25 February 1919
Location Established: Böblingen
Founding Commander: *Oblt.* Robert Holtzmann

Reported to: *Württembergische Freiwillige Abteilung* Haas
Bases: Böblingen; Grenzschutz Ost (25.2.19)*, Lager Lechfeld (24.4.19), Böblingen (14.6.1919).
Disbanded: *Fliegerhorst* Böblingen, 1 September 1919 per Württ. Kr. Min. Nr. 3936.K.19.A.1., 31.7.19

Commanding Officers: *Oblt.* Robert Holtzmann

Personnel known to have served with unit: 17 officers, 58 NCOs & 187 enlisted men, including:

Frw. Calestin Allgaier
Frw. Wilhelm Armbruster
Frw. Emil Hugo Arndt
Feldw.-Ltn. Ernst Assenheimer
Frw. Erwin Backse
Vzfw. Alfred Bäder - Driver
Frw. Wilhelm Friedrich Bader
Frw. Heinrich Bandele
Vzfw. Franz Banholzer
Frw. Wilhelm Bannier
Gefr. Willy Bauerle
Frw. Friedrich Baumgartner
Gefr. Eduard Bazlen
Gefr. Johann Beck
Gefr. Paul Robert Beck
Frw. Peter Beck
Frw. Karl Hermann Becker
Gefr. Ludwig Belz
Uffz. Heinrich Berger - *Korporalsschaftsführer* I (*2 Flugzeuge*)
Sergt. Joseph Bernhard
Uffz. Bernhardt - *Korporalsschaftsführer* III (*2 Flugzeuge*)
Frw. Friedrich Wilhelm Bidlingmayer
*Frw.*Max Paul Bihlmayer
*Frw.*Emil Rudolf Biselli
Johann Bitterle - Driver
Frw. Ernst Blessing
Frw. Johannes Blessing
Emil Bornhauser - Driver
*Frw.*Emil Bosch
Frw. Friedrich Walter Braun
*Frw.*Karl Braun
Off.-stv. Erich Brunst
Frw. Eugen Bubek
*Gefr.*Albert Michael Buchhorn
*Frw.*Emil August Buck
*Frw.*Richard Bürgin
Uffz. Erwin Burianeck
Uffz. Wilhelm Burkhardt
Feldw. Karl Carle
Frw. Adolf Ruppert Christ
*Frw.*Erwin Dambacher
Asst. Artzt d. R. Ernst Dame
Uffz. Johannes Diebold
Feldw. Emil Dippon
*Frw.*Gottlieb Friedrich Dittus
Hpt. Dreschel
Vzfw. Hans Drescher
*Frw.*Franz Xaver Dressler
Gefr. Hans Dukek
*Frw.*Hermann Dürr
Frw. Otto Eberle
Vzfw. Gottfried Ehmann
Anton Eichmayer - Driver
Frw. Karl Eifer
Frw. Hermann Eitel
Frw. Albert Ellensohn - Driver
Frw. August Eppler
Frw. August Erhardt
Frw. Paul Ernst Escher
Ltn. Feller
Frw. Karl Friedrich Femppel
Uffz. Karl Feucht
Frw. Paul Fiesel
Frw. Albert Fischer
Josef Fischer - Driver
Uffz. Kurt Frankl
August Karl Fritz
Frw. Max Fritzenschaft
Frw. Melchior Georg Fröhlich
*Sergt.*Paul Fröhlich - *Korporalsschaftsführer* 6 (*Kraftfahrer*)
Vzfw. Adam Funk - *Werkm.-stv.*
Vzfw. Josef Funk
*Uffz.*Paul Geserick
Gefr. Josef Gessler
Ltn. Glöcklen
Frw. Wilhelm Göbel
Xaver Göetz - Driver
*Frw.*Heinz Grad
Vzfw. Carl Graf
Frw. Klemens Gröber
Frw. Friedrich Gross
Friedrich Haas - Driver
Gefr. Alfred Habermann
Frw. Gustav Hafner
*Frw.*Georg Haller
*Frw.*Paul Adolf Hämmerle
Off.-stv. Paul Häussermann
Frw. Friedrich Häussler
Uffz. Friedrich Haussmann
Uffz. Friedrich Haussmann

(continued on page 2)

Aircraft:

6 two-seaters, including Halb. Cl.IV (Rol.) 9470 (or 9478)/18, shot down at Munich, 2 May 1919, & 3 single-seaters, including one Fokker D.VII.

A/C Unit Marking:

***Note:** Assigned to *Grenzschutz Ost* 25 February 1919, but was reassigned to the Munich operation.

Freiwillige Fliegerabteilung Haas
Personnel known to have served with unit (continued):

Frw. Karl Ludwig Haussmann
*Frw.*Eugen Albert Hermann
Anton Hitz - Driver
Karl Holch - Driver
Hpt. Robert Holtzmann - P, CO
Frw. Karl Hommel
Frw. Ludwig Hummel
Uffz. Erwin Jäger
Frw. Paul Jakob
Frw. Albert Jaser
Frw. Friedrich Joos
Vzfw. Eugen Christian Kasper
*Vzfw.*Kayser
Frw. Eberhardt Heinrich Keidel
Frw. Erwin Keinath
Frw. Eugen Friedrich Kemmler
Frw. Eugen Kerber
Gefr. Christian Kern
Frw. Otto Wilhelm Kientach
Frw. Albert Kiesecker
Hpt. Kirsch
Frw. Paul Klaile
Frw. Erwin Kleinath
Gefr. Wilhelm Kleinath
Frw. Ernst Klink
Oblt. Knöringer
Lt.d.R. Fritz Koehler - *Ordinnanz Offizier*
Frw. Matthaus Kohnle
Frw. Hermann Kollmar
Frw. Otto König
Frw. Karl Köpf
Uffz. Heinrich Georg Koppenhäfer - *Korporalsschaftsführer*, Driver
Frw. Erwin Karl Kost
Sergt. Johann Kraft
Frw. August Wilhelm Kramer
Sergt. Richard Krämer
Gefr. Karl Georg Kübler
Frw. Hermann Kuhn
Uffz. Karl Kupfer
Frw. Alfred Laib
Frw. Karl Lang
Vzfw. Johannes Lässle
Uffz. Andreas Lemmle
Frw. Otto Hermann Lidle
Frw. Oswald Link
Frw. Christian Lipp
Rittm. Logan
Frw. Wilhelm Karl Ludwig
Frw. Ernst Gottlob Lutz
Gefr. Joseph Maass
Frw. Friedrich Mack
Frw. Adolf Maier
Gefr. Ernst Maier
Vzfw. Friedrich Maier
Frw. Lorenz Maier
Frw. Johann Malsfey
Frw. Heinrich Otto Mammele
Gefr. Johannes Marx
Georg Maucher - Driver
Werkm.-stv. Eugen Otto Mauer
Werkm.-stv. Eugen Otto Mauer
Off.-stv. Mayer - II *Werkmeister*
Frw. Karl Meinhardt
Frw. Paul Mendele
Frw. Martin Menhofer
Frw. Albert Hans Mentele
Uffz. Firdolin(?) Mern
Lt. d. R. Hans Metzger
Gefr. Albert Meyer
Gefr. Karl Möhrle
Lt. d. R. Kurt Monnington - P
Frw. Albert Müller
Frw. Christian Müller
Frw. Max Nafe
Frw. Karl Nagel
Frw. Max Emil Nägele
Frw. Wilhelm Neidhardt
Frw. Wilhelm Niebling
Gefr. Karl Niederer
Frw. Friedrich Johann Niedmayer
Frw. Albert Pfänder
Frw. Friedrich Jakob Pfeil -- Radio Operator
Frw. Otto Pfister
Frw. Hermann Pflüger
Frw. Eugen Johann Polla
Uffz. Georg Raab
Frw. Gotthilf Max Rahm
Gefr. Stanislaus Raikowski
Hermann Rapp - Driver
Uffz. Richard Paul Rath
Gefr. Albert Jakob Rauscher
Frw. Heinrich Adolf Rehm
Frw. Eugen Reiber
Frw. Karl Reichardt
Frw. Michael Reicherzer
Frw. Richard Reichle
Frw. Herman Robert Reihling
Frw. Karl Christian Reinhardt
Frw. Konrad August Rempfer
Frw. Max Otto Renz
Frw. Robert Richard Reutter - Radio Operator
Sergt. Eugen Ripberger
Frw. Richard Rohn
Frw. Robert Römer
Frw. Heinrich Roth
Frw. Karl Eugen Rothfelder
Frw. Anton Rundel
Uffz. Pius Schach
Uffz. Schäppel - *Korporalsschaftsführer* IV (3 *Einsitzer*)
Robert Schaich - Driver
Sergt. Herrmann Christian Schelling
Frw. Otto Schirmer
Frw. Friedrich Schlekweis
Jakob Schlenker - Driver
Frw. Gotthold Johannes Schmid
Sergt. Jakob Schmid
Sergt. Karl Schmid
Off.-Stv. Karl Schmuckle
Otto Schnotz*
Uffz. Otto Schöppel
Frw. Joseph Schoy
Frw. Erhard Schrank
Frw. Wilhelm Schranz
Sergt. Klaus Willi Julius Schulz - Korporalsschaftsführer
Feldw. Paul Josef Schulz
Josef Schurr - Driver
Karl Schwab - Driver
Uffz. Wilhelm Schwenk - *Korporalsschaftsführer*
Frw. Albert Seifried - Driver
Uffz. Friedrich Paul Seitz - *Korporalsschaftsführer* II (2 *Flugzeuge*)
Frw. Adam Christian Silberzahn
Hpt. Willibald Spang - P, *Grufl.*, *Gruppe* West
Frw. Hermann Speidel
Frw. Leonhard Spiegel
Frw. Wilhelm Stähle
Uffz. Friedrich Steiner
Frw. Joseph Stengele
Uffz. Karl Strecker - Driver
Frw. Karl Strobel - Driver
Frw. Lorenz Gustav Strobel
Frw. Eugen Theurer
Frw. Karl Theurer
Frw. Eugen Uebelmesser
Frw. Alfons Karl Uhler
Gefr. Anton Benedict Vogel.
Frw. Karl Wacker
Frw. Wilhelm Wagner
Frw. Eugen Michael Wahl
Gefr. Fritz Louis Wahl
Frw. Irwin Wahl
Uffz. Wilhelm Waldbauer
Karl Walter - Driver
Frw. Wilhelm Walter
Frw. Lorenz Walz
Frw. Paul Walz
Gefr. Vitalis Weber
Hpt. Weingarth
Lt. d. R. Otto Weisshaar - P
Vzfw. Hermann Weller - I *Werkm.-stv.*
Frw. Otto Wenzler
Frw. Christian Wiedennauer
Frw. Wilhelm Wörmer
Frw. Albert Würner
Gefr. Ernst Zaisa
Uffz. Karl Zeller
Frw. Johann Ziefle
Jakob Ziegler -Driver
Frw. Lorenz Zielinsky -Driver

Appendix 3

Alternate Names for Flying Units

As mentioned in the text, most *Freikorps*/Provisional *Reichswehr* Aviation Units were known by alternate names in addition to their official designations. Those alternate names are listed in the profiles for each unit (Appendix 2).

To avoid the need for searching through the profile pages for a specific unit name, the alphabetized list below contains the alternate names and their corresponding official designation. A question mark indicates the name hasn't been positively tied to any recognized unit.

Alternate Name	Official Designation
1. *Baltische Fliegerabteilung*	*Flieger Abteilung* 433
2. *Freiwillige Flieger* Division	*Flieger Abteilung* 426
Aufklarungs-Staffel des Kampfgeschwader Sachsenberg	*Flieger Abteilung* 416
Aviation-*Abteilung*	*Flieger Abteilung* 405
Baltische Landeswehr Fliegerstaffel	*Flieger Abteilung* 433
Eiserne Flieger Abteilung 8	*Flieger Abteilung* 427
FA A2 *der Bayerische Schützenkorps*	*Flieger Abteilung* A2
FA der 7. *Garde-Kavallerie-Schutzen* Division	*Flieger Abteilung* 420
FA der Freiwilliges Landesjägerkorps Maercker	*Flieger Abteilung* 423
FA des Ostpreussisches-Freiwilligenkorps Memel	*Flieger Abteilung* 433
FA Kriegelsteiner	*Freiwillige Flieger Abteilung* Kriegelsteiner
FA Krocker *des Landesjäger-Korps*	*Flieger Abteilung* 423
FA Möhn *des Deutsche Schutzen* Division	*Flieger Abteilung* 421
Feld-Flieger-Abteilung Koch	*Flieger Abteilung* 410
Feldflieger-Abteilung 433	*Flieger Abteilung* 433
Fl. Abt. des Freikorps Lüttwitz	Possibly *FA* 419, 420, 421, 422 or 432
Fl. Abt. des Freikorps Schleswig-Holstein	*Flieger Abteilung* 422?
Fl. Abt. Keng	*Flieger Abteilung* Krug?
Fl. Abt. Miller	?
Flieger Abteilung 2	*Flieger Abteilung* 414 or 415?
Fliegerstaffel des Grenzschutz-Oberkommandos Nord	?
Flieger Abteilung A	*Flieger Abteilung* 400
Flieger Abteilung B	*Flieger Abteilung* 401
Flieger Abteilung bei der 117. I.-D.	*Flieger Abteilung* 401
Flieger Abteilung C	*Flieger Abteilung* 402
Flieger Abteilung D	*Flieger Abteilung* 403
Fliegerabteilung der Sturmabteilung Rossbach	*Schlachtstaffel der Sturmabteilung* Rossbach
Flieger Abteilung der Brigade z.v.B. 5	*Flieger Abteilung* 429?
Flieger Abteilung des 2. *Garde-Inf.* Division	*Flieger Abteilung* 401?
Flieger Abteilung des Freikorps Lichtschlag	*Flieger Abteilung* Löwenfeld?
Flieger Abteilung Goldingen	*Freiwillige Fliegerstaffel* Libau
Flieger Abteilung Grauert	*Flieger Abteilung* 420
Flieger Abteilung Häfner	*Flieger Abteilung* A2
Flieger Abteilung Kinkeln	*Kurierfliegerstaffel* Kinkeln

Flieger Abteilung Krausser	*Flieger Abteilung* A1
Flieger Abteilung Krug	*Freiwillige Flieger Abteilung* Fürth
Flieger Abteilung Löwe	*Flieger Abteilung* 422
Flieger Ersatz Abteilung Palmer	*Flieger Abteilung* 415?
Flieger Frei-Korps Horst Hannover	*Flieger Abteilung* Schlieben
Flieger-Kompanie Heintz	*Flieger Abteilung* 422?
Fliegerabteilung Behrendt	*Flieger Abteilung* 432
Fliegerstaffel Hassfeld	?
Fliegerstaffel Kimale	*Freiwillige Fliegerstaffel* Libau
Fliegerstaffel zur Befreiung der Ostmark	*Flieger Abteilung* 415
Flugpark Kurland	*Flieger Abteilung* 425, after absorbing *FA* 433
Freiwillige FA Weinschenck	*Flieger Abteilung* 426
Freiwillige FA Winkler	*Flieger Abteilung* 411
Freiwillige Festungs Flieger Abteilung Graudenz	*Flieger Abteilung* 435
Freiwillige Flieger Abteilung der X. *Armee*	*Flieger Abteilung* 425
Freiwillige Flieger Abteilung des Detachement Tüllmann	*Flieger Abteilung* 401?
Freiwillige Flieger Abteilung Näther	*Flieger Abteilung* 415
Freiwillige Flieger Abteilung Rühl	*Flieger Abteilung* 2, Fürth
Freiwillige Flieger Abteilung Schauer	*Kurierfliegerstation* Bamberg
Freiwillige Württemburgische FA Haas	*Freiwillige Fliegerabteilung* Haas
Freiwilligen Flieger-Geschwader Staffel 327	*Freiwilligen Fliegerstaffel* 18
Frw. Artilleriefliegerstaffel Grossenhain	*Grenzjäger Flieger Abteilung* 2?
Frw. Fl. Abt. Cadelke *der Landesschützenkorps*	*Flieger Abteilung* 419?
Frw. Fl. Abt. Lissa	*Flieger Abteilung* 407
Frw. Fl. Abt. von Mann	?
Frw.-Fliegergruppe 36	FA 408?, Truppen-Fliegerstaffel 38?
Gemischtes Geschwader Sachsenberg	*Kampfgeschwader* Sachsenberg
Gleiwitz *Grenzschutz Fliegerstaffel*	*Flieger Abteilung* 401
Grenzflieger Abt. I	*Grenzjäger Flieger Abteilung* 1
Grenzschutz FA bei den Truppen des General Haas	*Freiwillige Fliegerabteilung* Haas
Jagdgeschwader des Kampfgeschwader Sachsenberg	*Flieger Abteilung* 413
Jagdstaffel 202	?
Jagdstaffel der Landesschützenkorps	*Flieger Abteilung* 419?
Jagdstaffel Löwe	*Flieger Abteilung* 422
Kampfgeschwader von Bredow	*Flieger Abteilung* 431?
Kurierfliegerstaffel Bamberg	*Kurierfliegerstation* Bamberg
Kurierfliegerstaffel Schleissheim	*Kurierfliegerstation* Schleissheim
Schlachtstaffel des Kampfgeschwader Sachsenberg	*Flieger Abteilung* 417
Staffel Weimar	*Flieger Abteilung* 423
Truppenfliegerstaffel Bautzen	*Grenzjäger Flieger Abteilung* 1
Truppenfliegerstaffel Krug	*Kurierfliegerstation* Schleissheim

Appendix 4

Placename Cross Reference

As a rule, this account uses German spellings of place names in use in 1918 - 20. Following independence, Latvia, Lithuania, and to a lesser extent, Estonia, changed many of these names to conform to their own languages. The same happened in those areas of Germany that became part of Poland or Russia. For readers trying to locate these places on modern maps, this cross-reference list may prove helpful. We have included as many of the diacritical marks as the type font permits.

Former Name	Present Name
Aa River	Lielupe River, Latvia
Abau River	Abava River, Latvia
Alienhof	Alīnes muiža, Latvia
Allenstein	Olsztyn, Poland
Alt-Auz	Auce, Latvia
Alt Berun	Bieruń, Poland
Angernsee	Lake Engure, Latvia
Auciemi	Auciems, Latvia
Babitsee	Babītes ezers, Latvia
Baldon, Gut	Baldone, Latvia
Bartau River	Bartuva River, Lithuania, Barta River, Latvia
Bartenstein	Bartoszyce, Poland
Bausk	Bauska, Latvia
Bentschen	Zbąszyń, Poland
Bernstadt	Bierutów, Poland
Beuthen	Bytom, Poland
Bilderingshof	Bulduri, Latvia
Brieg	Brzeg, Poland
Briesen	Brzeźno, Poland
Bogutschütz	Bogucice, Poland
Breslau	Wrocław, Poland
Bromberg	Bydgoszcz, Poland
Culmsee	Chełmża, Poland
Czarnikau	Czarnków, Poland
Danzig	Gdańsk, Poland
Deuthen	Dajtki, Poland
Deutsch-Eylau	Iława, Poland
Dietrichsdorf, Gut	Falęcin, Poland
Doblen	Dobele, Latvia
Dorpat	Tartu, Estonia
Düna River	Daugava River, Latvia
Dünaburg	Daugavpils, Latvia
Eckau	Iecava, Latvia
Eichenau	Ławica, Poland
Elsenmühle	Dąbrówka Mała. Poland
Emanuelsegen	Murcki, Poland
Essern, Ezern	Ezere, Latvia

Former Name	Present Name
Faulbrücke bei Reichenbach	Moscisko, Poland
Freystadt	Kisielice, Poland
Friedrichstadt	Jaunjelgava, Latvia
Gandau	Gądów Mały, Poland
Getlingsmoor	Railway terminal in Riga, Latvia
Gieschewald	Giszowiec, Poland
Ginkunen, Gut	Ginkūnai , Lithuania
Gleiwitz	Gliwice, Poland
Glogau	Glogów, Poland
Görchen	Miejska Górka, Poland
Goldingen	Kuldīga, Latvia
Graudenz	Grudziądz, Poland
Gross Plowenz	Płowęż, Poland
Gross Roop	Straupe, Latvia
Grüssau	Krzeszów, Poland
Grundhofen	Grójec, Poland
Hasenpoth	Aizpute, Latvia
Heidekrug	Šilutė, Lithuania
Hohensalza	Inowrocław, Poland
Hundsfeld	Psie Pole, Poland
Illuxt	Ilūkste, Latvia
Imelin	Imielin, Poland
Insterburg	Chernyakhovsk, Russia
Jakobstadt	Jekabpils, Latvia
Jeseritz	Jezierzyce, Poland
Johnsdorf	Janów , Poland
Kedainius	Kėdainiai , Lithuania
Kalnzen	Area near Sloka, Latvia
Karlsbad	Melluži, Latvia
Karsakiski	Karsakiškis, Lithuania
Kalisch	Kalisz, Poland
Kielmy	Kelmė, Lithuania
Kimalen	Ķimale, Latvia
Königsberg	Kaliningrad, Russia
Köslin	Koszalin, Poland
Kolberg	Kołobrzeg, Poland
Kolmar	Chodzież, Poland
Kovno, Kowno	Kaunas, Lithuania
Kreuzburg, Gut	Krustpils, Latvia
Kreuzberg	Kluczbork, Poland
Kurland, Courland	Kurzeme, Latvia
Langfuhr	Wrzeszcz, Poland
Laurahütte	Huta Laury, Siemianowice, Poland
Lemsal	Limbaži, Latvia
Lendzin	Lędziny, Poland
Libau	Liepāja, Latvia
Lissa	Leszno, Poland
Livland	Vidzeme district, Latvia
Livländ. Aa River	Gauja River, Latvia

Former Name	Present Name
Löwen	Lewen Brzeski, Poland
Lublinitz	Lubliniec, Poland
Lutzausholm	Lucavas Island, Riga. Latvia
Lyck	Ełk, Poland
Majorienhof	Jūrmala, Latvia
Mariampol	Marijampolė, Lithuania
Memel	Klaipėda, Lithuania
Meseritz	Międzyrzecz, Poland
Mesoten	Mežotne, Lithuania
Mitau	Jelgava, Latvia
Murawjewo	Mažeikiai, Lithuania
Myslowitz	Mysłowice, Poland
Nakel	Nakło nad Notecią, Poland
Namslau	Namysłów, Poland
Nargö	Naissaar, Estonia
Neu Berun	Bieruń Nowy, Poland
Neu Gacz	Nowa Gać, Poland
Nikischschacht	Nikiszowiec-Katowice, Poland
Oger River	Ogre River, Latvia
Okmijany	Akmenė, Lithuania
Olai	Olaine, Latvia
Oppeln	Opole, Poland
Orany	Varėna, Latvia
Ostrowo	Ostrów Wielkopolski, Poland
Paulsgnade	Ozolnieki, Latvia
Peterfeld	Rail station between Auri & Ķirpēni , Latvia
Pillau	Baltiysk, Russia
Pitschen	Byczyna, Poland
Pless	Pszczyna, Poland
Popeljany	Papilė, Lithuania
Posen	Poznań, Poland
Pupe	Mārupe, Latvia
Radziwilischki	Radviliškis, Lithuania
Rawitsch	Rawicz, Poland
Rembata	Rembate, Latvia
Retowo	Rietavas, Lithuania
Reval	Tallinn, Estonia
Rojen	Roja, Latvia
Rudbaren	Rudbārži , Latvia
Rudelsdorf	Radzików, Poland
Sackenhausen	Sacka, Latvia
St. Olai	Olaine, Latvia
Schadow	Šeduva, Lithuania
Schaulen	Šiauliai, Lithuania
Schildberg	Ostrzeszów, Poland
Schlok	Sloka, Latvia
Schneidemühl	Piła, Poland
Schönberg	Skaiskalne, Latvia

Former Name	Present Name
Schoppinitz	Szopienice, Poland
Schrunden	Skrunda, Latvia
Schwed River	Šventė (Lithuania), Svēte (Latvia) River
Schwetz	Świecie, Poland
Seerappen	Liublino, Russia
Segewold	Sigulda, Latvia
Shagory	Žagarė, Lithuania
Siemianowitz	Siemianowice, Poland
Skaudwile	Skaudvilė, Lithuania
Stettin	Szczecin, Poland
Stockmannshof	Pļaviņas, Latvia
Stolp	Słupsk, Poland
Subocz	Subačius, Lithuania
Szaty	Svėdasai, Lithuania
Talssen	Talsi, Latvia
Tarnowitz	Tarnowskie Góry, Poland
Tauerkaln	Taurkalne, Latvia
Tauroggen	Tauragė, Lithuania
Telsche	Telšiai, Lithuania
Thorensberg	Torņakalns district, Riga, Latvia
Thorn	Toruń, Poland
Thüringshof	Tīriņmuiža, Latvia
Tilsit	Sovetsk, Russia
Tischau	Tychy, Poland
Tirul Swamp	Cenas tīrelis, Latvia
Tuckum	Tukums, Latvia
Üxküll	Ikšķile, Latvia
Vilna, Vilno	Vilnius, Lithuania
Wainoden	Vaiņode, Latvia
Walk	Valka, Latvia
Walkowitz	Walkowice, Poland
Weissenhof	Spilve, Latvia
Weitenfeld, Gut	Vītiņi, Latvia
Wenden	Cēsis, Latvia
Werro	Võru, Estonia
Wiebersholm	Vibera Island, Latvia
Wilkomierz	Ukmergė, Lithuania
Wilkowischken	Vilkaviškis, Lithuania
Windau	Ventspils, Latvia
Windau River	Venta River, Latvia & Lithuania
Wissek	Wysoka, Poland
Wormen	Vārme, Latvia
Yamburg	Kingisepp, Russia
Zarnikau	Carnikava, Latvia
Zirke	Sieraków, Poland

Appendix 5

Freikorps/Vorlaufige Reichswehr Aviation Personnel Deceased in Service

Records are not complete, but at least twenty-five members of the postwar *Freikorps* and Provisional *Reichswehr* aviation detachments perished in combat, accidents, or due to illness between January 1919 and March 1920. Inasmuch as the influenza pandemic was still ravaging the world in 1919, its safe to assume this list represents only a small portion of the flying personnel who died during this period.

	Date of death	Name & Rank	Unit	Location	Cause
1.	7 January 1919	*Ltn.* Walter Meuer	*Eiserne FA* 8	Riga	[1]
2.	8 January 1919	*Ltn.* Max Näther	*FA* 415	Czarnikau	KIA
3.	28 Jan (2 Feb?) 1919	*Ltn.* Gerhard Pape	FA 423	Luckenwalde	KIC
4.	8 February 1919	*Feldw.* Hawelik	Libau *Fliegerstaffel*	Libau	KIC[2]
5.	6 March 1919	*Lt. d. R.* Werner Preuss	*FA* 422	Rendsburg	KIC
6.	22 May 1919	*Ltn.* B. Hermann	*FA* 427	Riga	KIC
7.	22 May 1919	*Uffz.* Grund	*FA* 427	Riga	KIC
8.	19 June 1919	*Lt. d. R.* Wittemayer	*FA* 427	Livonia	KIA
9.	1 July 1919	Fritz Kalbreier	*FA* 406	Meseritz?	KIC
10.	7 July 1919	*Ltn.* Bovakht	*FA* 422	Ludwigslust	KIC
11.	19 August 1919	*Lt. d. R.* Weiss	*AFst* 129	Silesia	KIA
12.	28 August 1919	*Ltn.* Hellmuth Rüsche	*AFst* 129	Myslowitz	KIA
13.	28 August 1919	*Sgt.* Kieske	*AFst* 129	Myslowitz	KIA
14.	28 September 1919	*Oblt.* Fritz Schulz	Lithuanian AF	Aleksotas	KIC
15.	27-8 October 1919	*Vzfw.* Christian Mesch	2 *Plastun* Av. Div.	Rembata	KIA[3]
16.	4 November 1919	Unidentified pilot	3 *Plastun* Av. Div.?	Libau	KIA
17.	4 November 1919	Unidentified observer	3 *Plastun* Av. Div.?	Libau	KIA
18.	29 November 1919	*Oblt.* Hans-Helmut von Boddien	3 *Plastun* Av. Div.	Okmiany	KIA[3]
19.	2 December 1919	*Ltn.* von Sallerau	3 *Plastun* Av. Div.	Kielmy	KIA[3]
20.	2 December 1919	*Vzfw.* Weiss	3 *Plastun* Av. Div.	Kielmy	KIA[3]
21.	2 December 1919	Reich	3 *Plastun* Av. Div.	Kielmy	KIA[3]
22.	2 December 1919	Kretschmar	3 *Plastun* Av. Div.	Kielmy	KIA[3]
23.	18 March 1920	*Ltn.* Franz Büchner	*Fl'horst* Grossenhain[4]	Leipzig	KIA
24.	21 March 1920	Ltn. Kurt Scheibe	*Fl'horst* Halle	Halle a.d. Saale	KIA
25.		[5]	*FA* 424		

Key: KIA: Killed in Action. KIC: Killed in Crash

(1) Shot down over Riga, 2 January 1919, tried and executed as a spy 7 January 1919.
(2) Crashed after engine failure and subsequently died of his injuries. Date of death unknown.
(3) Killed on ground after being forced to land.
(4) *Flugabteilung der Reichswehr Fliegerhorst Grossenhain*
(5) Unknown member of *FA* 424. Based on photo of bough- and ribbon-draped coffin. Date, location and cause of death unknown.

Appendix 6

Glossary of German Military and Other Terms Appearing in this Book

German Term	English Translation
Abteilung	Department, Section or Detachment
Abteilungsführer	Detachment or Department Commander
a. D. (following a rank)	Retired
Armee	Army
AOK (Armee Oberkommando)	Army High Command
Artillerie	Artillery
Artillerie-Fliegerstaffel	Artillery Flying Squadron
Artzt	Medical Officer
Aufklärung(s)	Reconnaissance
Ballonzug	Balloon Section
Befehl	Order, Command
Beobachter	Observer
Bordschützen	Aerial Gunner
d. R. (following a rank)	Reserve
Detachement	Detachment
Eiserne	Iron
Feldflieger Abteilung (FFA)	Field Flying Detachment
Feldmarschall	Field Marshal
Feldwebel (Feldw.)	Sergeant Major (U. S.)
Fliegerbeobachterschule (Fliebsch.)	Aviation Observer School
Flieger	Aviation, Pilot, or Private rank in Aviation
Flieger Abteilung	Flying Detachment, Flying Section
Flieger Abteilung (A), FA(A)	Artillery Flying Detachment
Flieger Ersatz Abteilung (FEA)	Aviation Replacement Section
Fliegerhorst	Military Air Base
Fliegerschule	Aviation School
Fliegerstaffel	Aviation Squadron
Fliegertruppen	Aviation Troops
Flivo (Fliegerverbindungsoffizier)	Aviation Liaison Officer
Flugmaat	3rd Class Aviation Petty Officer
Flugpark	Aviation Park
Flugplatz	Airfield
Flugstation	Aviation Station
Flugzeug	Aircraft
Flugzeugmeisterei	Directorate of Aircraft Production
Freikorps, Freiwillige Korps	Volunteer Corps
Freiherr (Frhr.)	Baron
Freiwillige	Volunteer, or Private rank in some units
Führer	Commander or Leader
Garde	Guards
Gefreiter	Private First Class, Lance Corporal (U. S.)
Generalleutnant	Lieutenant General
Generalmajor	Major General

German Term	English Translation
Geschwader	Aviation Wing, Aviation Group
Grenzschutz	Border Protection
Grufl. (Gruppenflieger)	Group Aviation Commander
Gruppe	Group
Gruppenkommando	Group Command
Hauptmann (Hpt.)	Captain
Hauptquartier	Headquarters
Idflieg (Inspektor der Fliegertruppen)	Inspector of Flying Troops
Infanterie	Infantry
Jagdstaffel (Jasta)	Fighter Squadron, literally, Hunting Squadron
Iluft (Inspektor der Luftschiffertruppen)	Inspector of Airship Troops (Bavaria)
Kampf	Combat
Kampfgeschwader	Combat Aviation Wing or Group
Kampfstaffel (Kasta)	Combat Squadron
Kanone	Air ace (Literally, cannon)
Kogenluft (Kommandierender General der Luftstreitkräfte)	Commanding General of the Air Service
Kofl. (Kommandeur der Flieger)	Commander of Flying Troops
Korporatsschaft	Section or Department, literally, "Corporate body". Usage appears to be exclusive to Frw. FA Haas
Korps	Corps
Kreiegsgliederung	Order of battle
Kurierfliegerstaffel	Courier Aviation Squadron
Lager	Camp
Landesjäger	Gendarme
Landesschützen	Territorial Protection
Landeswehr, Landwehr	Territorial Army
Leutnant (Ltn.)	Second Lieutenant (U. S.)
Luftstreitkräfte	Air Service (during World War I)
Marine	Naval
Nord	North
Oberfehlshaber Ost (Ober Ost)	Supreme Commander East
Oberkommando	High Command
Oberleutnant (Oblt.)	First Lieutenant (U. S.)
Oberleutnant zur See (Oblt. z. S.)	Naval First Lieutenant
Oberst	Colonel
Oberstleutnant	Lieutenant Colonel
Offizier-stellvertreter (Off-stv.)	Acting Officer, Warrant Officer (U. S.)
Ost	East
Ostpreussisches-Freiwilligenkorps	East Prussian Volunteer Corps
O.z.b.V. (Offizier zur besondere Verwendung)	Adjutant
Polizei	Police
Reichswehr	Army, literally, Realm Defense

German Term	English Translation
Ritter	Knight
Rittmeister	Cavalry Captain
Rubild	Sequential strip aerial camera
Schlacht	Assault, Ground Attack
Schlachtstaffel (Schlasta)	Assault Squadron
Schule	School
Schützen	Protection, Defense
Schutzen	Infantry
Seeflugstation	Naval Air Station
Stab	Staff
Staffel	Squadron
Staffelführer	Squadron Commander
Stosstruppe	Shock Troops or Storm Troops
Sud	South
Tagsbefehl	Daily Order
Truppen-Fliegerstaffel	Troop Flying Squadron
Unteroffizier (Uffz.)	Corporal (U. S.)
USP (Unabhängige Sozialdemokratische Partei Deutschlands)	Independent Social Democratic Party of Germany
Vizefeldwebel (Vzfw., Vfw.)	Sergeant 1st Class (U. S.)
Vizeflugmeister	Aviation Vice Petty Officer
Vorlaufige Reichswehr	Provisional Reichs Army
Werkmeister	Foreman (of workshops, in flying units)
Zahlmeister	Paymaster
Zug	Platoon, Section

Below: Wrecked Halberstadt CL.II of *FA* 406/*Artillerie-Fliegerstaffel* 127 at Meseritz. This is likely the machine in which Fritz Kalbreier lost his life 1 July 1919. (Author's collection)

Appendix 7

Freikorps / Vorlaufige Reichswehr / Polizei aviation personnel who rose to the rank of General in the *Luftwaffe*

It shouldn't come as any surprise that leaders of the *Freikorps* and Provisional *Reichswehr* aviation units were quick to join the new *Luftwaffe* in 1935. It also shouldn't be a surprise that many of them rose to high ranks, given their experience and obvious dedication to German military aviation. In 1939, this must have been looked upon as a wise career move; in 1945, not so much. The list below contains summary details of 138 individuals known to have served in *Freikorps*, Provisional *Reichswehr* and *Polizei* aviation and who later became *Luftwaffe* Generals in World War II.

Primary source for this information was Hildebrand, K. F., *Die Generale der deutschen Luftwaffe* 1935-1945, Biblio Verlag, Osnabrück, 1990.

Name	*Freikorps/Vorl. Rw./ Polizei* Unit	Highest *Luftwaffe* Rank
Abernetty, *Ltn.* Otto	*FA* 409	*Generalmajor*
Alt, *Hpt.* Erich	*FA* 433	*Generalmajor*
Angerstein, *Oblt.* Karl	*FA* 420	*Generalleutnant*
Anton, *Ltn.* Werner	*Grenzjäger-Abt*.2	*Generalleutnant*
Ascheberg, Hpt. Percy *Baron* von	*Kofl*. XI, *Grufl*. 37	*Generalmajor*
Baerens, *Hpt.* Friedrich	*Fliegerhorst* Insterberg	*Generalmajor*
Banse, *Ltn.* Gustav	*Polizei F'staffel* Mecklenburg	*Generalmajor*
Barlen, *Oblt.* Karl	*FA* 419	*General der Flieger*
Bassenge, *Ltn.* Gerhard	*Fliegerhorst* Insterberg	*Generalmajor*
Baur de Betaz, *Hpt.* Wilhelm	*Kofl. Gkdo. XI A.K.*	*Generalleutnant*
Becker, *Ltn.* Wilhelm	*Beobachterschule* Thorn, *Fliegerhorst* Stolp	*Generalleutnant*
Behrendt, *Oblt.* Hans	*FA* 432 / *AFst*. 116, *Fliegerhorst* Paderborn	*Generalleutnant*
Berthold, *Hpt.* Friedrich	*Truppen-Fliegerstaffet* 21	*Generalmajor*
Bieneck, *Hpt.* Hellmuth	*Fliegerhorst* Brieg	*General der Flieger*
Bodenschatz, Oblt. Karl	*Freiwilligen-Detachement* Probstmayr	*General der Flieger*
Boehmer, *Rittm.* Hans	*Freikorps* Dohna	*Generalmajor*
Boenigk, *Oblt.* Oskar von	*FA* 418, *Jasta* 202	*Generalmajor*
Boettge, *Oblt.* Kurt	*FA* 433	*Generalmajor*
Bonatz, *Oblt.* Ernst	*FA* 420	*Generalmajor*
Breithaupt, *Kapitänleutnant* Joachim	Marine-Brigade Erhardt	*Generalmajor*
Buchholz, *Ltn.* Ulrich	*Grenzschutz Ost*	*Generalleutnant*
Bülow, *Hpt.* Hilmer *Frhr.* von	*Flieger Nord*	*Generalleutnant*
Bülowius, *Oblt.* Alfred	*FA* 420, *Polizei F'staffel Ostpreussen*	*General der Flieger*
Busch, *Hpt.* Hans	*Stab, Kogenluft*	*Generalmajor*
Christiansen, *Käpt.-Lt.*Friedrich	III. (Löwenfeld) Marine-Brigade	*General der Flieger*
Crailsheim, *Oblt.* Hanns *Frhr.*von	*Gruppe* Hierl	*Generalmajor*
Cranz, *Oblt.* Friedrich	*Fliegerhorst* Hannover, *Fliegerhorst* Paderborn	*Generalleutnant*

Name	*Freikorps/Vorl. Rw./ Polizei* Unit	Highest *Luftwaffe* Rank
Dahlmann, *Oblt.* Hermann	*Fliegerhorst* Gotha	*General der Flieger*
Deichmann, Ltn. Paul	*FA* 427, *Fl'staffel* Libau, *FA* 433, *AFP* Kurland, *TFst* 2	*General der Flieger*
Dessloch, *Oblt.* Otto	*Flugpark* Gersthofen, *FA* Dessloch	*Generaloberst*
Dieckhoff, *Lt. d. R.* Gerhard	*Fliegerhorst* Gotha	*Generalmajor*
Dörffler, *Hpt.* Ernst	*FA* 425, *Flugpark* Kurland	*Generalmajor*
Doerstling, *Hpt.* Egon	*Grenzflieger Abt.* 1	*General der Flieger*
Dommenger, *Hpt.* Erich	*Fliegerhorst* Elbing & Stolp	*Generalmajor*
Dransfeld, Hpt. Eduard	*Dilfer, Fliegerhorst* Braunschweig	*General der Flieger*
Drechsel, *Hpt.* Ernst	*Fliegerhorst* Böblingen	*Generalmajor*
Drum, *Oblt.* Karl	*Kofl*, XVII *A.K.*	*General der Flieger*
Durr, *Hpt.* Alfred	*Feld-Art. Rgt.* 29	*Generalmajor*
Erdmann, *Lt. d. R.* Hans	*Polizei F'staffel* Hamburg	*Generalmajor*
Exss, *Ltn.* Ernst	*FA* 423 / *AFst* 116, *Fliegerhorst* Paderborn	*Generalmajor*
Fahrig, *Rittm.* Hermann	*Kofl.* XII / XIX, *Fliegerhorst* Kowno	*Generalmajor*
Falkenhayn, *Hpt.* Erich	*FK* Faupel, *Inf.-Regt.* 145, *FK* Görlitz	*Generalmajor*
Felmy, *Oblt.* Gerhard	*FEA* 9, *Fliegerschule* Hagenau	*Generalmajor*
Felmy, *Oblt.* Hellmuth	*Fliegerschule* Lübeck	*General der Flieger*
Fink, *Hpt.* Carl	*FA* 407	*Generalmajor*
Fischer, *Ltn.* Eberhard	*FA* 401/*AFst* 129, *Fliegerhorst* Brieg	*Generalmajor*
Fischer, *Oblt.* Veit	*Flugplatz* Sonthofen, *FK* Schwaben	*General der Flieger*
Fisser, Oblt. Johann Volkmar	FA 432, *Fliegerhorste* Frankfurt / O, Döberitz	*Generalmajor*
Fitzau, Ltn. Erich-Erdmann	*FA* 423 / *AFst* 116	*Generalmajor*
Freyberg-Eisenberg-Allmendingen, *Hpt.* Egloff *Frhr. von*	*FK* Hülsen, *Fliegerhorst* Döberitz	*Generalmajor*
Fricke, *Oblt.* Hermann	*FA* 432, *FK* Hülsen	*Generalmajor*
Frommherz, *Lt. d. R.* Hermann	*Polizei-Fliegerstaffel* Hassfeld	*Generalmajor*
Fruhner, *Lt. d. R.* Otto	*FA* Schlieffen	*Generalmajor*
Fütterer, *Ltn.* Kuno Heribert	*FA* 420 / *AFst* 115	*Generalmajor*
Funck, *Hpt.* Werner	*AFP 2 & 17*	*Generalmajor*
Gablenz, *Ltn.* Carl-August *Frhr.*	*FK* Lüttwitz	*Generalmajor*
Gandert, *Oblt.* Hans-Eberhard	*Fliegerhorst* Neuruppin	*Generalmajor*
Gerlach, *Ltn.* Karl von	*Fliegerhorste* Königsberg, Döberitz, *Idflieg*	*Generalmajor*
Grauert, *Hpt.* Ulrich	*FA* 420	*Generaloberst*
Greim, *Oblt.* Robert *Ritter* von	*Korps* Oven	*Generalfeldmarschall*
Hanesse, *Oblt.* Friedrich-Carl	*FA* 400, *Fliegerhorste* Brieg & Paderborn	*General der Flieger*
Heidenreich, *Lt. d. R.* Anton	*Flugplatz* des *Gouvernments* Libau	*Generalmajor*
Heising, *Hpt.* Fritz	*Grenzschutz-Bombenstaffel Ost*	*Generalmajor*
Hempel, *Hpt.* Claus	*Idflieg, Fliegerhorst* Brieg	*Generalmajor*

Name	*Freikorps/Vorl. Rw./ Polizei* Unit	Highest *Luftwaffe* Rank
Hentschel, Sergeant Karl	*Fliegerhorst* Königsberg	*Generalmajor*
Heyde, *Ltn.* Heinrich-Sigismund von der	*Fliegerhorst* Neuruppin	*Generalmajor*
Homburg, *Oblt.* Erich	*Fliegerhorst* Freiburg i. Breisgau	*Generalleutnant*
Hückel, *Oblt.* Hans	*Fliegerhorst* Halle/Saale	*Generalleutnant*
Jeschonnek, *Ltn.* Hans	*Artillerie-Fliegerstaffel* 129	*Generaloberst*
Jordan, *Ltn.* Otto	*Fliegerhorst* Altenburg, *FA* 432/*AFst* 103	*Generalmajor*
Junck, *Ltn.* Werner	*Truppen-Fliegerstaffel* 8, *Fliegerhorst* Brieg	*Generalleutnant*
Kastner-Kirdorf, *Hpt.* Gustav	*Fliegerhorst* Neuruppin	*General der Flieger*
Kessler, *Lt. z. S.* Ulrich	*Garde-Kavallerie-Schützen* Division, *Gkdo.* Lüttwitz	*General der Flieger*
Klepke, *Hpt.* Waldemar	*Fliegerhorst* Schneidemühl, XX. *Armeekorps*	*General der Flieger*
Koehler, *Lt. d. R.* Fritz	*Truppen-Fliegerstaffel* 21b	*Generalmajor*
Kohlbach, *Ltn.* Wilhelm	*FA* 423/*AFst* 116	*Generalmajor*
Koller, *Vzfw.* Karl	*Polizei Fliegerstaffel* 1	*General der Flieger*
Kortüm, *Ltn.* Joachim	*AFst* 117, *Polizei-Fliegerstaffel* Königsberg	*Generalmajor*
Krahmer, *Oblt.* Eckhard	*Fliegerhorst* Jüterbog, *Polizei-F'staffel* Berlin	*Generalleutnant*
Kriegelsteiner, *Hpt.* Reinhard	*Frw. FA* Kriegelsteiner	*Generalmajor*
Krocker, *Hpt.* Victor	*FA* 423/*AFst* 116	*Generalleutnant*
Krüger, *Oblt.* Ernst	*Kofl.* X. *A. K.*	*Generalleutnant*
Krug, *Hpt.* Michael	*Kurierfliegerstationen* Schleissheim, *FA* Nürnberg-Fürth, *Truppen-Fliegerstation* 21a	*Generalleutnant*
Leesemann, *Lt. d. R.* Michael	*FA* 406/*AFst* 127, *Fliegerhorst* Brieg	*Generalmajor*
Loerzer, *Hpt.* Bruno	*FA* 427	*General der Flieger*
Lowenstern, *Rittm.* Elard Baron von	*Fliegerhorste* Altenburg, Halle/Saale, Braunschweig	*Generalmajor*
Lohmann, *Ltn.* Günther	*FA* 409, 427, 410, *Fliegerhorst* Königsberg	*Generalleutnant*
Lorenz, *Oblt.* Heinrich	*Fliegerhorst* Königsberg, *Hauptflugpark* 301, *A'flst* 12	*Generalmajor*
Maass, *Oblt.* Bruno	*Kofl.* XX. *A. K.*, *F'horst* Stolp	*Generalleutnant*
Mahnke, Hpt. Alfred	*Fliegerhorst* Königsberg	*General der Flieger*
Maier, Lt. d. R. Nikolaus	*FA* Miller, *TFst.* Krug	*Generalmajor*
Manteuffel, *Ltn.* Eitel-Friedrich Roediger von	*FEA* Darmstadt, *FA* 432	*Generalmajor*
Markert, *Lt. d. R.* Wilhelm	*FA* Sorg, *Gkdo. II.* Bayr. *A. K.*, *Fliegerhorst* Fürth	*Generalarzt*
Meister, *Ltn.* Rudolf	*Fliebsch.* Stolp, *FA* 420	*General der Flieger*

Name	*Freikorps/Vorl. Rw./ Polizei* Unit	Highest *Luftwaffe* Rank
Milch, *Hpt.* Erhard	*Kofl.* 1, *FA* 429, 412/*AFst* 117 *Polizei Flst.* Königsberg	*Generalfeldmarschall*
Missfelder, *Hpt.* Johannes	*FA* 408	*Generalmajor*
Morzik, *Off.-Stv.* Friedrich-Wilhelm	*FA* 418/*TFst.* 27, *TFst* 8, *Polizei Flst.* Königsberg	*Generalmajor*
Mühlig-Hofmann, *Hpt.* Albert	*FA* 414/*AFst* 102	*Generalleutnant*
Müller, *Ltn.* Gottlob	*FA* A2, *AFst* 121	*Generalleutnant*
Müller-Arles, *Hpt.* Aribert	*Grenzflieger-Abt* I, *Fliegerhorst* Grossenhain	*Generalmajor*
Müller-Kahle, *Oblt.* Albert	*FA* 409, *Fliegerhorst* Stolp	*Generalmajor*
Muggenthaler, *Oblt.* Hermann	*FA* 21b	*Generalleutnant*
Nordt, *Hpt.* Gustav	*Grenzschutz Ober-Kdo Nordostpreussen*	*Generalmajor*
Ossig, *Off-stv* Richard	*FA* 403/*TFst* 6	*Generalmajor*
Osterkamp, *Lt. d. R.* Theodor	*FA* 413/*TFst.* 34	*Generalleutnant*
Pfeiffer, *Ltn.* Herbert	*FA* 421	*Generalmajor*
Ploch, *Ltn.* August-Albert	*F'horste* Hannover, Cottbus & Paderborn	*Generalmajor*
Putzier, *Oblt.* Richard	*Fliegerhorst* Bromberg	*General der Flieger*
Quade, *Hpt.* Erich	*Truppen-Fliegerstaffel l 9*	*General der Flieger*
Raithel, *Ltn.* Johann	*Kurierfliegerstation* Bamberg	*Generalleutnant*
Rauch, *Ltn.* Heinrich	*FA* 410, 433, 426	*Generalleutnant*
Richthofen, *Ltn.* Wolfram *Frhr.* von	*Fliegerhorst* Gotha	*Generalmajor*
Rieckhoff, *Ltn.* Herbert	*FA* 414, 400	*Generalleutnant*
Riecke, *Ltn.* Georg	*Fliegerhorst* Neuruppin	*General der Flieger*
Sattler, *Ltn.* Xaver	*Fliegerhorst* Döberitz	*Generalleutnant*
Schauer, *Hpt.* Ludwig	*F'horste* Bamberg, Schleissheim	*Generalmajor*
Schleich, *Hpt.* Eduard *Ritter* von	*Bayr. Iluft*	*Generalleutnant*
Schoenebeck, *Ltn.* Carl-August	*FA* 424	*Generalmajor*
Schroeder, *Vzfw.* Walter	*F'horst.* Elbing, *TFst.* 20	*Generalmajor*
Schroth, *Ltn.* Günther	*FEA* 2, *FA* 415, *F'horst.* Stolp	*Generalmajor*
Schubert, *Hpt.* Theodor	*Kofl.* XVII *A.K.*, *FA* 429	*Generalleutnant*
Schüz, *Oblt.* Hans	*F'horst.* Freiburg/Breisgau	*Generalmajor*
Schultheiss, *Oblt.* Paul	*Fliegerhorst* Böblingen	*Generalleutnant*
Schulz, *Hpt.* Julius	*FA* 409, *Eiserne* Div.	*General der Flieger*
Schwartzkopff, *Ltn.* Günter	*FA* 406	*Generalmajor*
Seeboth, Oblt. Joseph	*Gkdo.* X. *A.K.*	*Generalrichter*
Seywald, *Ltn.* Heinrich	*FA* A2	*Generalmajor*
Spang, *Hpt.* Willibald	*FA* Haas	*Generalleutnant*

Name	Freikorps/Vorl. Rw./ *Polizei* Unit	Highest *Luftwaffe* Rank
Sperrle, *Hpt.* Hugo	*Kofl. Korps* Lüttwitz	*Generalfeldmarschall*
Stahr, *Hpt.* Walter	*Kofl.* XX. *A. K.*	*Generalmajor*
Steinmann, Heinrich	*FA z. b. V.* A/*FA* 400	*Generalmajor*
Treindl, *Oblt.* Theodor	*Artillerie-Fliegerstaffel* 121	*Generalleutnant*
Wadehn, *Ltn.* Walther	*Kampfgeschwader* Sachsenberg	*Generalmajor*
Walz, *Hpt.* Franz Josef	*F'horst.* Schleissheim	*Generalleutnant*
Weiner, *Ltn.* Georg	*AFst.* Grossenhain	*Generalmajor*
Wieland, *Ltn.* Günther	*FA* 428, *F'horst* Stolp	*Generalmajor*

Above: Halberstadt C.V(Av) 6902/18 of *FA* 417/*KG* Sachsenberg, crashed by Averdam and Ruppricht at Vaiņode, probably in May 1919. (Peter M. Grosz collection/STDB)

Bibliography

Books

6. Rīgas *Kājnieku Pulks, Rīgas Kājnieku Pulka Vēsture 1919 – 1929,* Rīgā, 1929.

Angiola, John R. & Clint R. Hackney, Jr., *The Pour le Mérite and Germany's First Aces,* Hackney Publishing Co., Friendswood, 1984.

Awaloff, General *Fürst* Pawel, *Im Kampf gegen Bolschevismus, Druck und Verlag von,* J. J. Augustin, Glückstadt & Hamburg, 1925.

Belcarz, Bartolomiej & Robert Peczkowski, *White Eagles: The Aircraft, Men and Operations of the Polish Air Force, 1918 - 1939,* Hikoki Publications, Ottingham, 2001.

Bennett, Geoffrey, *Freeing the Baltic,* Birlin Limited, Edinburgh, 2002.

Benoist-Méchin, Jacques, Taylor, Eileen R. (tr.), *History of the German Army since the Armistice Vol. I, From the Imperial Army to the Reichswehr,* Howard Fertig, New York, 1988.

Bischoff, Major Josef, *Die letzte Front: Geschichte der Eisernen Division im Baltikum,* 1919, *Buch- und Teifdruck G. m. b. H. Abteilung Buchverlag,* Berlin, 1935.

Borlinghaus, H., *Die deutsche Feldpost im Ersten Weltkrieg* 1914–1918 / *Stempelhandbuch* (*Band* II), Jülich, 2006.

Brzenk, *Lt. d. R.* Hans, *Im Kampf wider Kokarde und weissen Adler, Germanen-Verlag,* Berlin-Lichterfelde, 1924.

Bucholtz, Hansgeorg, *Der Flieger Thom,* Gräfe & Unzer, Königsberg, 1937.

Cowin, Hugh, *Profile 187. The Junkers Monoplanes,* Profile Publications, Leatherhead, 1967.

von Engelhardt, Eugen *Freiherr, Der Ritt nach Riga, Volk und Reich Verlag,* Berlin, 1938.

Franke, *Generalmajor* Hermann, *Handbuch der neuzeitlichen Wehrwissenschaften, Dritter Band, 2. Die Luftwaffe, Verlag von* Walter de Gruyter & Co., Berlin, 1939.

Franks, Norman, Frank Bailey & Rick Duiven, *Casualties of the German Air Service, 1914–1920,* Grub Street, London, 1999.

Franks, Norman, Frank Bailey, & Rick Duiven, *The Jasta Pilots,* Grub Street, London, 1996.

Franks, Norman & Greg VanWyngarden, *Fokker D VII Aces of World War I, OspreyAircraft of the Aces 53,* Osprey Publishing Ltd., Botley, 2003.

Gerste, City Councillor Max, *Der Münchener Räte-Republik, Verlag der Politischen Zeitragen,* Munich, 1919, quoted in Weiss, Dr. G., *Kernpunkte der Schulorganisation,* Hermann Bayer & *Söhne* (Bayer *und* Mann), Herz., Langensalza.

von der Goltz, General *Graf* Rüdiger, *Meine Sendung in Finnland und in Baltikum, Verlag von* K. F. Koebler, Leipzig, 1920.

Grimm, Claus, *Vor dem Toren Europas* 1918–1920, *Geschichte des Baltischen Landeswehr,* August Friedrich Velmede *Verlag,* Hamburg, 1963.

Grosz, P. M., Windsock Datafile 25. Fokker D. VII, Albatros Productions Ltd., Berkhamsted, 1991.

Herris, Jack, *Pfalz Aircraft of World War I (Great War Aircraft in Profile)*, Flying Machines Press, Boulder, 2001.

Herris, Jack, *Halberstadt Aircraft of World War I, Volume 2: CL.IV-CLS.I & Fighters, A Centennial Perspective on Great War Airplanes*, Aeronaut Books, Reno, 2020.

Hildebrand, K. F., *Die Generale der deutschen Luftwaffe* 1935–1945, Biblio Verlag, Osnabrück, 1990.

von Hoeppner, General *der Kavallerie* Ernst, *Deutschlands Krieg in der Luft, Verlag von* K. F. Koehler, Leipzig, 1921.

von Hösslin, *Rittmeister a.D.* Hubert, *Die Organisation der K. B. Flieger-Truppe 1912–1919, Verlag des Bayerischen Kriegsarchivs*, München, 1924.

Imrie, Alex, German Naval Air Service, Vintage Aviation Fotofax, Arms & Armour Press, London, 1989.

Imrie, Alex, *Pictorial History of the German Army Air Service 1914–1918*, Henry Regnery Company, Chicago, 1971.

Inspektion der Fliegertruppen, *Ehrentafel, der in Flugdienst während des Weltkrieges gefallenen Offiziere der deutsche Fliegerverbuande*, E. S. Mittler und Sohn, Berlin, 1920.

Jones, Nigel H., *Hitler's Heralds. The Story of the Freikorps 1918–1923*, John Murray, London, 1987. Jones' very readable book covers much the same ground as Waite's (see below), but in a less academic style. It contains a useful appendix of basic information about many of the *Freikorps* units. Like Waite's book, however, it makes little or no mention of aviation units or their role.

Kilduff, Peter, *Iron Man: Rudolf Berthold: Germany's Indomitable Fighter Ace of World War I*, Grub Street, London, 2012.

Kitvel, Toivo, & Johannes Tik, *Eesti lennukroonika: tekste ja pilte aastani 1940*, Aviopol, 2003.

von Langsdorff, W., *Flieger am Feind*, Bertelsmann, Gütersloh, 1942.

Maercker, General Ludwig R. G., *Vom Kaiserheer zur Reichswehr, Ein Beitrag zur Geschichte der deutschen Revolution, Verlag von* K. F. Koehler, Leipzig, 1921.

Matius, Dr. J., Military Archive no. XI, 1939.

Mückler, Jörg, & Rainer Absmeier, *Die Geschichte der Jagdstaffel* 14 – 1916–1918, *Luftfahrtverlag-Start*, Bad Zwischenhahn, 2021.

Musicano, Walter A., *Eagles of the Black Cross*, Ivan Obolensky, New York, 1965.

Neumann, Major *a. D.* Georg Paul, *In der Luft unbesiegt*, J. F. Lehmanns *Verlag*, München, 1923.

Nowarra, Heinz J., *Die verboten Flugzeuge 1921–1935, Motorbuch Verlag*, Stuttgart, 1985.

Ockel, Heinz, *Revolutionäre Volkswehr 1918–19, Militärverlag*, Berlin, 1968.

O'Connor, Neal, *Aviation Awards of Imperial Germany in WW I, Volume I -- The Aviation Awards of the Kingdom of Bavaria*, Foundation for Aviation World War I, Princeton, 1989.

O'Connor, Neal, *Aviation Awards of Imperial Germany in WW I, Volume II -- The Aviation Awards of the Kingdom of Prussia*, Foundation for Aviation World War I, Princeton, 1990.

O'Connor, Neal, *Aviation Awards of Imperial Germany in WW I, Volume III -- The Aviation Awards of the Kingdom of Saxony*, Foundation for Aviation World War I, Princeton, 1993.

O'Connor, Neal, *Aviation Awards of Imperial Germany in WW I, Volume IV – The Aviation Awards of the Kingdom of Württemburg,* Foundation_for Aviation World War I & Flying Machines Press, Princeton, 1995.

O'Connor, Neal, *Aviation Awards of Imperial Germany in WW I, Volume V – The Aviation Awards of the Eight Thüringian States and the Duchy of Anhalt,* Foundation for Aviation World War I, Princeton, 1998.

O'Connor, Neal, *Aviation Awards of Imperial Germany in WW I, Volume VI – The Aviation Awards of the Grand Duchies of Baden and Oldenburg,* Foundation for Aviation World War I & Flying Machines Press,_Princeton, 1999.

O'Connor, Neal, *Aviation Awards of Imperial Germany in WW I, Volume VII -- The Aviation Awards of Eight German States and the Three Free Cities*, Schiffer Military History, Atglen, 2002.

von Oertzen, Friedrich Wilhelm, *Die deutschen Freikorps 1918–1923,* F. Bruckmann *Verlag,* München, 1937.

Osterkamp, Theo, *Du oder ich, deutsche Jagdflieger in Höhen und Tiefen,* Albert Nauck & Co., Berlin-Charlottenburg, 1938.

Owers, Colin A. & Jack Herris, *Hannover Aircraft of WWI, A Centennial Perspective on Great War Airplanes,* Aeronaut Books, Las Vegas, 2020.

Owers, Colin A., *Junkers Airplanes of WWI Volume 2: J.5–J.11 A Centennial Perspective on Great War Airplanes,* Aeronaut Books, Reno, 2018.

Pitrof, *Oberst d. Landes-Polizei a. D.* & Daniel Ritter, *Gegen Spartikus in München und im Allgäu, Buchdruckerei und Verlagsanstalt* Carl Gerber, München, 1937.

Pletschacher, Peter, *"Das Ende der bayerischen Fliegertruppen* 1919", *Die Königlich Bayerischen Fliegertruppen* 1912–1919, 2nd Edition, Aviatik *Verlag*, Planegg, 1992.

Pilawski, Wladislaw., *Ziemia czarnkowska w walce o wolnosc 1918-1919,* Studio Prograf- A. J. Nowiccy, Czarnków, 2006.

von Pölnitz, Götz, *Emir - Das tapfere Leben des Freiherrn Marschall von Bieberstein, Verlag* Georg D. Callwey, München, 1938

Pyragaius, J., *Musu sparnai,* Kaunas, 1929

Reichskriegsministerium, Darstellungen aus den Nachkriegskämpfen deutscher Truppen und Freikorps, E. S. Mittler & Sohn, Berlin. These are the official *Reichswehr* histories of the German army and *Freikorps* from the Armistice until 1923. They were published in six volumes between 1936 and 1940:

1. *Die Rückführung des Ostheeres*
2. *Der Feldzug im Baltikum bis zur zweiten Einnahme von Riga. Januar bis Mai* 1919
3. *Die Kämpfe im Baltikum nach der zweiten Einnahme von Riga. Juni bis Dezember* 1919
4. *Die Niederwerfung der Räterschaft in Bayern* 1919
5. *Die Kämpfe in Südwestdeutschland* 1919–1923
6. *Die Wirren in der Reichshauptstadt und im nördlichen Deutschland* 1918–1920

These accounts are a mixed bag -- on one hand, they are a valuable source of factual information with their mind-numbing detail of the various campaigns and battles. On the other, they overlay those actions with the twisted Nazi view of right and wrong.

Rempe, Paul, *From German Cavalry Officer to Reconnaissance Pilot: The World War I History, Memories, and Photographs of Leonhard Rempe, 1914–1921,* Savas Beatie, El Dorado Hills, 2016.

Rimell, Raymond Laurence, *Vintage Warbirds No. 2. The German Army Air Service in World War One,* Arms & Armour Press, London,1985.

von Salomon, Ernst, ed., *Das Buch des deutschen Freikorpskämpfer*, Wilhelm Limpert, Berlin, 1938. This book is primarily a collection of articles that appeared previously in *Der Reiter gen Osten*, a periodical for *Freikorps* veterans, edited by von Salomon.

Schmäling, Bruno, Reinhard Kastner & Jörn Leckscheid, *Royal Bavarian Jagdstaffel 76,* Aeronaut Books, Monee, 2011.

Schroeder, Generalmajor a. D., *Die Sächsischen Freiwilligen-Truppen in Litauen 1919, Buchdruckerei der* Wilhelm *und* Bertha von Baensch *Stiftung,* Leipzig, 1933.

Suchenwirth, Richard, *The Development of the German Air Force, 1919–1939*, USAF Historical Division, Aerospace Studies Institute, Air University, Arno Press, New York, 1968.

Supf, Peter, *Das Buch der deutschen Fluggeschichte,* Stuttgart, 1926.

Tessin, Georg, *Deutsche Verbände und Truppen 1918–1939*, Biblio *Verlag*, Osnabrück, 1974. Tessin's book is indispensible for understanding the structure and reporting relationships of the *Freikorps* and Provisional *Reichswehr*, along with all their component forces, except, alas, the aviation units.

Thielhaber, Felix A., *Jüdische Flieger im Weltkrieg, Verlag der Schild*, Berlin, 1924.

Wagener, *Hauptmann a. D., Von der Heimat geächtet*, Chr. Belsersche *Verlagsbuchhandlung,* Stuttgart, 1920.

Waite, Robert G. L., *Vanguard of Nazism. The Free Corps Movement in Postwar Germany 1918–1923*, Harvard University Press, Cambridge, 1952, reissued 1970.

Waite's book is probably the best general study of the *Freikorps* era in English. Comprehensive and thoroughly footnoted, it's an essential reference to anyone researching this topic. I have relied on this source, along with Jones' *Hitler's Heralds,* for most of the background narrative for events, especially within Germany.

VanWyngarden, Greg, *Aces of Jagdstaffel 17*, Osprey Aircraft of the Aces 118, Osprey Publishing Ltd., Oxford, 2013.

Wagener, *Hauptmann a. D., Von der Heimat geächtet*, Chr. Belsersche *Verlagsbuchhandlung,* Stuttgart, 1920.

Wenzl, Richard, Richthofen-*Flieger, Badische Zeitung*, Karlsruhe, 1930.

Woolley, Charles, *German Uniforms, Insignia & Equipment 1918–1923*, Schiffer Military History, Atglen, 2002.

Zuerl, Walter, *Pour le Mérite-Flieger*, Curt Pechstein *Verlag,* München, 1938.

Periodicals

Air Enthusiast
Arbeitsgemeinschaft Dt. Luftfahrthistorik
Arbeitskreis Sächsische Militärgeschichte e. V.,
Berliner Volks-Zeitung
Blätter zur Geschichte der Deutschen Luft- und Raumfahrt
Cross & Cockade Journal (U.S.)
Cross & Cockade International Journal, formerly *Cross & Cockade Great Britain Journal*
Dobeles Balss
Die Freiheit, organ of the *USPD* (Independent Socialist party of Germany), a daily newspaper published from 1918 till 1922.
Jägerblatt, *Mitteilungsblatt der Gemeinschaft der Jagdflieger e.V.*, published from 1975 to 2005.
Jet & Prop, journal of the German Aviation History Study Group

Journal of Petroleum and Gas Engineering
Kariškių Žodis
Der Landesknecht
Latvijas Sargs
Libauische Zeitung
Kurjer Polski
Kurzemes Vārds
Luftpost
Over the Front, Journal of the League of World War I Aviation Historians, 1986–present
Plieno Sparnai
Das Propellerblatt
Reiter gen Osten, monthly newsletter for *Freikorps* veterans, published from 1929 to 1944.
Rigasche Rundschau
Die Rote Fahne, newspaper published intermittently by the German Socialist Workers' Party, the Spartacus League and the Communist party of Germany.
Rundbrief der Gemeinschaft Alte Adler
Der Schild
Die Trommel, newspaper for German troops in the West Russian Army, published from 8 October to 20 November 1919.

Websites
Aerodrome Forum (www.the aerodrome.com / forum)
Axis History (www.axishistory.com)
Flieger und Luftschiffer (www.buddeke.de)
Feldgrau (www.feldgrau.net / forum)
Forum-der-Wehrmacht (forum-der-wehrmacht.de)
Frontflieger (www.frontflieger.de)
Gentleman's Military Interest Club (www.gmic.co.uk)
Lexicon der Wehrmacht (lexicon-der-wehrmacht.de)
Wikipedia

Right: Fokker D.VII of *FA* 423 /*Artillerie-Fliegerstaffel* 116 over the unit's airfield in Weimar. (Paul Rempe)

Index

Aircraft

Persons

Units

Above: Air-to-air shot of one of *FA* 423 / *Artillerie-Fliegerstaffel* 116's Fokker D.VIIs. (Paul Rempe)

Above: *Polizei Fliegerstaffel* Hamburg's Halberstadt C.V(Av) 6930/18 with its Benz Bz.IVü removed for maintenance. (Peter M. Grosz collection/STDB)

Made in United States
Orlando, FL
04 November 2026

90137849R00190